Water Quality

Claude E. Boyd

Water Quality

An Introduction

Second Edition

Springer

Claude E. Boyd
School of Fisheries, Aquaculture and Aquatic Sciences
Auburn University
Auburn, AL, USA

ISBN 978-3-319-33062-4 ISBN 978-3-319-17446-4 (eBook)
DOI 10.1007/978-3-319-17446-4

Springer Cham Heidelberg New York Dordrecht London

Softcover reprint of the hardcover 2nd edition 2015

Printed on acid-free paper

Springer International Publishing AG Switzerland is part of Springer Science+Business Media (www.springer.com)

Foreword

Claude Boyd is a name that is well known in the field of aquaculture and aquatic sciences, and for good reason. His research and teaching are marked with clarity and pragmatism that have addressed gaps in knowledge and brought about leaps in understanding. Much of the knowledge that we take for granted today, with regards to water quality and its management, is built upon the work that he has shared. This book represents the continuation of his efforts to advance the field. While Dr. Boyd's work is typically associated with aquaculture, the principles and theories elaborated upon in this book are fundamental to all aquatic sciences with widespread application to our aquatic environments and water resources.

Water Quality: *An Introduction* bridges a gap. Traditionally, books on this topic fall on two ends of a spectrum. Books on one end of the spectrum touch upon the subject matter and lack depth. On the other end of the spectrum are highly specialized references intended for practitioners or researchers with years of experience in the field. Neither serves to provide a strong foundation from which to build expertise. *Water Quality*: *An Introduction* creates that opportunity, and indeed has for countless professionals. Furthermore, the information contained within these pages can be found scattered among books on a range of topics from chemistry to environmental policy. This single concise volume efficiently organizes these topics and provides students, researchers, and educators with a comprehensive introduction to the discipline.

The second edition is more than an update to the original. Years of teaching and research have informed this edition, resulting in expanded chapters, additional figures, and illustrative problems. Notable changes include an additional chapter on basic chemistry, a more comprehensive chapter on hydrology, and an updated chapter on regulations and standards. This renewal will allow *Water Quality*: *An Introduction* to serve as the foundation for the discipline for years to come.

Colby-Sawyer College
New London, NH, USA
December 1, 2014

Harvey J. Pine Ph.D

Preface

Water is a familiar substance; it is as commonplace as air, soil, and concrete. Like the other three substances—except possibly concrete—water is essential to human and other forms of life. There is a lot of water; two-thirds of the world is covered by the ocean, while roughly 3.5 % of the global land mass is inundated permanently with water. Water exists in the hydrosphere in a continuous cycle—it is evaporated from the surface of the earth but subsequently condenses in the atmosphere and returns as liquid water. The world is not going to run out of water; there is as much as ever or is ever going to be.

Despite the rosy scenario expressed above, water can be and often is in short supply—a trend that will intensify as the global population increases. This is because all places on the earth's land mass are not watered equally. Some places always have little water, while many other localities are water deficient during droughts. The quality of water also varies from place to place and time to time. Most of the world's water is too saline for most human uses, and pollution from anthropogenic sources degrades the quality of freshwater, lessening its usefulness. As implied above, evaporation is a water purification process, but salts and pollutants left behind on the land and in water bodies remain to contaminate the returning rainwater.

Availability of freshwater has always been and continues to be an important factor affecting human population. But, as the world's population has grown, water quality has become an equally important issue. Water quality is a critical consideration in domestic, agricultural, and industrial water supply, fisheries and aquaculture production, aquatic recreation, and the health of ecosystems. Professionals in many disciplines need a basic understanding of the factors controlling concentrations of water quality variables as well as the effect of water quality of aquatic life and water use by humans. These individuals also need to know how to manage the resources or activities for which they are responsible in a manner that is protective of water quality.

Water quality is a complex subject, and unfortunately, the teaching of this important topic is not well organized. In most colleges and universities, water quality instruction is given in certain engineering curricula. The classes emphasize specific aspects of water quality management and tend to focus on water treatment. Such classes typically have prerequisites or other conditions that prohibit students from other disciplines from enrolling in them. As a graduate student, I attempted to enroll in such a class with the instructor's consent, but the engineering dean refused to allow me to take the class on the grounds that I would possibly in the future try to do engineering work without an engineering degree. I doubt that the attitude has changed much in the intervening 50 years. Those of us in academics tend to stay in our boxes and to defend them. But, our boxes are interconnected in the real world just as are the compartments of the hydrosphere.

In agriculture, forestry, fisheries and aquaculture, biology, chemistry, physics, geology, nutrition, science education, and a few other disciplines, specific aspects of water quality are taught in some classes. The coverage of water quality in these classes focuses on specific issues and is incomplete. Of course, a class in general environmental science considers water quality along with several other major environmental issues in a general way. But, there is not time in such a class to provide much detail on water quality.

It is interesting to compare the teaching of water quality with that of general soil science. Many universities teach an excellent class on the fundamentals of soil science that contains a wealth of information on the geology, physics, chemistry, and biology of soils. This class is taken by students from a variety of disciplines. In the author's opinion, colleges and universities should teach a class providing an easily understandable but comprehensive overview of water quality modeled somewhat along the lines of a general soil science class.

In 1971, my immediate boss at Auburn University asked me to consider developing a class in water quality management for fishery science majors—mostly graduate students—many of whom were interested primarily in aquaculture. I apparently was chosen for this task because, as a student, I had taken numerous classes in chemistry, physics, mathematics, plant physiology, soils, and toxicology. I really knew very little about water quality per se, but I had been exposed to a lot of its component parts. Being a young faculty member and eager to please, I accepted. I spent a lot of time studying in order to arrange the material that I presented in the lectures. It became obvious that my students had such a poor knowledge of basic chemistry and the fundamentals of water quality that they had trouble understanding the material on water quality management. Thus, I convinced my boss to allow me to develop a class on the principles of water quality to serve as a prerequisite to the course in water quality management.

My students had relatively little background in physics, chemistry, and mathematics, and I had to provide a review of the basic principles of chemistry and physics using simple mathematics—nothing beyond algebra. Moreover, I could not find a suitable water quality or water chemistry book for the class—all texts were far too complicated for my students.

I wrote a manual about water quality management that consisted of a review of basic chemistry followed by a discussion on water quality; the remainder—about half of the text—dealt with water quality management. The College of Agriculture at Auburn University published this book primarily for use in my class. I finally prepared a book manuscript on the fundamentals of water quality by expanding the section on water quality in the manual; this became the first edition of

Water Quality: An Introduction. I believe that the first edition provided a reasonable overview of water quality. The discussion was slanted toward physical and chemical aspects of water quality, but an effort was made to illustrate the interactions with biological components of water bodies. The chemistry was presented at a basic level, and as a result, some of the explanations and solutions to problem examples were only approximate. Nevertheless, I felt that simplification would allow students to grasp salient points with relatively little "weeping, wailing, and gnashing of the teeth."

I want to emphasize that chemistry usually is taught to "non-major" students in such a formidable way that many of them have a life-long aversion to the subject. Some appear to actually abhor chemistry—it seems to terrify them. And, I blame this phobia on freshman chemistry teachers who are doing a great disservice by teaching chemistry in such a rigorous and theoretical manner that even the students who pass the class do not understand the importance of chemistry in everyday life. Many of the students who get a bad attitude about chemistry will, during their professional career, be called upon to make decisions requiring knowledge of chemistry. As a society, we would benefit from instilling a better practical understanding of chemistry in such students rather than trying to demonstrate the difficulties of the subject to "non-major" students. I cannot help saying this—classes should be held to teach students rather than to demonstrate the difficulty of the subject or the brilliance of the professor.

The second edition of *Water Quality: An Introduction* has been revised heavily. Numerous errors found in the first edition were corrected—but, there no doubt are sufficient errors in the second edition to satisfy everyone. The contents of the first edition were rearranged, and many sections were updated or improved. I believe that the second edition of *Water Quality: An Introduction* has been greatly enhanced for use as a text for a general water quality class or for self-study of the subject. I hope that you will agree.

Preparation of this book would have been impossible without the excellent assistance of June Burns in typing the manuscript and proofing the tables, references, and examples. Three of my graduate students, Rawee Viriyatum, Lauren Jescovitch, and Zhou Li, assisted with the drawings.

Auburn, AL, USA Claude E. Boyd

Introduction

Water is a common substance that life cannot exist without. Water is a major component of living things; humans are about two-thirds water, and most other animals contain equal or even greater proportions. Woody plants are at least 50 % water, and the water content of herbaceous plants usually is 80–90 %. Bacteria and other microorganisms normally contain 90–95 % water. Water is important physiologically. It plays an essential role in temperature control of organisms. It is a solvent in which gases, minerals, organic nutrients, and metabolic wastes dissolve. Substances move among cells and within the bodies of organisms via fluids comprised mostly of water. Water is a reactant in biochemical reactions, the turgidity of cells depends upon water, and water is essential in excretory functions.

Water is important ecologically for it is the medium in which many organisms live. The distribution of vegetation over the earth's surface is controlled more by the availability of water than by any other factor. Well-watered areas have abundant vegetation, while vegetation is scarce in arid regions. Water plays a major role in shaping the earth's surface through the processes of dissolution, erosion, and deposition. Large water bodies exert considerable control over air temperature of surrounding land masses. This is especially true for coastal areas that may have cooler climates than expected because of cold ocean current offshore or vice versa. Thus, ecosystems have a great dependence upon water.

In addition to requiring water to maintain bodily functions, humans use water for domestic purposes such as food preparation, washing clothes, and sanitation. Many aquatic animals and some aquatic plants have long been an essential part of mankind's food supply.

Early human settlements developed in areas with dependable supplies of water from lakes or streams. Gradually, humans learned to tap underground water supplies, store water, convey water, and irrigate crops. This permitted humans to spread into previously dry and uninhabitable areas. Even today, population growth in a region depends upon water availability.

Water is a key ingredient of industry for power generation through direct use of the energy of flowing water to turn water wheels or turbines, steam generation, cooling, and processing. In processing, water may be used as an ingredient, solvent, or reagent. It also may be used for washing or conveying substances, and wastes from processing often are disposed of in water.

Water bodies afford a convenient means of transportation. Humans find water bodies to be aesthetically pleasing, and many recreational activities are conducted in and around them. Much of the world's commerce depends upon maritime shipping that allows relatively inexpensive transport of raw materials and industrial products among continents and countries. Inland waterways also are important in both international and domestic shipping. For example, in the USA, huge amounts of cargo are moved along routes such as the Mississippi and Ohio Rivers and the Tennessee–Tombigbee waterway.

Water Quality

At the same time that humans were learning to exert some control over the quantity of water available to them, they found different waters varied in qualities such as temperature, color, taste, and odor. They noted that these qualities influenced the suitability of water for certain purposes. Salty water was not suitable for human and livestock consumption or irrigation. Clear water was superior over turbid water for domestic use. Some waters could cause illness or even death when consumed by humans. The concepts of water quantity and water quality were probably developed simultaneously, but throughout most of human history, there were few ways for evaluating water quality beyond sensory perception and observations of the effects that certain waters had on living things or other water uses.

Any physical, chemical, or biological property that influences the suitability of water for natural ecological systems or the use by humans is a water quality variable, and the term water quality refers to the suitability of water for a particular purpose. There are literally hundreds of water quality variables, but for a particular purpose, only a few variables usually are of interest. Water quality standards have been developed to serve as guidelines for selecting water supplies for various activities or for protecting water bodies from pollution. These standards will be discussed later in this book, but a few introductory comments will be made here. The quality of drinking water is a health consideration. Drinking water must not have excessive concentrations of minerals, must be free of toxins, and must not contain disease organisms. People prefer their drinking water to be clear and without bad odor or taste. Water quality standards also are established for bathing and recreational waters and for waters from which shellfish are cultured or captured. Diseases can be spread through contact with contaminated water. Oysters and some other shellfish can accumulate pathogens or toxic compounds from water, making them dangerous for human consumption. Water for livestock does not have to be as high in quality as water for human consumption, but it must not cause sickness or death in animals. Excessive concentrations of minerals in irrigation water have adverse osmotic effects on plants, and irrigation water also must be free of phytotoxic substances. Water for industry also must be of adequate quality. Extremely high quality water may be needed for some processes, and even boiler feed water must not contain excessive suspended solids or a high concentration of carbonate hardness. Solids can settle in plumbing systems and calcium carbonate can precipitate to form scale. Acidic waters can cause severe corrosion of metal objects with which it comes in contact.

Water quality effects the survival and growth of plants and animals in aquatic ecosystems. Water often deteriorates in quality as a result of its use by humans, and much of the water used for domestic, industrial, or agricultural purpose is discharged into natural bodies of water. In most countries, attempts are made to maintain the quality of natural waters within limits suitable for fish and other aquatic life. Water quality standards may be recommended for natural bodies of water, and effluents have to meet certain requirements to prevent pollution and adverse effects on the flora and fauna.

Aquaculture, the farming of aquatic plants and animals, now supplies about half of the world's fisheries production for human consumption, because the capture fisheries have been exploited to their sustainable limit. Water quality is a critical issue in cultivation of aquatic organisms.

Factors Controlling Water Quality

Pure water that contains only hydrogen and oxygen is rarely found in nature. Rainwater contains dissolved gases and traces of mineral and organic substances originating from gases, dust, and other substances in the atmosphere. When raindrops fall on the land, their impact dislodges soil particles and flowing water erodes and suspends soil particles. Water also dissolves mineral and organic matter from the soil and underlying formations. There is a continuous exchange of gases between water and air, and when water stands in contact with sediment in the bottoms of water bodies, there is an exchange of substances until equilibrium is reached. Biological activity has a tremendous effect upon pH and concentrations of dissolved gases, nutrients, and organic matter. In general, natural bodies of water approach an equilibrium state with regard to water quality that depends upon climatic, hydrologic, geologic, and biologic factors.

Human activities also strongly influence water quality, and they can upset the natural status quo. The most common human influence for many years was the introduction of disease organisms via disposal of human wastes into water supplies. Until the past century, waterborne diseases were a leading cause of sickness and death throughout the world. We have greatly reduced the problems of waterborne diseases in most countries, but because of the growing population and increasing agricultural and industrial effort necessary to support mankind, surface waters and groundwaters are becoming increasingly contaminated. Contaminants include suspended soil particles from erosion that cause turbidity and sedimentation in water bodies; inputs of nutrients that promote eutrophication and depletion of dissolved oxygen; toxic substances such as heavy metals, pesticides, and industrial chemicals; and heated water from cooling of industrial processes.

Purpose of Book

Water quality is a complex topic of importance in many scientific and practical endeavors. It is a key issue in water supply, wastewater treatment, industry, agriculture, aquaculture, aquatic ecology, human and animal health, and many other areas. People in many different occupations need information on water quality. Although the principles of water quality are covered in specialized classes dealing with environmental sciences and engineering, many who need a general understanding of water quality are not exposed to the basic principles of water quality in a formal way. The purpose of this book is to present the basic aspects of water quality with emphasis on physical, chemical, and biological factors controlling the quality of surface waters. However, there will be brief discussions of groundwater and marine water quality as well as water pollution, water treatment, and water quality standards.

It is impossible to provide a meaningful discussion of water quality without considerable use of chemistry and physics. Many water quality books are available, in which the level of chemistry and physics is far above the ability of the average readers to understand. In this book, I have attempted to use only first-year, college-level chemistry and physics in a very basic way. Thus, most of the discussion of water quality hopefully will be understandable even to readers with only rudimentary formal training in chemistry and physics.

Contents

Physical Properties of Water

1

Abstract

This chapter discusses the physical properties of water, but by necessity, it contains an explanation of both atmospheric and hydrostatic pressure. The water molecule is dipolar—there are negatively and positively charged sites on opposite sides of the molecule. As a result, water molecules form hydrogen bonds with other water molecules. Hydrogen bonds are stronger than van der Waals attractions that occur among molecules, and impart unique physical properties to water. The dipolar molecule and hydrogen bonding greatly influence the density, vapor pressure, freezing point, boiling point, surface tension, phase changes, specific heat, dielectric constant, viscosity, cohesion, adhesion, and capillarity of water. Water has its maximum density at about 4 °C; this results in ice floating and the potential for thermal stratification of natural water bodies. The physical properties of water are of intrinsic interest, but some of these properties influence water quality as will be seen in other chapters.

Keywords

Water molecule • Atmospheric pressure • Physical characteristics • Albedo of water • Thermal stratification

Introduction

Water is a simple molecule with two hydrogen atoms and an oxygen atom. Its molecular weight is 18. Water has a unique feature that causes it to behave differently than other compounds of similar molecular weight. The water molecule is dipolar because it has a negatively charged side and a positively charged one. This polarity leads to water having high freezing and boiling points, large latent heat requirements for phase changes between ice and liquid and between liquid and

C.E. Boyd, *Water Quality*, DOI 10.1007/978-3-319-17446-4_1

vapor, temperature dependent density, a large capacity to hold heat, and good solvent action. These physical properties influence the behavior of water in nature, and the student of water quality should be familiar with them.

Structure of Water Molecule

The water molecule consists of two hydrogen atoms covalently bonded to one oxygen atom (Fig. 1.1). The angle formed by lines through the centers of the hydrogen nuclei and the oxygen nucleus in water is 105 °. The distance between hydrogen and oxygen nuclei is 0.96×10^{-8} cm. The oxygen nucleus is heavier than the hydrogen nucleus, so electrons are pulled relatively closer to the oxygen nucleus. This gives the oxygen atom a small negative charge and each of the hydrogen atoms a slight positive charge resulting in a separation of electrical charge on the water molecule or polarity of the molecule (Fig. 1.1).

The molecules of any substance attract each other as well as molecules of other substances through van der Waals forces. Negatively charged electrons of one molecule and positively charged nuclei of another molecule attract. This attraction is almost, but not completely, neutralized by repulsion of electrons by electrons and nuclei by nuclei, causing van der Waals forces to be weak. Electrostatic van der

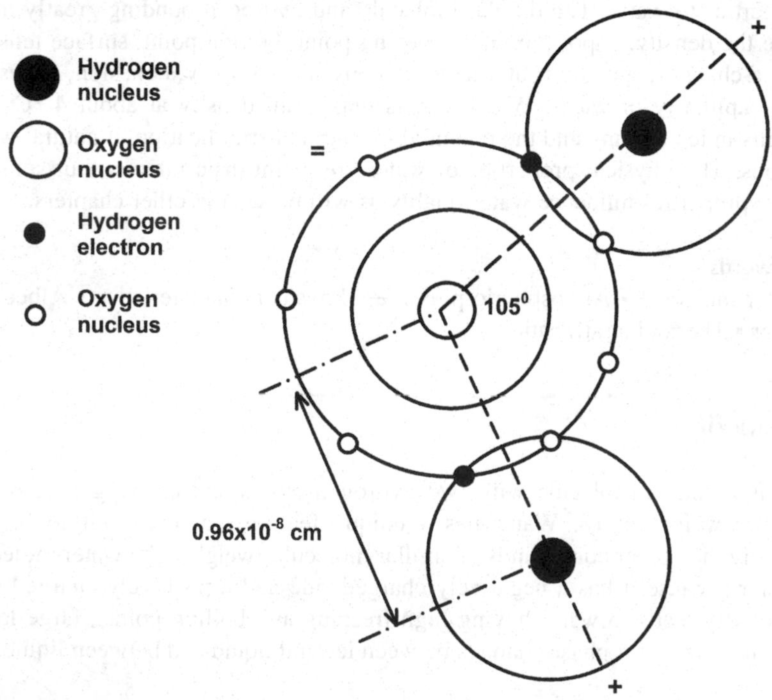

Fig. 1.1 The water molecule

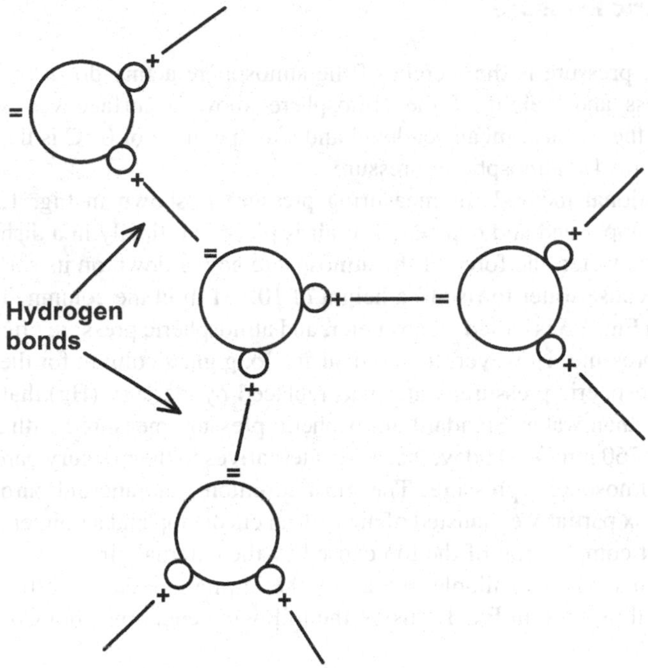

Fig. 1.2 Hydrogen bonds between water molecules

Waals attractions between molecules increase in intensity with increasing molecular weight, because the number of electrons and nuclei increases as molecular weight increases.

The positively charged side of a water molecule attracts the negatively charged side of another to form hydrogen bonds (Fig. 1.2). Hydrogen bonding is illustrated here in two dimensions, but hydrogen bonding actually is three-dimensional. One water molecule can be bonded to another with the axis of attraction extending in any direction. Hydrogen bonds are not as strong as covalent bonds or ionic bonds, but they are much stronger than van der Waals attractions. The degree of molecular attraction in water is much greater than in nonpolar substances. Because of hydrogen bonding, water actually has the structure $(H_2O)_n$ rather than H_2O. The number of associated molecules (n) is greatest in ice and decreases as temperature increases. All hydrogen bonds are broken in vapor, and each molecule exists as a separate entity (H_2O).

Properties of Water

The physical properties of water have a profound influence on the behavior of water in nature. In explaining those properties, reference often will be made to standard atmospheric pressure. Thus, it is necessary to discuss the concept of standard atmospheric pressure before proceeding.

Atmospheric Pressure

Atmospheric pressure is the weight of the atmosphere acting down on a surface. The thickness and weight of the atmosphere above a surface varies with the elevation of the surface; mean sea level and a temperature of 0 °C is the reference point (standard) for atmospheric pressure.

The traditional method for measuring pressure is shown in Fig. 1.3. A tube closed at its upper end and evacuated of air is placed vertically in a dish of liquid. If the liquid is water, the force of the atmosphere acting down on its surface at sea level would cause water to rise to a height of 10.331 m in the column. The device illustrated in Fig. 1.3 is called a barometer, and atmospheric pressure often is called barometric pressure. However, to avoid such a long glass column for the measurement of atmospheric pressure, water was replaced by mercury (Hg) that is 13.594 times denser than water. Standard atmospheric pressure measured with a mercury barometer is 760 mmHg. Today, there are alternatives to the mercury barometer for measuring atmospheric pressure. The most common—the aneroid barometer—is basically a box partially exhausted of air with an elastic top and a pointer to indicate the degree of compression of the top caused by the external air.

If a barometer is unavailable, the atmospheric pressure can be estimated from elevation as illustrated in Ex. 1.1 using the following equation from Colt (2012):

$$\log 10\ \mathrm{BP} = 2.880814 - \frac{\mathrm{h}}{19,748.2} \tag{1.1}$$

where BP = barometric pressure (mmHg) and h = elevation (m).

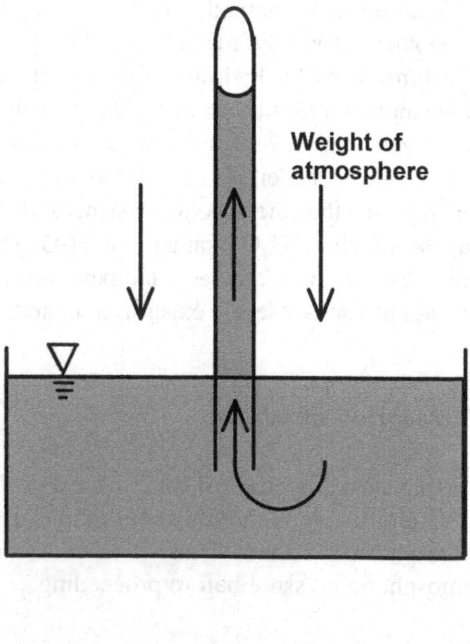

Fig. 1.3 A schematic view of a traditional mercury barometer

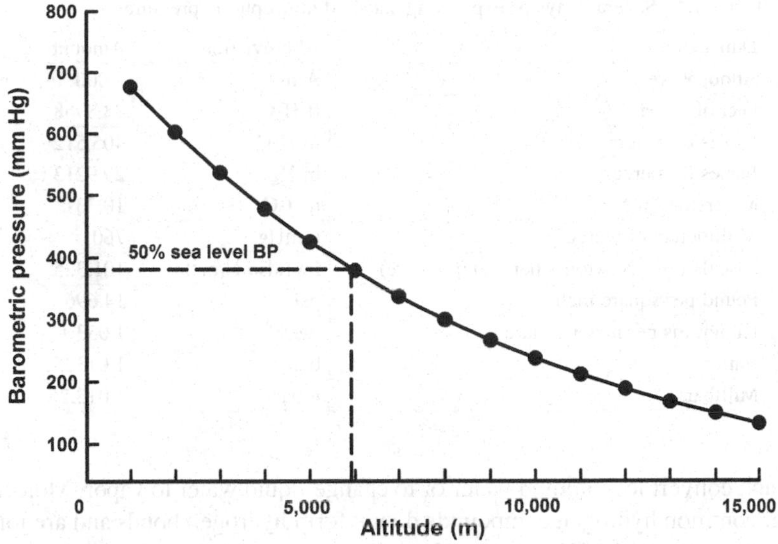

Fig. 1.4 Standard barometric pressures at different elevations

Ex. 1.1: *The atmospheric pressure at 500 m will be estimated using (1.1).*

Solution:

$$log_{10} BP = 2.880814 - \frac{500 \text{ m}}{19,748.2}$$

$$= 2.880814 - 0.025319 = 2.855495$$

$$BP = antilog\ 2.855495 = 717 \text{ mm Hg}.$$

The effect of increasing altitude on barometric pressure is illustrated (Fig. 1.4) with estimated values calculated with (1.1). The atmospheric pressure is roughly 50 % of that at sea level at an elevation of 5,945 m.

Atmospheric pressure may be expressed in a variety of dimensions. The most common are presented (Table 1.1).

Thermal Characteristics

Water is a liquid between 0 and 100 °C at standard atmospheric pressure. Freezing and boiling points of water, 0 and 100 °C, respectively, are much higher than those of other hydrogen compounds of low molecular weight, e.g., methane (CH_4), ammonia (NH_3), phosgene (PH_3), and hydrogen sulfide (H_2S), that are gases at ordinary temperatures on the earth's surface. The aberrant behavior of water results from hydrogen bonding. Considerable thermal energy is required to break hydrogen

Table 1.1 Several ways of expressing standard atmospheric pressure

Dimension	Abbreviation	Amount
Atmosphere	Atm	1.000
Feet of water	ft H_2O	33.8958
Inches of water	in H_2O	405.512
Inches of mercury	in Hg	29.9213
Meters of water	mH_2O	10.331
Millimeters of mercury	mmHg	760
Pascals (also Newton's per meter square)	Pa (also N/m^2)	101.325
Pound per square inch	psi	14.696
Kilograms per meter square	kg/m^2	1.033
Bars	bar	1.01325
Millibars	mbar	1,013.25

bonds and convert ice to liquid water or to change liquid water to vapor. Molecules of other common hydrogen compounds do not form hydrogen bonds and are joined only by weaker van der Waals attractions.

Depending on its internal energy content, water exists in solid, liquid, or gaseous phase. In ice, all hydrogen atoms are bonded; in liquid phase, a portion of the hydrogen atoms is bonded; in vapor, there are no hydrogen bonds. An increase in the internal energy content of water agitates its molecules, causes hydrogen bonds to stretch and break, and temperature to rise. The opposite effect occurs when the energy content of water declines.

The amount of energy (or heat) in calories (cal) required to raise the temperature of a substance by 1 °C is the specific heat. The specific heat of ice is about 0.50 cal/g/°C, the specific heat of water is 1.0 cal/g/°C, and steam has a specific heat of 0.48 cal/g/°C. Water has a high specific heat compared with most other substances. Thus, the specific heat of many common substances in nature depends on their water content. Dry mineral soils have specific heats of 0.18–0.24 cal/g/°C, but most soils may have specific heats of 0.5 cal/g/°C or more. Wood typically has a specific heat of 0.60 cal/g/°C or more, but the specific heat of succulent leaves is around 0.9 cal/g/°C.

Water freezes when its energy content declines and molecular motion slows so that hydrogen bonds form to produce ice. Ice melts when its energy content rises and molecular motion increases and too few hydrogen bonds are present to maintain the crystalline structure of ice. If the temperature of water falls to 0 °C, 80 cal of heat must be removed from each gram of water to cause it to freeze with no change in temperature. It follows that to melt 1 g of ice at 0 °C with no change in temperature requires 80 cal. The energy necessary to cause the phase change between liquid water and ice (80 cal/g) is called the latent heat of fusion.

Water changes from liquid to vapor when it attains enough internal energy and molecular motion to break all hydrogen bonds. Water vapor condenses to form liquid water when it loses energy and molecular motion decreases to permit formation of hydrogen bonds. The amount of energy necessary to cause the liquid to vapor phase change is 540 cal/g. This quantity of energy is termed the latent heat of vaporization.

The total amount of heat required to convert ice to liquid water and finally to water vapor can be calculated as illustrated in Ex. 1.2.

Ex. 1.2: *The heat required to convert 10 g of ice at −15 °C to water vapor at 100 °C will be calculated.*

Solution:
To raise temperature to 0 °C: 10 g × 15 °C × 0.50 cal/g/°C = 75 cal.

(i) *To melt ice at 0 °C: 10 g × 80 cal/g = 800 cal.*
(ii) *To raise temperature to 100 °C: 10 g × 100 °C × 1.0 cal/g/°C = 1,000 cal.*
(iii) *To convert water to vapor at 100 °C: 10 g × 540 cal/g = 5,400 cal.*
(iv) *The required heat is: 75 + 800 + 1,000 + 5,400 cal = 7,275 cal.*

Ice also can go from a solid to a vapor without going through the liquid phase. This is what happens when wet clothes suspended on a line outdoors in freezing weather dry. The process is called sublimation, and the latent heat of sublimation is 680 cal/g. Of course, water vapor also can change from vapor to ice without going through the liquid phase. This process is known as deposition for which the latent heat also is 680 cal/g.

Vapor Pressure

Vapor pressure is the pressure exerted by a substance in equilibrium with its own vapor. In a beaker of water inside a sealed chamber containing dry air, molecules from the water in the beaker enter the air until equilibrium is reached. At equilibrium, the same number of water molecules enters the air from the water as enters the water from the air, and there is no net movement of water molecules. The pressure of the water molecules in the air (water vapor) acting down on the water surface in the beaker is the vapor pressure of water. Vapor pressure increases as temperature rises (Table 1.2). The vapor pressure of pure water reaches atmospheric pressure at 100 °C. Bubbles then form and push back the atmosphere to break the water surface. This is the boiling point. Atmospheric pressure varies with altitude and weather conditions, so the boiling point of water is not always exactly 100 °C. At low enough pressure, water will boil at room temperature.

Density

Ice molecules form a tetrahedral lattice through hydrogen bonding. Spacing of molecules in the lattice creates voids, so a volume of ice weighs less than the same volume of liquid water. Because ice is lighter (density = 0.917 g/cm^3) than water (density = 1 g/cm^3), it floats.

Table 1.2 Vapor pressure of water in millimeters of mercury (mmHg) at different temperatures (°C)

°C	mmHg	°C	mmHg	°C	mmHg
0	4.579	35	42.175	70	233.7
5	6.543	40	55.324	75	289.1
10	9.209	45	71.88	80	355.1
15	12.788	50	92.51	85	433.6
20	17.535	55	118.04	90	525.8
25	23.756	60	149.38	95	633.9
30	31.824	65	187.54	100	760.0

Table 1.3 Density of freshwater (g/m^3) at different temperatures between 0 and 40 °C

°C	g/cm^3	°C	g/cm^3	°C	g/cm^3
0	0.99984	14	0.99925	28	0.99624
1	0.99990	15	0.99910	29	0.99595
2	0.99994	16	0.99895	30	0.99565
3	0.99997	17	0.99878	31	0.99534
4	0.99998	18	0.99860	32	0.99503
5	0.99997	19	0.99841	33	0.99471
6	0.99994	20	0.99821	34	0.99437
7	0.99990	21	0.99800	35	0.99403
8	0.99985	22	0.99777	36	0.99309
9	9.99978	23	0.99754	37	0.99333
10	0.99970	24	0.99730	38	0.99297
11	0.99961	25	0.99705	39	0.99260
12	0.99950	26	0.99678	40	0.99222
13	0.99938	27	0.99652		

The density of liquid water increases as temperature rises until its maximum density of 1.000 g/cm^3 is attained at 3.98 °C. Further warming decreases the density of water. Two processes influence density as water warms above 0 °C. Remnants of the crystalline lattice of ice break up and increase density, and bonds stretch to decrease density. From 0 to 3.98 °C, destruction of the remnants of the lattice has the greatest influence on density, but further warming causes density to decrease through bond stretching. Densities of water at different temperatures are given in Table 1.3. The change in density caused by temperature does not greatly affect the weight of a unit volume of water: a cubic meter of water at 10 °C will weigh 999.70 kg; the same volume weighs 995.65 kg at 30 °C. The effect of temperature on density of water, however, allows water to stratify thermally as will be discussed later.

The dissolved mineral content or salinity of water also influences density (Table 1.4). Water at 20 °C with 30 g/L salinity has a density of 1.0210 g/cm^3 as compared to 0.99821 g/cm^3 for freshwater at the same temperature. Thus, water at 20 °C with 30 g/L salinity weighs 22.79 kg more per cubic meter than freshwater at 20 °C. The more important phenomenon in nature, however, is that small, salinity-related weight differences can lead to stratification of water bodies.

Table 1.4 The density of water (g/cm^3) of different salinities at selected temperatures between 0 and 40 °C

	Salinity (g/L)				
°C	0	10	20	30	40
0	0.99984	1.0080	1.0160	1.0241	1.0321
5	0.99997	1.0079	1.0158	1.0237	1.0316
10	0.99970	1.0075	1.0153	1.0231	1.0309
15	0.99910	1.0068	1.0144	1.0221	1.0298
20	0.99821	1.0058	1.0134	1.0210	1.0286
25	0.99705	1.0046	1.0121	1.0196	1.0271
30	0.99565	1.0031	1.0105	1.0180	1.0255
35	0.99403	1.0014	1.0088	1.0162	1.0237
40	0.99222	0.9996	1.0069	1.0143	1.0217

Surface Phenomena

The rise of water in small-bore tubes or soil pores is called capillary action. To explain capillary action, one must consider cohesion, adhesion, and surface tension. Cohesive forces result from attraction between like molecules. Water molecules are cohesive because they form hydrogen bonds with each other. Adhesive forces result from attraction between unlike molecules. Water adheres to a solid surface if molecules of the solid surface form hydrogen bonds with water. Such a surface is called hydrophilic because it wets easily. Adhesive forces between water and the solid surface are greater than cohesive forces among water molecules. Water will bead on a hydrophobic surface and run off, for cohesion is stronger than adhesion. For example, raw wood wets readily, but a coat of paint causes wood to shed water.

The net cohesive force on molecules within a mass of water is zero, but cohesive forces cannot act above the surface. Molecules of the surface layer are subjected to an inward cohesive force from molecules below the surface. Surface molecules act as a skin and cause the phenomenon known as surface tension. Surface tension is strong enough to permit certain insects and spiders to walk over the surface of water, and to allow needles and razor blades to float. The strength of the surface film decreases with increasing temperature, and increases when electrolytes are added to water. Soap and most dissolved organic substances decrease surface tension.

Water will rise in small-bore glass tubes when inserted vertically into a beaker of water (Fig. 1.5). This phenomenon, called capillary action, is the combined effects of surface tension, adhesion, and cohesion. Water adheres to the walls of the tube and spreads upward as much as possible. Water moving up the wall is attached to the surface film, and molecules in the surface film are joined by cohesion to molecules below. As adhesion drags the surface film upward, it pulls water up the tube against the force of gravity. The column of water is under tension because water pressure is less than atmospheric pressure. Capillary rise is inversely proportional to tube diameter.

Capillary action occurs in soils and other porous media. Space exists among soil particles because they do not fit together perfectly. Pore space in soil is

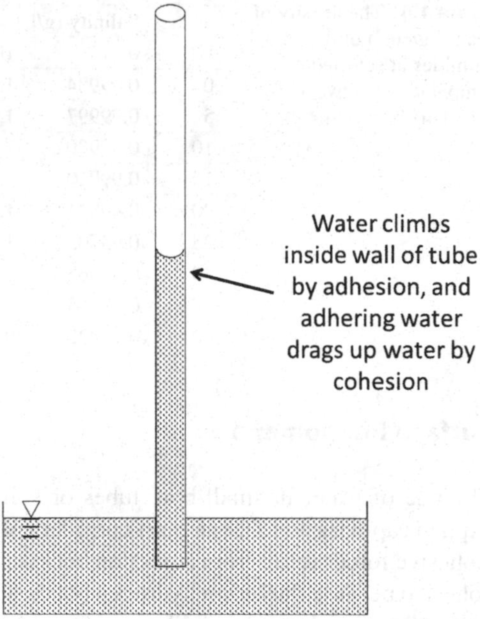
Water climbs
inside wall of tube
by adhesion, and
adhering water
drags up water by
cohesion

Fig. 1.5 Illustration of capillary action

interconnected, and it can function in the same manner as a thin glass tube. A good example is the capillary rise in water above the water table. In fine-grained soils, groundwater may rise a meter or more above the top of the water table creating a capillary fringe from which deep-rooted ponds may obtain water.

Viscosity

Water, like other fluids, has viscosity or internal resistance to flow. Viscosity represents the capacity of a fluid to convert kinetic energy to heat energy. Viscosity results from cohesion between fluid particles, interchange of particles between layers of different velocities, and friction between the fluid and the walls of the conduit. In laminar flow, water moves in layers with little exchange of molecules among layers. During laminar flow in a pipe, the water molecules in the layer in contact with the pipe often adhere to the wall and do not flow. There is friction between the pipe wall and those molecules that do flow. The influence of the pipe wall on the flow of molecules declines with distance from the wall. Nevertheless, there is still friction between the layers of the flowing water. When flow becomes turbulent, the molecules no longer flow in layers and the movement of water becomes more complex.

Water has a greater viscosity than would be expected for its molecular weight because of hydrogen bonding. Viscosity often is reported popularly in a unit of force called the centipoise; water with a viscosity of 0.89 cP at 25 °C is considerably more viscous than acetone (0.306 cP), methanol (0.544 cP), and many other low-molecular weight liquids. More complex liquids have greater viscosities, e.g., olive oil (81 cP), SAE40 motor oil (319 cP), and corn syrup (1,381 cP). In scientific applications viscosity usually is reported in Newton second per square meter (N s/m^2); 1 cP = 0.001 N s/m^2. The viscosity of water increases with decreasing temperature (0.00178 N s/m^2 at 0 °C) and decreases with rising temperature (0.00055 N s/m^2 at 50 °C). Water flow in pipes and channels, seepage through porous media, and capillary rise in soil are favored slightly by warmth, because viscous shear losses decrease with decreasing viscosity.

Elasticity and Compressibility

Water, like other fluids, has little or no elasticity of form and conforms to the shape of its container, i.e., it does not resist shearing forces. Unless completely confined, a liquid has a free surface that is always horizontal except at the edges.

The ideal liquid is incompressible, and water often is said to be incompressible. But, this statement is not actually correct—water is slightly compressible. Its coefficient of compressibility at 20 °C is 4.59×10^{-10} Pa^{-1}. At 4,000 m depth in the ocean, water would be compressed by about 1.8 % as compared to water at the surface.

Water Pressure

The pressure of water at any particular depth below the surface water bodies is equal to the weight of the water column above that depth (Fig. 1.6). The water column of height h pressing down on a small area ΔA has a volume of hΔA. The weight of water or force (F) is

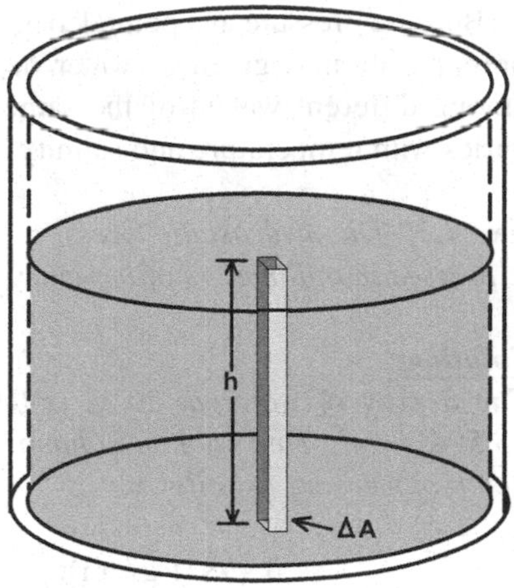

Fig. 1.6 Pressure of water on a surface (ΔA) beneath water of depth (h)

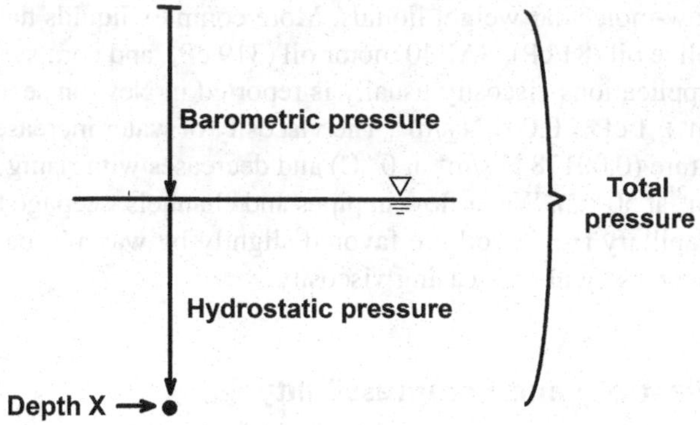

Fig. 1.7 Total pressure at a point below a water surface

$$F = \gamma h \Delta A \tag{1.2}$$

where γ = weight of water per unit volume.

Pressure (P) is a force acting over a unit area

$$P = \frac{F}{\Delta A}. \tag{1.3}$$

Thus, the pressure on the area denoted as ΔA (Fig. 1.6) may be expressed as

$$P = \frac{\gamma h \Delta A}{\Delta A} = \gamma h. \tag{1.4}$$

The pressure is for the water only, i.e., the hydrostatic pressure, and it is the product of the height of the water column and the unit weight of water (Ex. 1.3). To obtain the absolute pressure, atmospheric pressure must be added to the hydrostatic pressure (Fig. 1.7) by multiplying by the area over which the pressure is acting. This force is always normal to the surface.

Because pressure at a point depends mainly upon depth of water above the point, pressure often is given as water depth. The actual pressure could vary slightly among different waters of the same depth because the specific weight of water varies with temperature and salinity.

Ex. 1.3: *The hydrostatic pressure of a 1-m depth of water at 20 °C will be expressed in millimeters of mercury.*

Solution:

The density of water at 20 °C is 0.99821 g/cm³ and the density of mercury is 13.594 g/cm³. Thus, a 1 m column of water may be converted to an equivalent depth of mercury as follows:

$$\frac{0.99821\ \text{g/cm}^3}{13.594\ \text{g/cm}^3} \times 1\ \text{m} = 0.0734\ \text{m}\ \textit{or}\ 73.4\ \text{mm}.$$

Thus, freshwater at 20 °C has a hydrostatic pressure of 73.4 mmHg/m.

When barometric pressure is 760 mmHg, a point at 1 m depth in a water body has a total pressure of 760 mmHg + 73.4 mmHg = 833.4 mmHg.

Pressure also can result from elevation of water above a reference plane, water velocity, or pressure applied by a pump. In hydrology and engineering applications, the term head expresses the energy of water at one point relative to another point or reference plane. Head often is expressed as depth of water.

Dielectric Constant

In the electric field of a condenser, water molecules orient themselves by pointing their positive ends toward the negative plate and their negative ends toward the positive plate (Fig. 1.8). The orientation of molecules neutralizes part of the charge applied to the condenser plates. The voltage required to produce a given voltage on condenser plates is a measure of the dielectric constant of the substance surrounding the condenser plates. In a vacuum, 1 V applied from a battery to the plates produces a voltage of 1 V on the condenser plates. In contrast, the dielectric constants of air and water are 1.0006 and 81 V, respectively.

A crystal of salt maintains its structure in air because of electrical attraction between anions and cations, e.g., $Na^+ + Cl^- = NaCl$. Salt dissolves readily in water because the attractive forces between anions and cations are 81 times less in water than in air. Water insulates ions from each other because water molecules are attracted to dissolved ions. Each anion attracts the positive sides of several water molecules, and each cation attracts the negative sides of other water molecules (Fig. 1.9). Ions can each attract several water molecules because ionic charges are much stronger than the charges on opposite sides of a water molecule. Water is said

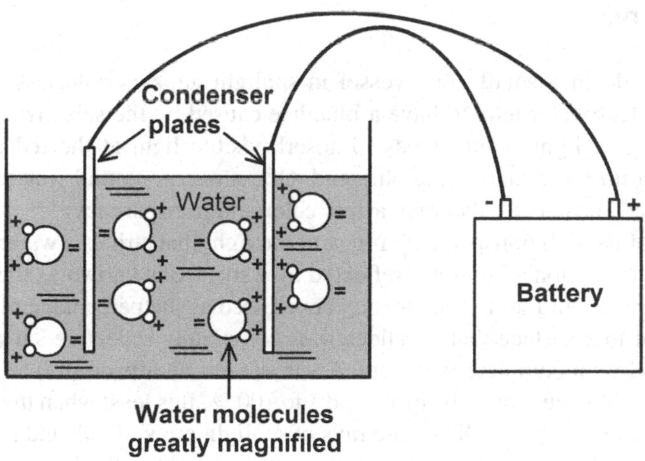

Fig. 1.8 Orientation of water molecules in an electric field

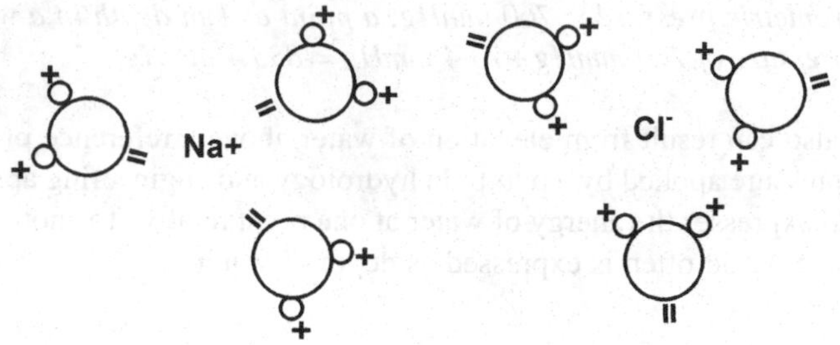

Fig. 1.9 Hydration of dissolved ions by water molecules

to hydrate ions, and hydration neutralizes charges on ions just as the water molecules neutralize the charge on a condenser plate. Because of its high dielectric constant and polar properties, water is an excellent solvent for most inorganic and many organic substances.

Conductivity

Conductivity is the ability of a substance to convey an electrical current. In water, electrical current is conveyed by the dissolved ions. Pure water contains only a low concentration of hydrogen and hydroxyl ions; it is a poor conductor. Natural waters, however, contain a greater concentration of dissolved ions and are better conductors than pure water. Conductivity increases in direct proportion to dissolved ion concentrations, and the conductivity of water is an important water quality variable that will be discussed in Chap. 4.

Transparency

Pure water held in a small glass vessel in sunlight appears colorless, but larger amounts of clear water tend to have a blue hue caused by the selective absorption and scattering of light. Water tends to absorb visible light at the red end of the visible spectrum more than at the blue end. Of course, in natural water, dissolved and suspended substances also may affect color and transparency.

Because of its high transparency, much of the light that strikes a water surface is absorbed. The fraction of sunlight reflected by a surface is known as the albedo—from albus (white) in Latin. The albedo is expressed as the percentage of incoming light incident to a surface that is reflected. A completely reflective surface has an albedo of 100 %; a completely absorptive surface has an albedo of 0 %.

The albedo of water varies from around 1 to 100 %. It is least when the surface of a very clear water body is still and the rays of sunlight are vertical, and it is greatest when the sun is below the horizon. A mirror reflects light at the same angle as the incident light as shown in Fig. 1.10. This is known as specular reflection, and in

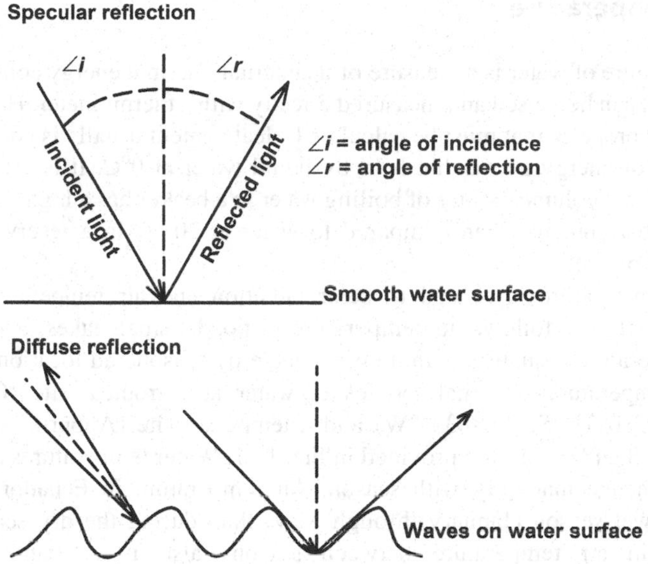

Fig. 1.10 Illustration of specular reflection from a still water surface (*upper*) and diffuse reflection from a water surface with waves (*lower*)

nature, the angle of incidence of the sun's rays will vary with time of day and with the progression of the seasons. Seldom will the angle of incidence be exactly perpendicular with the water surface. However, water surfaces seldom are completely smooth—there usually are waves. This results in the angle of incidence varying and the angle of the reflected light also varying (Fig. 1.10). This type of reflection results in light being reflected at many different angles and is known as diffuse reflection.

In spite of the various factors affecting reflectivity, when the angle of incidence is 60 ° or less, e.g., the sun's rays at 30 ° or more above the horizon, the albedo of water usually is less than 10 %. When the angle of incidence is 0 %—the sun's rays are vertical—the albedo of clear water usually would be around 1–3 %. According to Cogley (1979), depending upon the method of calculation, the average annual albedos for open water surfaces range from 4.8 to 6.5 % at the equator and from 11.5 to 12.0 % at 60 ° latitude. On a monthly basis, albedos varied from 4.5 % in March and September to 5.0 % in June and December at the equator. The corresponding values at 60 ° latitude were 7.0 % in June and 54.2 % in December.

Water absorbs an average of about 90 % of the radiation that strikes its surface. Thus, water with its high specific heat is the major means of detaining incoming solar energy at the earth's surface. Of course, water bodies like land masses re-radiate long wave radiation continuously, and the input of solar energy is balanced by outgoing radiation.

The light that penetrates the water surface is absorbed and scattered as it passes through the water column. This phenomenon has many implications in the study of water quality, and will be discussed several places in this book—especially in Chap. 5.

Water Temperature

The temperature of water is a measure of its internal, thermal energy content. It is a property that can be sensed and measured directly with a thermometer. Heat content is a capacity property that must be calculated. Heat content usually is considered as the amount of energy above that held by liquid water at 0°C. It is a function of temperature and volume. A liter of boiling water in a beaker has a high temperature but small heat content when compared to water at 20 °C in a reservoir of five million m^3 volume.

Water temperature is related to solar radiation and air temperature. Water temperature closely follows air temperature in ponds, small lakes, and streams. Water temperatures usually are quite predictable by season and location. Average monthly temperatures in small bodies of water at a tropical site (Guayaquil, Ecuador at 2.1833 °S; 79.8833 °W) and a temperate site (Auburn, Alabama at 32.5977 °N; 85.4808 °W) are provided in Fig. 1.11. Water temperatures at Auburn, Alabama, change markedly with season. Air temperature in Ecuador is higher during the wet season (January through May) than during the dry season. This difference in air temperature between seasons also is reflected in water temperatures. Thus, water temperatures vary with season even in the tropics, but the variation is less than in the temperate zone. Air temperatures at a given locality may deviate from normal for a particular period causing deviation in water temperatures.

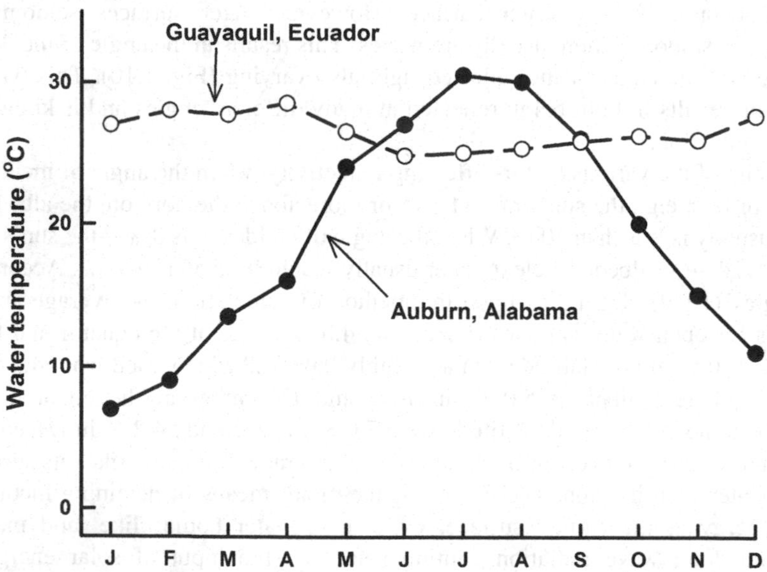

Fig. 1.11 Average monthly water temperatures at Auburn, Alabama (32.5 °N latitude) and Guayaquil, Ecuador (2.1 °S latitude)

Thermal Stratification

Light energy is absorbed exponentially with depth so much heat is absorbed within the upper layer of water. This is particularly true in eutrophic water bodies where high concentrations of dissolved particulate organic matter greatly increase the absorption of energy as compared to less turbid waters. The transfer of heat from upper to lower layers of water depends largely upon mixing by wind.

Density of water is dependent upon water temperature (Table 1.3). Ponds and lakes may stratify thermally, because heat is absorbed more rapidly near the surface making the upper waters warmer and less dense than the deeper waters. Stratification occurs when differences in density between upper and lower strata become so great that the two strata cannot be mixed by wind. At the spring thaw, or at the end of winter in a lake or pond without ice cover, the water column has a relatively uniform temperature. Although heat is absorbed at the surface on sunny days, there is little resistance to mixing by wind and the entire volume circulates and warms. As spring progresses, the upper stratum heats more rapidly than heat is distributed from the upper stratum to the lower stratum by mixing. Waters of the upper stratum become considerably warmer than those of the lower stratum. Winds that often decrease in velocity as weather warms no longer are powerful enough to mix the two strata. The upper stratum is called the epilimnion and the lower stratum the hypolimnion (Fig. 1.12). The stratum between the epilimnion and the hypolimnion has a marked temperature differential. This layer is termed metalimnion or thermocline. In lakes, a thermocline is defined as a layer across which the temperature drops at a rate of at least 1°C/m. The depth of the thermocline below the surface may fluctuate depending upon weather conditions, but most large lakes do not destratify until autumn when air temperatures decline and surface waters cool. The

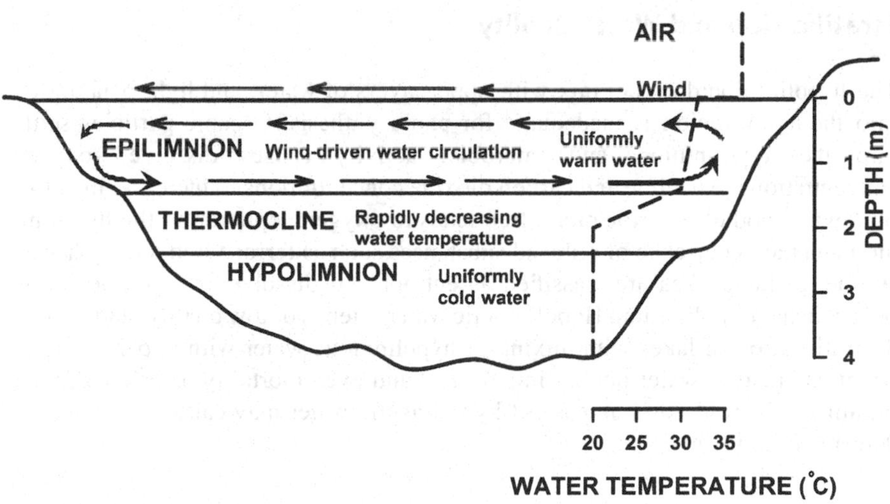

Fig. 1.12 Thermal stratification in a small lake

difference in density between upper and lower strata decreases until wind mixing causes the entire volume of water in a lake or pond to circulate and destratify.

Small water bodies are shallower and less affected by wind than lakes. For example, the ordinary warmwater fish pond seldom has an average depth of more than 2 m and a surface area of more than a few hectares. Marked thermal stratification can develop in fish ponds because of rapid heating of surface waters on calm, sunny days.

The stability of stratification is related to the amount of energy required to mix the entire volume of a water body to a uniform temperature. The greater the energy required, the more stable is stratification. Small water bodies with average depths of 1 m or less and maximum depths of 1–2 m often thermally stratify during daylight hours. They destratify at night when the upper layers cool by conduction. Larger, deeper water bodies stratify for long periods.

Some large deep tropical lakes tend to be permanently stratified, but most occasionally destratify as the result of weather events. Smaller water bodies in the tropics typically destratify during the rainy season and stratify again in the dry season. Events that can lead to sudden destratification of water bodies—in any climate—are strong winds that supply enough energy to cause complete circulation, cold, dense rain falling on the surface that sinks through the warm epilimnion causing upwelling and destratification, and disappearance of heavy plankton blooms allowing heating to a greater depth to cause mixing.

The density of water varies with salinity, so stratification may occur in areas where waters of different salt contents converge. Where rivers discharge into the ocean, freshwater will tend to float above the salt water because it is less dense. Density wedges often extend upstream for considerable distances beneath the freshwater in the final coastal reaches of rivers.

Stratification and Water Quality

The hypolimnion does not mix with upper layers of water, and light penetration into the hypolimnion is inadequate for photosynthesis. Organic particles settle into the hypolimnion, and microbial activity causes dissolved oxygen concentrations to decline and carbon dioxide concentrations to increase. In eutrophic water bodies, stratification often leads to oxygen depletion in the hypolimnion and the occurrence of reduced substances such as ferrous iron and hydrogen sulfide. In fact, lakes are classified as eutrophic if dissolved oxygen depletion occurs in the hypolimnion. Hypolimnetic water often is of low quality, and sudden destratification of lakes with mixing of hypolimnetic water with upper layers of water can lead to water quality impairment and even mortality of fish and other organisms. In reservoirs, release of hypolimnetic water may cause water quality deterioration downstream.

Ice Cover

Ice cover can have an important influence on water quality, because ice cover prevents contact between air and water. Gases cannot be exchanged between an ice-covered water body and air. Moreover, ice cover reduces light penetration into water, and snow cover over ice blocks light penetration entirely. Thus, photosynthetic production of oxygen is greatly reduced or prevented by ice and snow cover. Winter kills of fish because of low dissolved oxygen concentration beneath the ice can occur in eutrophic water bodies.

References

Cogley JG (1979) The albedo of water as a function of latitude. Monthly Weather Rev 107:775–781

Colt J (2012) Dissolved gas concentration in water. Elsevier, Amsterdam

An Overview of Hydrology and Water Supply

2

Abstract

An overview of the hydrological cycle, how this cycle is influenced by local conditions, and the distribution of water within the hydrosphere is presented. Water quality investigations often require an assessment of water quantity. This is because studies of water quality may necessitate information on amounts of dissolved or particulate substances added to or discharged from aquatic systems as a result of inflowing or outflowing water of natural or anthropogenic origin (pollution). A discussion of simple water velocity and volume measurements and their use in the preparation of water budgets is included.

Keywords

Hydrologic cycle • Evaporation • Rainfall • Runoff • Groundwater

Introduction

This book is about water quality, but a fundamental knowledge of hydrology is highly beneficial to students of water quality. For example, the basic composition of water differs among components of the hydrosphere, e.g., rainwater is less concentrated in minerals than runoff, and runoff usually does not contain as much dissolved mineral matter as groundwater. One needs to know about the movement of water through the hydrologic cycle to appreciate why such differences occur. In most discussions, water quality cannot be separated entirely from water quantity. A specific illustration is that the amount of nutrient entering a water body in an industrial effluent depends on the volume or flow rate of the effluent in addition to its nutrient concentration.

This chapter provides a brief discussion of the hydrologic cycle and some simple methods of water quantity measurement.

C.E. Boyd, *Water Quality*, DOI 10.1007/978-3-319-17446-4_2

The Earth's Water

The hydrosphere consists of the gaseous, liquid, and solid water of the earth. The ocean has a volume of around 1.3 billion km^3 and comprises 97.40 % of the earth's water. The remaining water—about 36 million km^3—is contained in several freshwater compartments of the hydrosphere (Table 2.1). The largest proportion of the earth's freshwater is mainly unavailable for direct use by mankind, because it is either bound in ice or occurs as deep groundwater. The amount of freshwater available for human use at any particular time is shallow groundwater and water in lakes, manmade reservoirs, and rivers and other streams—about 4.6 million km^3 or 12 % of the freshwater in the hydrosphere. But, as will be seen later, not all of this water is accessible or sustainable.

The volumes of water in the different compartments of the hydrosphere are not directly related to their comparative importance as sources of water for ecological systems or humans because of differences in renewal times (Table 2.1). For example, the volume of water in the atmosphere is around 14,400 km^3 at a particular time, but the residence time of atmospheric water vapor is only 9 days. Water in the atmosphere can be recycled 1,500,000 times during the 37,000 years necessary to recycle the volume of water held by the ocean. The amount of water cycling through the atmosphere is 21,600,000,000 km^3 in 37,000 years, or roughly 16 times the volume of the ocean.

Table 2.1 Volumes of ocean water and different compartments of freshwater

Compartment	Volume (km^3)	Proportion of total (%)	Renewal time
Oceans	1,348,000,000	97.40	37,000 years
Freshwater			
Polar ice, icebergs, and glaciers	27,818,000	2.01	16,000 years
Groundwater (800–4,000 m depth)	4,447,000	0.32	–
Groundwater (to 800 m depth)	3,551,000	0.26	300 years
Lakes	126,000	0.009	1–100 years
Soil moisture	61,100	0.004	280 days
Atmosphere (water vapor)	14,400	0.001	9 days
Rivers	1,070	0.00008	12–20 days
Plants, animals, humans	1,070	0.00008	–
Hydrated minerals	360	0.00002	–
Total freshwater	(36,020,000)	(2.60)	–

Renewal times are provided for selected compartments (Baumgartner and Reichel 1975; Wetzel 2001)

The Hydrologic Cycle

The hydrologic or water cycle is the processes whereby water moves among the different compartments of the hydrosphere. The familiar and continuous motion of water in the hydrologic cycle is depicted in Fig. 2.1. Water evaporates from oceans, lakes, ponds, and streams and from moist soil; water also is transpired by plants. The sun provides energy to change liquid water to water vapor, so solar radiation is the engine that drives the hydrologic cycle. Water vapor is condensed and returned to the earth's surface as dew, frost, sleet, snow, hail, or rain. A portion of the water reaching the earth's surface evaporates back into the atmosphere almost immediately, but another part runs over the land surface as storm runoff, collects in streams, rivers, and lakes, and finally enters the oceans. Water is continually evaporating as the runoff flows toward the oceans, and it evaporates from the oceans. Some water infiltrates into the ground and becomes soil water or reaches saturated formations (aquifers) to become groundwater. Soil water can be returned to the atmosphere by transpiration of plants. Groundwater seeps into streams, lakes, and oceans, and it may be removed by wells for use by mankind.

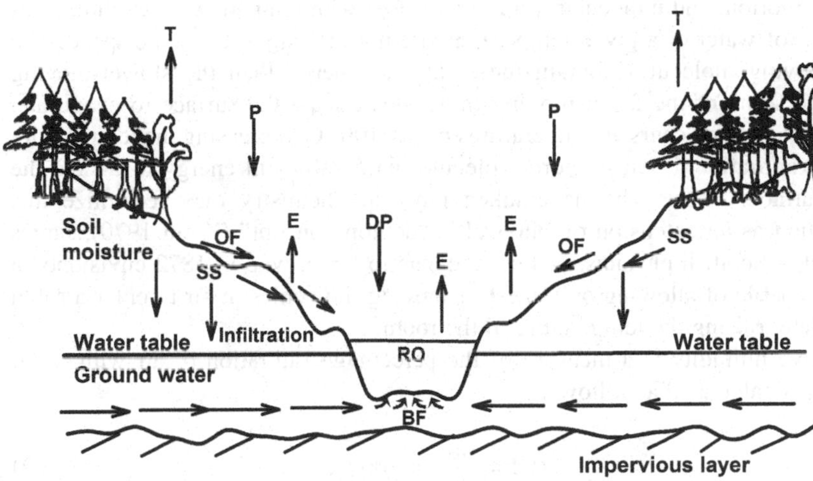

P = Precipitation
DP = Direct precipitation into water body
E = Evaporation } Evapotranspiration
T = Transpiration }
SS = Subsurface stream drainage
OF = Overland flow
BF = Base flow
RO = Runoff (stream flow) = DP+OF+SS+BF
ET = P - RO

Fig. 2.1 The hydrologic cycle or water cycle

Evaporation

The capacity of air to hold water vapor depends on its temperature (Table 1.2). The maximum amount of moisture that air can hold at a given temperature is the saturation vapor pressure. If unsaturated air is brought in contact with a water surface, molecules of water bounce from the surface into the air until the vapor pressure in air is equal to the pressure of water molecules escaping the surface, and evaporation ceases. Water molecules continue to move back and forth across the surface, but there is no net movement in either direction. The driving force for evaporation is the vapor pressure deficit (VPD)

$$VPD = e_s - e_a \quad (2.1)$$

where e_s = saturation vapor pressure and e_a = actual vapor pressure. The greater the vapor pressure deficit, the greater is the potential for evaporation. When $e_s - e_a = 0$, evaporation stops.

Water must change from the liquid state to the vapor state to evaporate, and the energy relationship for evaporation is the same as occurs when a mass of water in a container is heated to 100 °C and additional heat added to cause vaporization. Some may have difficulty reconciling this fact with the observation that water evaporates readily at temperatures well below 100 °C. Molecules in a mass of water are in constant motion, and molecular motion increases with temperature. All molecules in a mass of water of a given temperature are not moving at the same speed. The faster-moving molecules contain more thermal energy than the slower-moving ones, and some of the faster-moving molecules escape the surface to evaporate. Thus, evaporation occurs at temperatures below 100 °C. Increasing the temperature favors evaporation, because more molecules gain sufficient energy to escape the water surface. Those who have taken physical chemistry may recognize this explanation as an extension of Maxwell's "demon concept" (Klein 1970). James Maxwell, a Scottish physicist of thermodynamic fame, who in 1872 envisioned a demon capable of allowing only the fast-moving molecules in air to enter a room and thereby raising the temperature in the room.

Relative humidity is a measure of the percentage saturation of air with water vapor. It is calculated as follows:

$$RH = \frac{e_a}{e_s} \times 100 \quad (2.2)$$

where RH = relative humidity (%), e_a = actual vapor pressure, and e_s = saturation vapor pressure. Saturation vapor pressure (Table 1.2) increases with warmth, so warm air has the capacity to hold more water vapor than does cool air. Air with vapor pressure equal to the saturation vapor pressure has a relative humidity of 100 %. Evaporation rate obviously is favored by low relative humidity.

Wind velocity also influences evaporation. Wind blowing over the surface replaces humid air with drier air to favor evaporation. When there is no air movement over a water surface, evaporation quickly saturates the layer of air above the surface and evaporation ceases or its rate diminishes. The influence of wind on evaporation is especially pronounced in arid regions where air contains little moisture.

Dissolved salts decrease the vapor pressure of water, and evaporation rate under the same conditions is about 5 % greater from a freshwater surface than from an ocean water one. Dissolved salt concentrations do not vary enough among freshwater bodies to influence evaporation rates appreciably. Turbid waters heat faster than clear waters, so an increase in turbidity enhances evaporation. Changes in atmospheric pressure also slightly affect evaporation.

Temperature has the greatest influence on evaporation, and the temperature of a locality is closely related to the amount of incoming solar radiation. The temperature difference between air and water also affects evaporation. A cold water surface cannot generate a high vapor pressure, and cold air does not hold much water vapor. Cold air over cold water, cold air over warm water, and warm air over cold water are not as favorable for evaporation as warm air over warm water. Evaporation rates tend to increase from cooler regions to warmer regions. Relative humidity must be considered, because cold, dry air may take on more moisture than warm, moist air if the greater moisture-holding capacity of the warm air has already been filled.

Evaporation from land areas usually occurs from moist soil surfaces and from the leaves of plants (transpiration). The combined water loss by evaporation and transpiration is called evapotranspiration. Unlike evaporation from the surface of a water body, evapotranspiration often is limited by a lack of moisture.

Measurement

Three types of evaporation instruments were recommended by the World Meteorological Organization (Hounam 1973): the 3,000-cm^2 sunken tank; the 20-m^2 sunken tank; and the Class A evaporation pan. The Class A pan is widely used in the United States. The Class A pan is made of stainless steel and is 120 cm in diameter by 25 cm deep. The pan is mounted on a wooden platform with the pan base 5–10 cm above the ground. The pan is filled within 5 cm of its rim with clear water. Water depth in an evaporation pan is measured with a stilling well and hook gauge or by other devices. The stilling well is a vertical pipe with a small hole in its side to allow the water level in the well to equilibrate with the water level in the pan. Water inside the stilling well has a smooth surface. The top of the stilling well supports the hook gauge. The hook gauge consists of a pointed hook that can be moved up and down with a micrometer. With the point below the water surface, the micrometer is turned to move the hook upward until its point makes a pimple in the surface film. The hook is not permitted to break the surface. The micrometer is read to the nearest tenth of a millimeter, and the value recorded. The procedure is repeated after 24 h. The difference in the two hook gauge readings is the water loss by evaporation. A rain gauge positioned beside the evaporation pan allows correction for rain falling into the pan.

The evaporation rate from a pan is not the same as the evaporation rate from an adjacent body of water. A pan has a smaller volume than a lake or pond and a different exposure to the elements. There is considerable discrepancy between water temperature patterns in an evaporation pan and in a lake or pond. The water surface in an evaporation pan is smaller and smoother than the surface of a lake or pond. Thus, a coefficient must be used to adjust pan evaporation data. Pan coefficients have been developed by estimating evaporation from lakes by mass transfer, energy budget, or water budget techniques and relating these values to evaporation from an adjacent Class A pan. The pan coefficient is

$$C_P = \frac{E_L}{E_P} \tag{2.3}$$

where E_L = evaporation from lake, E_P = evaporation from pan, and C_P = pan coefficient. Coefficients for Class A pans range from 0.6 to 0.8. A factor of 0.7 is recommended for general use in estimating lake evaporation. Boyd (1985) found that a factor of 0.8 was more reliable for ponds.

Potential evapotranspiration (PET) can be measured in lysimeters—small, isolated soil-water systems in which PET can be assessed by the water budget method. Empirical methods such as the Thornthwaite equation or Penman equation may be used to estimate PET based on air temperature and other variables (Yoo and Boyd 1994). Of course, for a watershed where both precipitation and runoff are known, annual evapotranspiration (ET) is

$$\text{ET} = \text{Precipitation} - \text{Runoff}. \tag{2.4}$$

The equation is valid on an annual basis because there is little change in soil water and groundwater storage from a particular month this year to the same month in the next year, and any water falling on the watershed that does not become stream flow must be evaporated or transpired.

Rainfall

Water vapor enters the atmospheric circulation, and remains there until it returns to the earth's surface as precipitation. Rising air is necessary for precipitation. As air rises, the pressure on it decreases, because atmospheric pressure declines with increasing elevation, and the air expands. The temperature of an air parcel, like the temperature of any substance, results from the thermal energy of its molecules. The air parcel contains the same number of molecules after expansion as before, but after expansion, its molecules occupy a larger volume. Thus, the temperature of an air parcel decreases upon expansion.

The rate of cooling (adiabatic lapse rate) is 1 °C/100 m for air at less than 100 % relative humidity. The word adiabatic implies that no energy is gained or lost by the rising air mass. If air rises high enough, it will cool until 100 % relative humidity is reached and moisture will condense to form water droplets. The elevation at which air reaches the dew point and begins to condense moisture is called the lifting condensation level. Once rising air begins to condense moisture, it gains heat, for water vapor must release latent heat to condense. The heat from condensation counteracts some of the cooling from expansion, and the adiabatic lapse rate for rising air with 100 % relative humidity is 0.6 °C/100 m. This is the wet adiabatic lapse rate.

Clouds form when air cools to 100 % humidity and moisture condenses around hydroscopic particles of dust, salts, and acids to form tiny droplets or particles of ice. Droplets grow because they bump together and coalesce when the temperature is above freezing. At freezing temperatures, supercooled water and ice crystals exist in clouds. The vapor pressure of supercooled water is greater than that of ice. Ice particles grow at the expense of water droplets because of their lower vapor pressure. Precipitation occurs when water droplets or ice particles grow too large to remain suspended by air turbulence. The size of droplets or particles at the initiation of precipitation depends on the degree of turbulence. In the tops of some thunderheads, turbulence and freezing temperature can lead to ice particles several centimeters in diameter known as hail. Precipitation may begin its descent as sleet, snow, or hail, only to melt while falling through warm air. Sometimes ice particles may be so large that they reach the ground as hail even in summer.

The major factors causing air to rise are topographic barriers such as mountain ranges, warm air masses rising above heavier, cold air masses, and convectional heating of air masses. These facts may be combined to provide some basic rules for the amounts of precipitation at a location:

- Precipitation is normally greater in areas where air tends to rise than in areas where air tends to fall.
- Amounts of precipitation generally decrease with increasing latitude.
- The amount of precipitation usually declines from the coast to the interior of a large land mass.
- A warm ocean offshore favors high precipitation while a cold ocean favors low precipitation.

Amounts of annual precipitation vary greatly from location to location, with some arid regions receiving <5 cm/year and some humid regions having >200 cm/year. The world average annual precipitation on the land masses is around 70 cm/year. Precipitation totals vary greatly from year to year, and some months typically have greater precipitation than others.

Measurement

The amount of precipitation falling on the earth's surface at a particular place is determined by capturing precipitation in a container—melting it if it is frozen—and measuring the water depth. The standard US National Weather Service rain gauge consists of a brass bucket or overflow can into which a brass collector tube is placed. A removable, brass collector funnel is mounted on top of the overflow can to direct water into the collector tube. The rain gauge is mounted in a support attached to a wooden or concrete base. Rain is concentrated ten times in the collector tube, because the area of the collector funnel is ten times the area of the collector tube. Concentration of the rainfall facilitates measurement with a calibrated dipstick on which the wetted distance indicates rainfall depth. If rainfall exceeds the capacity of the collector tube, it overflows into the overflow bucket. Water from the overflow can is poured into the collector tube for measurement. Several other types of rain gauges including recording rain gauges are available.

Soil Water

Some of the water falling on the ground enters the soil. This water enters pore spaces of soil or is adsorbed to the surface of soil particles. A soil completely saturated with water is said to be at its maximum retentive capacity. Water will drain from the saturated soil by gravity, and the remaining water is soil moisture. At maximum soil moisture content, a soil is said to be at field capacity. Soil water can be removed by plants and lost to the air through transpiration by plants or evaporation from soil surfaces. When the water content of a soil drops so low that plants remain wilted at night, the soil is at its permanent wilting percentage. The water remaining at the permanent wilting percentage is held in very small pores and around the particles. The moisture status at this point is known as the hydroscopic coefficient. Water remaining at the hydroscopic coefficient is tightly attached to colloidal soil particles and not biologically available.

The maximum retentive capacity of soil depends on the volume of pore space alone. Fine-textured soils have higher maximum retentive capacities than coarse-textured soils. The proportion of biologically available water decreases in fine-textured soils, but they contain more available water than coarse-textured ones because they have greater pore space.

Groundwater

At some depth beneath most places on the land, soils and rocks are saturated with water that has percolated down from the surface. These saturated geological formations are called aquifers, and the top of the saturated formation is known as the water table. There are two kinds of aquifers, confined and unconfined (Fig. 2.2). The top of an unconfined aquifer is open to the atmosphere through voids in the rocks and soils above; an unconfined aquifer is known as a water table aquifer. Water in an unpumped well in an unconfined aquifer stands at the level of the water table.

In a confined aquifer, water is trapped between two impervious layers of rock. The land area between outcrops of the two confining strata of a confined aquifer is the recharge area, because rain falling here can infiltrate into the confined aquifer.

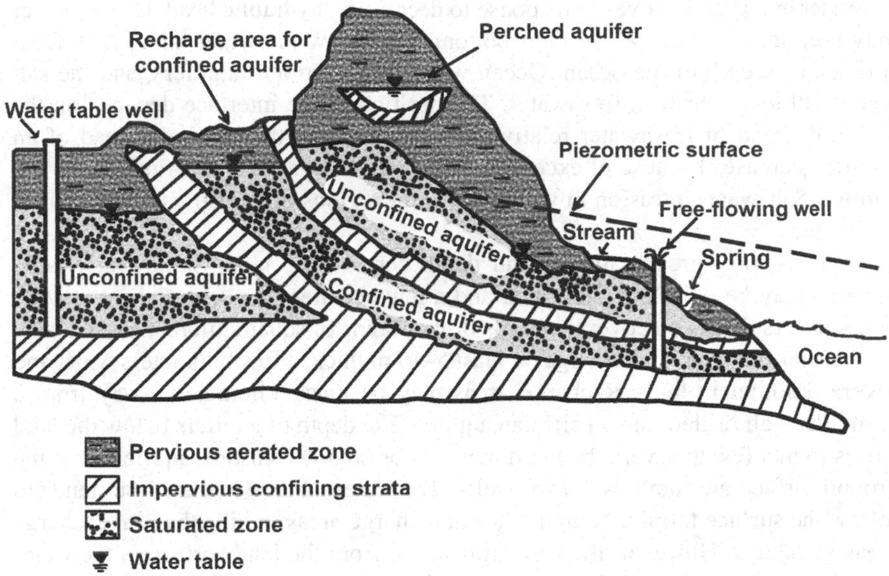

Fig. 2.2 A cross-sectional area of the earth's surface illustrating different kinds of aquifers

Water at any point in a confined aquifer is under pressure (Fig. 2.2) because the confining strata act like a pipe with one end higher than the other. Water pressure is a function of elevation difference between a point in the aquifer and the level of the water table beneath the recharge area. Water often stands above the upper confining stratum when a well is drilled through the upper confining stratum into a confined aquifer. This phenomenon results from the bore hole of the well releasing pressure on the aquifer allowing the water to expand. This is a case where the slight compressibility of water plays an important role.

The level at which water stands in an unpumped well is the piezometric level. Piezometric levels mapped over the entire aquifer form a piezometric surface. Water rises above the land surface from a well drilled into a confined aquifer when the piezometric level is above the land surface. Confined aquifers often are called artesian aquifers, and free-flowing wells frequently are said to exhibit artesian flow. The piezometric surface slopes downward, and water in a well drilled into the aquifer does not rise to the same elevation as that of the water table in the recharge area between the two confining layers (Fig. 2.2). This results primarily from reduction in hydraulic head caused by friction as water infiltrates through the aquifer. A well drilled from the recharge area into a confined aquifer will behave like a water table well.

As illustrated in Fig. 2.2, several water tables may occur beneath a particular place on the land surface. Small perched aquifers are common on hills where water stands above a small hardpan or impervious layer. Water may leak from an artesian aquifer into aquifers above or below. One or more artesian aquifers may occur below the same point, or artesian aquifers may be absent.

Water in aquifers moves in response to decreasing hydraulic head. Groundwater may seep into streams where their bottoms cut below the water table, flow from springs, or seep into the ocean. Ocean water may seep into aquifers, and the salt water will interface with freshwater. The position of the interface depends on the hydraulic head of freshwater relative to salt water. As the hydraulic head of an aquifer decreases because of excessive pumping, salt water moves further into the aquifer. Salt water intrusion in wells of coastal regions often causes water quality problems.

Aquifers vary greatly in size and depth beneath the land. Some water table aquifers may be only 2 or 3 m thick and a few hectares in extent. Others may be many meters thick and cover hundreds or thousands of square kilometers. Artesian aquifers normally are rather large, at least 5–10 m thick, a few kilometers wide, and several kilometers long. Recharge areas may be many kilometers away from a particular well drilled into an artesian aquifer. The depth of aquifers below the land varies from a few to several hundred meters. For obvious reasons, aquifers near the ground surface are more useful for wells. The water table is not level but tends to follow the surface terrain, being higher in recharge areas in hills than in discharge areas in valleys. However, the vertical distance from the land surface to the water table is often less under valleys than under hills. Water table depth also changes in response to rainfall; it declines during prolonged periods of dry weather and rises following heavy rains.

Runoff

Runoff from a watershed or other catchment usually exits as stream flow that consists of both storm flow (overland flow) and subsurface sources of water to the stream. Storm flow is the portion of rain falling on a watershed that is not retained on or beneath the watershed surface or lost from the watershed by evapotranspiration, It is the water that is ultimately transported from the watershed to streams of other water bodies by flow over the land surface after rainstorms.

Storm flow begins after there has been sufficient rainfall to exceed the capacity of watershed surfaces to detain water by absorption or in depressions, and the rate of rainfall exceeds the rate of infiltration of water into the soil. Features of the watershed, duration and intensity of rainfall, season, and climate influence the amount of runoff generated by a watershed. Factors favoring large amounts of storm flow are intense rainfall, prolonged rainfall, impervious soil, frozen or moist soil, low air temperature, high proportion of paved surface, steep slope, little surface storage capacity on land surface, sparse vegetative cover, soils with a low moisture-holding capacity, and a shallow water table. Storm flow moves downslope in response to gravity, it erodes the soil creating channels for water flow and ultimately forming a channel that becomes an intermittent or permanent stream.

An ephemeral stream only flows after rains and does not have a well-defined channel. An intermittent stream flows only during wet seasons and after heavy rains and has a well-defined channel. A perennial stream normally flows year-round and has a well-defined channel. Perennial streams receive groundwater inflow when their bottoms are below the water table; this inflow—called base flow—sustains stream flow in dry weather.

Streams in an area form a drainage pattern that can be seen on a topographic map. A tree-shaped pattern, called a dendritic pattern, forms where the land erodes uniformly and streams randomly branch and advance upslope. Where faulting is prevalent, streams follow the faults to form a rectangular pattern. Where the land surface is folded or is a broad, gently sloping plain, streams form a trellis or lattice pattern.

Streams are classified according to the number of tributaries. A stream without tributaries is a first-order stream, and its basin is a first-order basin. Two, first-order streams combine to form a second-order stream. A stream with a second-order branch is a third-order stream, and so on. The stream system that contributes to the discharge at a specified point in a higher-order stream is called a drainage network. The number of streams decreases and stream length increases as stream order increases.

Stream discharge is the amount of water flowing through the stream's cross section in a given time and place. A stream hydrograph is a plot of discharge versus time. The hydrograph in Fig. 2.3 shows the discharge of a small stream before, during, and after a storm. During dry weather, the hydrograph represents only groundwater seepage into the stream (base flow). During the initial phase of a rainstorm, only rain falling directly into the channel (channel precipitation) contributes to the hydrograph. Channel precipitation seldom is great enough to cause an appreciable increase in discharge. During a fairly large rain, stream discharge rises sharply as storm flow enters causing the rising limb of the hydrograph. Finally, peak discharge is reached causing the crest segment of the hydrograph. Discharge then declines resulting in the falling limb of the hydrograph. Subsurface storm drainage—water flowing downslope through the soil—often is blocked from entering streams by high water levels in the

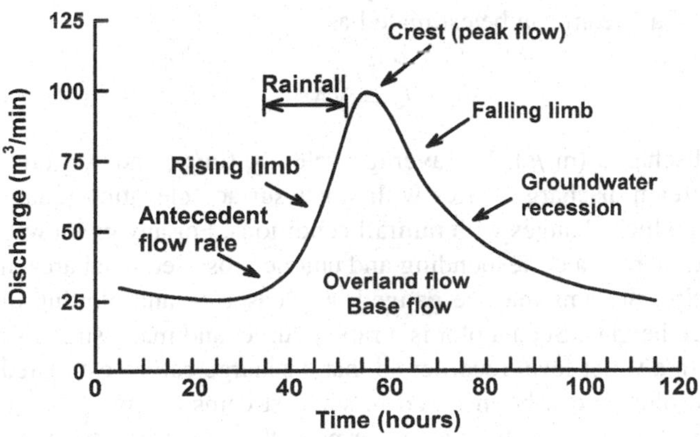

Fig. 2.3 A stream hydrograph

stream until the overland flow component has passed downstream. Subsurface storm drainage cannot be separated from base flow on the hydrograph. After a heavy rain, the water table usually rises because of infiltration, and base flow increases. Because of subsurface storm drainage and greater base flow, it may be days or weeks before discharge declines to the pre-storm rate after a period of greater than normal rainfall. The plot of discharge during this time is the groundwater recession segment of the hydrograph.

Watershed characteristics influence hydrograph shape. Watershed features that favor high rates of storm flow result in steep, triangular-shaped hydrographs. Streams on such watersheds are often said to be "flashy." Basins with permeable soils, good vegetative cover, or appreciable storage capacity tend to have trapezoidal hydrographs.

Streams are classified as young, mature, or old. Young streams flow rapidly and continually cut their channels. Their sediment loads are transported with no deposition. In mature streams, slopes are less and there is no down cutting of channels. Flows are adequate to transport most of the sediment load. Old streams have gentle slopes and sluggish flows. They have broad floodplains, and their channels meander. Sediment deposition near the area where streams discharge into large water bodies leads to delta formation. Larger streams usually may be classified as young near their sources, mature along middle reaches, and old near their mouths.

Measurement of Runoff and Stream Flow

The amount of overland flow can be assessed by various methods, but the most common procedure is the curve number method. In this procedure, an assessment of soil infiltration rate, antecedent rainfall, and extent and type of watershed cover (vegetation, paving, building, etc.) is used to obtain a curve number for a particular watershed area. A nomograph is used to estimate overland flow from rainfall amount and curve number (Yoo and Boyd 1994).

The flow of a stream can be estimated as

$$Q = VA \tag{2.5}$$

where Q = discharge (m^3/s), V = average velocity (m/s), and A = cross-sectional area (m^2). Stream discharge varies with water surface elevation (gauge height or stream stage) which changes with rainfall conditions. For any given water level, or gauge height, there is a corresponding and unique cross-sectional area and average stream velocity. Streams may be gauged by measuring and plotting discharge at different stage heights. Such a plot is a rating curve, and many streams are gauged and fitted with a water level recorder so that discharge can be estimated.

Weirs and plumes can be installed in small streams or other open channels to estimate flow. A weir consists of a barrier plate that constricts the flow of an open channel and directs it through a fixed-shape opening. Common shapes of weirs are

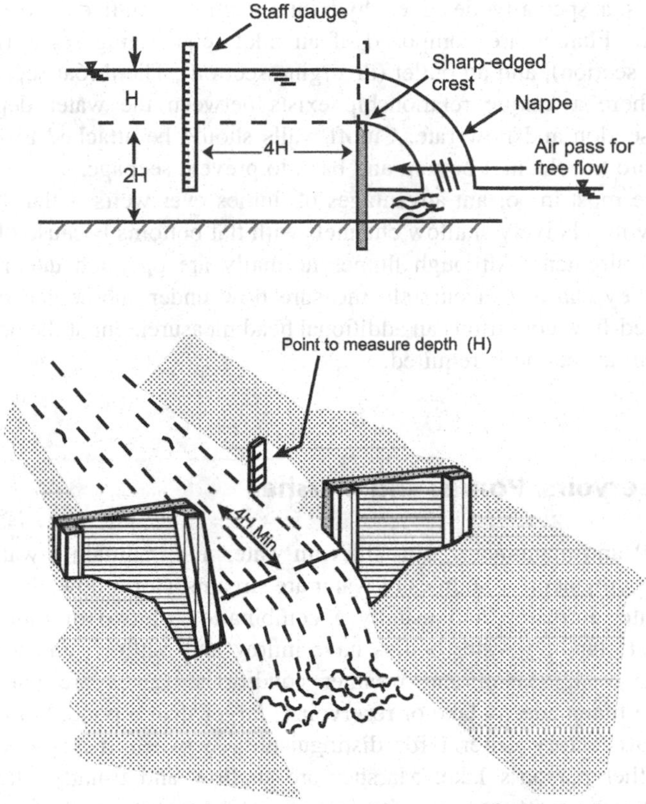

Fig. 2.4 An example with an end-contracted rectangular weir

rectangular, trapezoidal, and triangular. The profile of a sharp-crested rectangular weir is shown in Fig. 2.4. The bottom edge of the barrier plate is the weir crest, and flow depth over the crest is measured upstream from the crest and is the effective head. The overflowing stream of water is the nappe. The effective head and the shape and size of a weir crest determine the flow rate. The flow through a weir must maintain a free discharge for an accurate measurement, and water level on the downstream side must be low enough to maintain a free-flow. This requires some amount of head loss and limits the use of weirs in channels with very little slope.

Equations are used to estimate the discharge of streams over weirs. For the weir illustrated in Fig. 2.4, the weir equation is

$$Q = 0.0184(L_w - 0.2H)H^{1.5} \tag{2.6}$$

where Q = discharge (L/s), L_w = length of weir notch (width of crest) (cm), and H = head (cm).

A flume is a specially designed hydraulic section installed in a channel to measure flow. Flumes are composed of an inlet (converging section), a throat (constricting section), and an outlet (diverging section). The throat section acts as a control where a unique relationship exists between the water depth in the converging section and flow rate. Cutoff walls should be attached to the flumes and driven into the channel bottom and bank to prevent seepage.

One of the most important advantages of flumes over weirs is that flumes can measure flow in relatively shallow channels with flat bottoms because of their low head loss requirement. Although flumes normally are operated under free-flow conditions, they can also accurately measure flow under submerged conditions. For submerged-flow conditions an additional head measurement at the downstream side of the throat section is required.

Lakes, Reservoirs, Ponds, and Marshes

Many natural and manmade basins fill with water and maintain a water surface more or less permanently. Sources of water are streams, storm flow, direct rainfall, or groundwater intrusion and usually a combination of two or more of these sources. Most standing water bodies have inflow and outflow, and water levels that fluctuate with season and weather conditions. Lakes are natural while reservoirs are manmade. A lake or reservoir is larger than a pond, but no one has developed satisfactory criteria for distinguishing between them—one person's pond is another person's lake. Marshes are shallow and usually infested with aquatic plants, and they may be quite extensive. These water bodies usually hold water in temporary storage and can often be thought of as enlargements within streams. Because the water is in residence for a longer period, water quality dynamics and biological systems are much different in standing waters of lakes, reservoirs, marshes, and ponds than in streams.

Measurement of Volumes of Standing Water Bodies

The volumes of standing water bodies can be estimated as average depth times area

$$V = Ad \tag{2.7}$$

where V = volume, A = surface area, and d = average depth. For small bodies of water, the average depth can be estimated from random soundings (Ex. 2.1). Areas have traditionally been determined from aerial photographs and maps or by surveying techniques. Today, it is possible to locate standing water bodies on satellite imagery such as Google Earth Pro and use a tool provided with the software to estimate area.

Ex. 2.1: *A rectangular pond is 40 m long and 15 m wide. Random soundings in meters of depth from smallest to greatest were 0.1, 0.2, 0.5, 0.6, 0.7, 1.1, 1.3, 1.4, 1.5, 1.6, 1.8, 1.8, 1.9, 2.0, 2.1, 2.1. The volume will be estimated.*

Solution:

$$Area = 40\ \text{m} \times 15\ \text{m} = 600\ \text{m}^2.$$

$$Average\ depth = \frac{\sum soundings}{n} = \frac{20.7}{16} = 1.29\ \text{m}.$$

$$Volume = 600\ \text{m}^2 \times 1.29\ \text{m} = 774\ \text{m}^3.$$

Oceans and Estuaries

Most of the world's water is contained in the ocean. The ocean system is much larger than all other water bodies combined and contains salt water instead of freshwater. Water movements in the ocean are complex because of the large expanses of open water, great depths, wind driven currents, coriolis effect, tides, and other factors, but predictable and well-defined currents exist.

Estuaries are the entrances of streams into the ocean. Estuaries contain brackish water because the inflow of freshwater from streams dilutes ocean water. Because of tidal influence there are continuous changes in salinity at any given location in an estuary. When the tide recedes, the freshwater influence will extend farther into the estuary than when the tide advances. Salt or brackishwater is heavier than freshwater and a salt water wedge often extends into rivers beneath their freshwater flow. This can result in stratification of salinity at any given point in an estuary. Inflow of freshwater from streams increases after heavy rains and reduces the salinity of estuaries. Estuaries with poor connection to the ocean may become hypersaline in arid areas or during the dry season. The tide flushes estuaries, and the combination of freshwater inflow, size of the connection with the ocean, and tidal amplitude determines the water retention time in estuaries. Water quality in estuaries is highly dependent upon water residence time. Many estuaries are greatly influenced by human activities, and pollution problems are much more likely in estuaries where water exchange with the ocean is incomplete and slow.

Water Budgets

Water budgets often are made for the purposes of describing the hydrology of catchments or watersheds, natural water bodies, or water supply systems. Water budgets can be made by the general hydrologic equation

$$\text{Inflow} = \text{Outflow} \pm \text{Change in storage.} \tag{2.8}$$

This equation appears simple, but it often becomes complex when expanded to take into account all inflows and outflows. For example, the expanded equation for a small pond made by damming a watershed might be

$$\begin{aligned}\text{Rainfall} + \text{Seepage in} + \text{Runoff} =& (\text{Evaporation} + \text{Seepage out}\\ &+ \text{Overflow} + \text{Consumptive use})\\ &\pm \text{Change in storage.}\end{aligned} \tag{2.9}$$

Measurements of some of the terms are difficult to make. In (2.9), rainfall and change in storage can be measured easily. Evaporation can be estimated by applying a coefficient to Class A pan evaporation data. A measuring device would need to be installed in the pond to obtain overflow values. Runoff would have to be estimated by some appropriate procedure that takes into account rainfall amount and runoff-producing characteristics of the watershed. Consumptive use measurements could be made, but one would have to know the activities that withdraw water and quantify them. Seepage in and seepage out probably could not be measured independently, but if all other terms in (2.9) were known, net seepage could be estimated by difference.

The hydrologic equation allows the estimation of net seepage in a small pond as shown in Ex. 2.2.

Ex. 2.2: *A small pond has a water level of 3.61 m on day 1, and after 5 days, the water level is 3.50 m. There is no rainfall or consumptive water use over the 5-day period, and Class A pan evaporation is 6.0 cm. Net seepage will be estimated.*

Solution:

$$\textit{Pond evaporation} = \textit{Class A pan evaporation} \times 0.8$$

$$\textit{Pond evaporation} = 6\text{ cm} \times 0.8 = 4.8\text{ cm}$$

The water budget equation is

$$\textit{Inflow} = \textit{Outflow} \pm \Delta\ \textit{Storage}.$$

There was no inflow, outflows were seepage and evaporation, and storage change was 3.61–3.50 m = 0.11 m or 11 cm. The seepage was outward—lost from the pond. Thus,

$$0.0\text{ cm} = (\textit{seepage} + 4.8\text{ cm}) - 11\text{ cm}$$

$$(\textit{Seepage} + 4.8\text{ cm}) - 11\text{ cm} = 0$$

$$\textit{Seepage} + 4.8\text{ cm} = 11\text{ cm}$$

$$\textit{Seepage} = 11 - 4.8\text{ cm} = 6.2\text{ cm} = 1.24\text{ cm/day}.$$

Chemical Mass Balance

In studies of water quality, it often is necessary to calculate the amounts of substances contained in inputs, outputs, transport, or storage for specific masses of water. For example, it may be desired to calculate the amount of total suspended solids transported by a stream or the amount of phosphorus retained within a wetland. These calculations illustrated in Ex. 2.3 and 2.4 basically involve expanding the hydrologic equation by multiplying the amounts of water by the concentration of the variable of interest

$$\begin{aligned}\text{Inputs (concentration)} &= \text{Outputs (concentration)}\\ &\pm \Delta\ \text{Storage (concentration)}. \qquad (2.10)\end{aligned}$$

Ex. 2.3: *Suppose rainfall is 120 cm/year and evapotranspiration is 85 cm/year for a watershed. The stream flowing from this watershed has an average total nitrogen concentration of 2 mg/L. The annual nitrogen loss per hectare will be estimated.*

Solution:

$$\textit{Runoff} = \textit{Rainfall} - \textit{Evapotranspiration}$$

$$\textit{Runoff} = 120\ \text{cm/year} - 85\ \text{cm/year} = 35\ \text{cm/year}$$

$$\textit{Runoff volume} = 0.35\ \text{m/year}\ \times 10{,}000\ \text{m}^2/\text{ha} = 3{,}500\ \text{m}^3/\text{ha/year}$$

$$\begin{aligned}\textit{Nitrogen loss} &= 3{,}500\ \text{m}^3/\text{ha/year} \times 2\ \text{g N/ m}^3\\ &= 7{,}000\ \text{g N/ha/year}\ \textit{or}\ 7\ \text{kg N/ha/year}.\end{aligned}$$

Ex. 2.4: *A plastic lined pond (2,000 m^3) in an arid region is filled and supplied with city water containing 500 mg/L TDS. Over a year, 1,000 m^3 of water must be added to the pond to replace evaporation. What would be the approximate concentration of TDS after 3 years?*

Solution:
Quantities of dissolved solids added to the pond are

$$\textit{Filling}\ 2{,}000\ \text{m}^3\ \times 500\ \text{g/ m}^3 = 1{,}000{,}000\ \text{g}$$

$$\textit{Maintenance}\ 1{,}000\ \text{m}^3/\text{year}\ \times 3\ \text{years}\ \times 500\ \text{g/m}^3 = 1{,}500{,}000\ \text{g}$$

$$Total = 2,500,000\ \text{g}$$

$$Volume\ after\ 3\ years\ is\ still\ 2,000\ \text{m}^3.$$

$$Concentration\ TDS\ after\ 3\ years = \frac{2,500,000\ \text{g}}{2,000\ \text{m}^3} = 1{,}250\ \text{mg/L}.$$

Estimates of volumes, flow rates, or both almost always are needed in efforts related to water quality. Water quantity variables will be more difficult to measure in most instances than will be the concentrations of water quality variables.

Water Supply

Precipitation on the earth's land masses totals about 110,000 km^3 annually. Evapotranspiration from the land returns about 70,000 km^3 water to the atmosphere each year, and the rest—an estimated 40,000 km^3 becomes runoff and flows into the ocean (Baumgartner and Reichel 1975). Of course, if the ocean is included, evapotranspiration from land plus evaporation from the ocean must equal to precipitation onto the entire earth's surface.

Runoff consists of overland flow and base flow, but overland flow (storm flow) passes downstream quickly and cannot be used effectively by humans. Removing groundwater for water supply potentially can cause a decrease in the base flow of streams. Thus, groundwater use often is not considered sustainable. It often is assumed that about 30 % of runoff or 12,000 km^3 is stream base flow, and another 6,000 km^3 is captured in impoundments (Table 2.2). However, most authorities believe that only about 70 % of runoff is actually spatially available for extraction for human use. The annual supply of water for human use is around 12,600 km^3 (Table 2.2).

The water footprint—defined as the total amount of water used for a particular purpose—is becoming a popular calculation. Water footprints have been calculated for individuals, countries, the world, and for specific goods and services. The average water footprint for humans was estimated to be 1,385 m^3/cap/year (Hoekstra and Mekonnen 2012); water footprints in cubic meters per capita

Table 2.2 Renewable, accessible water for humankind

Rainfall onto land masses	110,000
Evapotranspiration from land masses	70,000
Renewable: total stream flow	40,000
Available: base stream flow (12,000 km^3) and reservoirs (6,000 km^3)	18,000
Accessible: about 70 % of available	12,600

annually for several countries were: China, 1,071; South Africa, 1,255; France, 1,786; Brazil, 2,027; United States, 2,842. Water footprints were calculated to include rainwater that evaporates from agricultural fields. This water is called "green" water and it would have evaporated whether or not it had fallen on cropland.

The global water footprint for agriculture given by Hoekstra and Mekonnen (2012) was 8,363 km^3/year while those of industrial and domestic use were around 400 km^3/year and 324 km^3/year, respectively—a total of 9,087 km^3/year. However, this includes 6,684 km^3/year of "green" water use in agriculture. A more reasonable estimate of water use is to combine water use for irrigation of 3,200 km^3/year (Wisser et al. 2008), animal water supply of 46 km^3/year, and industrial (400 km^3/ear) and domestic (324 km^3/year) purposes—a total of 3,970 km^3/year.

When compared to the amount of runoff that is renewable and accessible to humans, the total amount of water available for human use appears quite adequate for the future if green water is omitted. This unfortunately is not the case because of several reasons. The water is not uniformly distributed with population (Table 2.3), and many countries or areas of countries with rapidly growing populations have a natural scarcity of water. Moreover, the water supply varies from year to year with respect to weather rainfall patterns, and places that normally have high rainfall may suffer water shortages during droughts. For example, Birmingham, Alabama, has high rainfall normally, but it suffered water shortages during a drought in 2006 and 2007. The growth of population makes water shortage more likely during drought years. Water pollution associated with growing population also degrades the quality of water supplies making them more expensive to treat for human uses. Add to these issues the facts that many countries do not have adequate water supply infrastructure, suffer political instability, or are embroiled in armed conflicts, and it is easy to understand why water shortage is a frequent problem facing mankind that will become of even greater concern in the future.

Table 2.3 Distribution of global runoff by continent and population

Continent	Total runoff (km^3/year)	Population in 2013 (millions)	Runoff (m^3/capita/year)	Runoff (as % world)	Population (as % world)
Africa	4,320	1,110.6	3,890	10.6	15.5
Asia	14,550	4,298.7	3,385	35.7	60.0
Europe	3,240	742.5	4,364	8.0	10.4
North America (including Central America and Caribbean)	6,200	565.3	10,968	15.2	7.9
Oceania (including Australia)	1,970	38.3	51,436	4.8	0.5
South America	10,420	406.7	25,621	25.6	5.7
World	40,700	7,162.1	5,683	100.0	100.0

References

Baumgartner A, Reichel E (1975) The world water balance. Elsevier, Amsterdam

Boyd CE (1985) Pond evaporation. Trans Am Fish Soc 114:299–303

Hoekstra AY, Mekonnen MM (2012) The water footprint of humanity. Proc Natl Acad Sci U S A 109:3,232–3,237

Hounam CE (1973) Comparisons between pan and lake evaporation. Technical Note 126. World Meteorological Organization, Geneva

Klein MJ (1970) Maxwell, his demon and the second law of thermodynamics. Am Sci 58:84–97

Wetzel RG (2001) Limnology, 3rd edn. Academic, New York

Wisser D, Frolking S, Douglas EM, Fekete BM, Vöoösmarty CJ, Schumann AH (2008) Global irrigation water demand: variability and uncertainties arising from agricultural and climate data sets. Geophys Res Let 35(24), L24408. doi:10.1029/2008GL035296

Yoo KH, Boyd CE (1994) Hydrology and water supply for pond aquaculture. Chapman & Hall, New York

Review of Basic Chemistry, Solubility, and Chemical Equilibrium

3

Abstract

Water quality is governed by basic chemical principles taught in general chemistry classes. Nevertheless, some readers will no doubt have only vague recollection of these principles. This chapter reviews some basic topics of chemistry to include atomic structure, atomic number, and atomic mass (weight), valence, solubility, reactions, the equilibrium principle and its relationship to thermodynamics, and electrostatic interactions among ions.

Keywords

Atomic structure • Solubility principles • Le Chatelier's principle • Equilibrium and thermodynamics • Ionic interactions

Introduction

This book was written with the intent of providing a text on water quality in which the water chemistry aspects are at a level understandable to those with relatively little formal training in chemistry. Chemical concepts and methods for computations used in the book are covered in a freshman-level college course in inorganic chemistry. Moreover, even those with only a high school-level chemistry background should be able to understand most of the contents of this book.

The discussions of most variables are based on chemistry. Despite the effort at a simple presentation, it is anticipated that many of those using this book will benefit from a review of basic chemical principles and calculations—especially those related to solubility and equilibrium. This chapter has been included to provide a rapid review of the chemical knowledge necessary for understanding later chapters. One important concept—that of pH—was deferred to Chap. 8 where it is an inseparable aspect of the discussion of carbon dioxide, alkalinity, and acid-base relationships in natural waters.

C.E. Boyd, *Water Quality*, DOI 10.1007/978-3-319-17446-4_3

Review of Some Chemical Principles

Anything that has mass (or weight) and occupies space is matter made up of chemical elements such as silicon, aluminum, iron, oxygen, sulfur, copper, etc. Elements are the basic substances of which more complex substances are composed. Substances more complex than atoms consist of molecules. A molecule is the smallest entity of a substance that has all the properties of that substance.

Elements are classified broadly as metals, nonmetals, or metalloids. Metals have a metallic luster, are malleable (but often hard), and can conduct electricity and heat. Some examples are iron, zinc, copper, silver, and gold. Nonmetals lack metallic luster, they are not malleable (but tend to be brittle), some are gases, and they do conduct heat and electricity. Metalloids have one or more properties of both metals and nonmetals. Because metalloids can both insulate and conduct heat and electricity, they are known as semiconductors. The best examples of metalloids are boron and silicon, but there are several others.

There also are less inclusive groupings of elements than metals, nonmetals, and metalloids. Alkali metals such as sodium and potassium are highly reactive and have an ionic valence of +1. Alkaline earth metals include calcium, magnesium, and other elements that are moderately reactive and have an ionic valence of +2. Halogens illustrated by chlorine and iodine are highly reactive nonmetals, and they have an ionic valence of −1. This group is unique in that its members can exist in solid, liquid, and gaseous form at temperatures and pressures found on the earth's surface. Because of their toxicity, halogens often are used as disinfectants. Noble gases such as helium, argon, and neon are unreactive except under very special conditions. In all there are 18 groups and subgroups of elements. Most elements are included in the groups referred to as alkali metals, alkaline earth metals, transitional metals, halogens, noble gases, and chalcogens (oxygen family). The elements are grouped based on similar properties and reactions. However, in reality, each element has one or more unique properties and reactions.

The most fundamental entity in chemistry and the smallest unit of an element is the atom. An atom consists of a nucleus surrounded by at least one electron. Electrons revolve around the nucleus in one or more orbitals or shells (Fig. 3.1). The nucleus is made up of one or more protons, and with the sole exception of hydrogen, one or more neutrons. The protons and neutrons do not necessarily occur in the nucleus in equal numbers. For example, the oxygen nucleus has eight protons and neutrons, sodium has 11 protons and 12 neutrons, potassium has 19 protons and 20 neutrons, copper has 29 electrons and 34 neutrons, and silver has 47 electrons and 61 neutrons.

Protons are positively charged and assigned a charge value of +1 each, while electrons possess a negative charge and are assigned a charge value of −1 each. In their normal state, atoms have equal numbers of protons and electrons resulting in them being charge neutral. Neutrons are charge neutral.

Atoms are classified according to the numbers of protons. All atoms with the same number of protons are considered to be the same element. For example, oxygen atoms always have eight protons while chlorine atoms always have

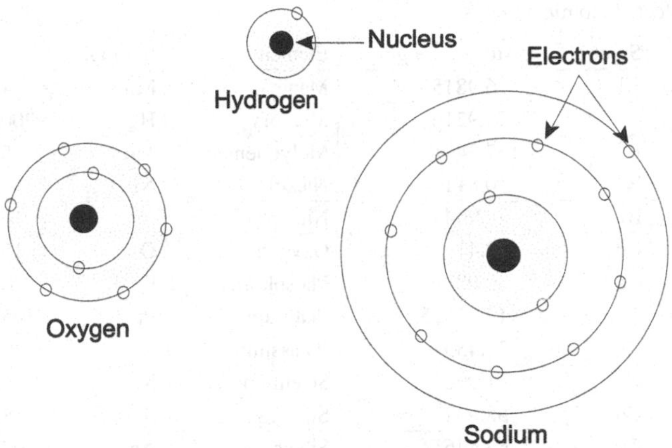

Fig. 3.1 The structure of oxygen, hydrogen, and sodium atoms

17 protons. The atomic number of an element is the same as the number of protons, e.g., the atomic number of oxygen is 8 while that of chlorine is 17. There are over 100 elements each with an atomic number assigned to it according to the number of protons in its nucleus.

Some atoms of the same element may have one to several neutrons more than do other atoms of the particular element, e.g., carbon atoms may have 6, 7, or 8 neutrons but only 6 protons. These different varieties of the same element are known as isotopes. Moreover, all atoms of the same element may not have the same number of electrons. When uncharged atoms come close together, one or more electrons may be lost from one atom and gained by the other. This phenomenon results in an imbalance between electrons and protons in each of the two interacting atoms that imposes a negative charge on the one that gained the electron(s) and a positive charge on the one that lost the electron(s). The charge on the atom is equal to the number of electrons gained or lost (-1 or $+1$ charge per electron). Charged atoms are called ions, but the normal atom is uncharged. In the periodic table of chemical elements, an element is assumed uncharged and to have equal numbers of protons and electrons.

The mass of atoms results almost entirely from their neutrons and protons. The masses of the two entities are almost identical; 1.6726×10^{-24} g for a proton and 1.6749×10^{-24} g for a neutron. Thus, their atomic masses usually are considered unity in determining relative atomic masses of elements. An electron has a mass of 9.1×10^{-28} g—nearly 2,000 times less than the masses of protons and neutrons. The mass of electrons is omitted in atomic mass calculations for elements. The atomic mass of elements increases as the atomic number increases, because the number of protons and neutrons in the nucleus increases with greater atomic number. The atomic mass of a given element differs among its isotopes because some isotopes have more neutrons than others, while all isotopes of an element have the same number of protons. The loss or gain of electrons by atoms forming ions does not affect atomic mass.

Table 3.1 Selected atomic weights

Element	Symbol	Atomic weight	Element	Symbol	Atomic weight
Aluminum	Al	26.9815	Manganese	Mn	54.9380
Arsenic	As	74.9216	Mercury	Hg	200.59
Barium	Ba	137.34	Molybdenum	Mo	95.96
Boron	B	10.811	Nickel	Ni	58.69
Bromine	Br	79.904	Nitrogen	N	14.0067
Cadmium	Cd	112.41	Oxygen	O	15.9994
Calcium	Ca	40.08	Phosphorus	P	30.9738
Carbon	C	12.01115	Platinum	Pt	195.09
Chlorine	Cl	35.453	Potassium	K	39.102
Chromium	Cr	51.996	Selenium	Se	78.96
Cobalt	Co	58.9332	Silicon	Si	28.086
Copper	Cu	63.546	Silver	Ag	107.868
Fluorine	F	18.9984	Sodium	Na	22.9898
Gold	Au	196.967	Strontium	Sr	87.62
Helium	He	4.0026	Sulfur	S	32.064
Hydrogen	H	1.00797	Thallium	Tl	204.37
Iodine	I	126.9044	Tin	Sn	118.71
Iron	Fe	55.847	Tungsten	W	183.85
Lead	Pb	207.19	Uranium	U	238.03
Lithium	Li	6.939	Vanadium	V	50.942
Magnesium	Mg	24.312	Zinc	Zn	65.38

Because of the different natural isotopes of atoms of a particular element, the atomic mass typically listed in the periodic table for elements does not equal to the sum of the masses of the protons and neutrons contained in atoms of these elements. This results because the atomic masses typically reported for elements represent the average atomic masses of their isotopes. For example, copper typically has 29 protons and 34 neutrons, and the atomic mass of the most common isotope would be 63 based on addition of neutrons and protons. However, the atomic mass reported in the periodic table of copper is 63.546. The additional mass results from the effects of averaging the atomic masses of the copper isotopes. Atomic masses of some common elements are provided (Table 3.1).

The atomic mass is very important in stoichiometric relationships in reactions of atoms and molecules. The relative molecular mass (or weight) of a molecule is the sum of the atomic masses of the atoms contained in the molecule. Thus, when sodium atoms (relative atomic mass of 22.99) react with chlorine atoms (relative atomic mass of 35.45) to form sodium chloride, the reaction will always be in the proportion of 22.99 sodium to 35.45 chlorine, and the molecular mass of sodium will be the sum of the atomic masses of sodium and chloride or 58.44. Atomic and molecular masses may be reported in any unit of mass (or weight), but the most common is the gram. Thus, the mass or weight of atoms often is referred to as the gram atomic mass and the molecular mass of molecules usually is referred to as the gram molecular weight.

Each element is assigned a symbol, e.g., H for hydrogen, O for oxygen, N for nitrogen, S for sulfur, C for carbon, and Ca for calcium. But, because of the large number of elements, it was not possible to have symbols suggestive of the English name for all elements. For example, sodium is Na, tin is Sn, iron is Fe, and gold is Au. The symbols must be memorized or found in reference material. The periodic table of the elements is a convenient listing of the elements in a way that elements with similar chemical properties are grouped together. The periodic table is presented in various formats, but most presentations include, at minimum, each element's symbol, atomic number, and relative atomic mass and indicate the group of elements to which it belongs.

The maximum number of electrons that can occur in a shell of an atom usually is $2n^2$ where n is the number of the shell starting at the shell nearest the nucleus, i.e., two electrons in the first shell, eight in the second shell, etc. Elements with a large atomic number have several shells and some shells may not contain the maximum number of electrons. Chemical reactions among atoms involve only the electrons in the outermost shell of atoms.

The laws of thermodynamics dictate that substances spontaneously change towards their most stable states possible under existing conditions. To be stable, an atom needs two electrons in the inner shell, and at least eight electrons in the outermost shell. Thus, chemical combinations (reactions) occur so that atoms gain or lose electrons to attain stable outer shells.

Nonmetals tend to have outer shells nearly full of electrons. For example, chlorine has seven electrons in its outer shell. If it gains one electron, it will have a stable outer shell, but the acquisition of this electron will give the chlorine atom a charge of −1. Gaining the electron causes a loss of energy and the chlorine ion is more stable than the free atom. Nonmetals tend to capture electrons. The source of electrons for nonmetals is metals which tend to have only a few electrons in their outer shells. Sodium is a metal with one electron in its outer shell. It can lose the single electron and attain a charge of +1 and lose energy to become more stable.

If sodium and chlorine are brought together, each sodium atom will give up an electron to each chlorine atom (Fig. 3.2). The chlorine atoms will now have fewer protons than electrons and acquire a negative charge while the opposite will be true of sodium atoms. Because opposite charges attract, sodium ions and chlorine ions will combine because of the attraction of unlike charges forming sodium chloride or common salt (Fig. 3.2). Bonds between sodium and chlorine in sodium chloride are called ionic bonds. The number of electrons lost or gained by an atom or the charge it acquires when it becomes an ion is the valence.

Some elements like carbon are unable to gain or lose electrons to attain stable outer shells. These elements may share electrons with other atoms. For example, carbon atoms have four electrons in their outer shells and can share electrons with four hydrogen atoms as shown in Fig. 3.3. This provides stable outer shells for both the carbon and hydrogen atoms. The resulting compound (CH_4) is methane. The chemical bond that connects each hydrogen to carbon in methane is known as a covalent bond.

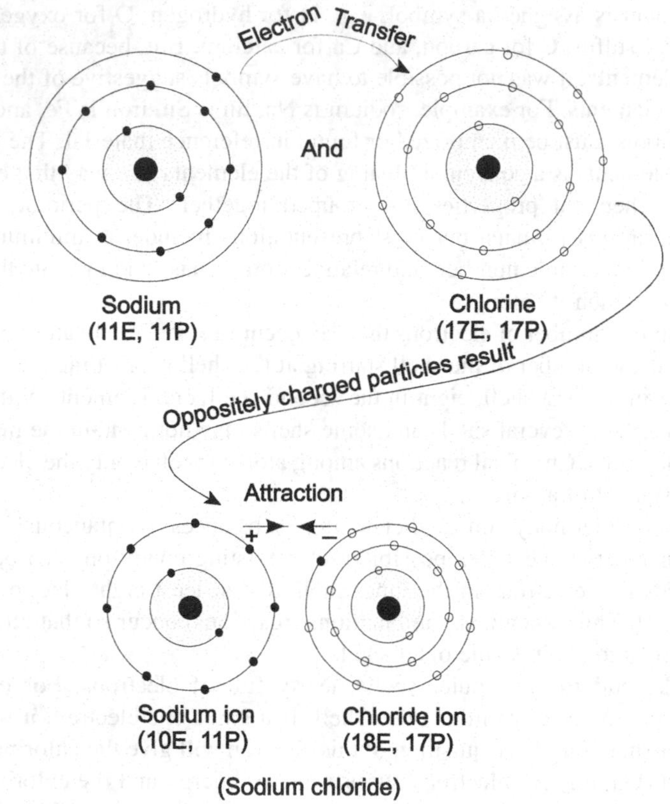

Fig. 3.2 Transfer of electron from a sodium atom to a chlorine atom to form an ionic bond in sodium chloride

When two or more elementary substances bond together, the resulting substance is called a compound. A compound has a characteristic composition and unique properties that set it aside from all other compounds. The Law of Definite Proportions holds that all samples of a given compound substance contain the same elements in the same proportions by weight. The smallest part of a substance with all of the properties of that substance is a molecule. A molecule may be composed of a single element as in the case of an elemental substance such as oxygen, sulfur, or iron or it may be composed of two or more elements as in acetic acid (CH_3COOH) which contains carbon, hydrogen, and oxygen. The gram molecular weight or gram atomic mass of a substance contains as many molecules as there are oxygen atoms in 15.9994 g of oxygen. This quantity is Avogadro's number of molecules (6.02×10^{23}), and it is known as a mole.

Elements are represented by symbols, but to represent molecules, the elemental symbols are given numerical subscripts to represent their proportions. Such notation is called a formula. For example, molecular oxygen and nitrogen have the

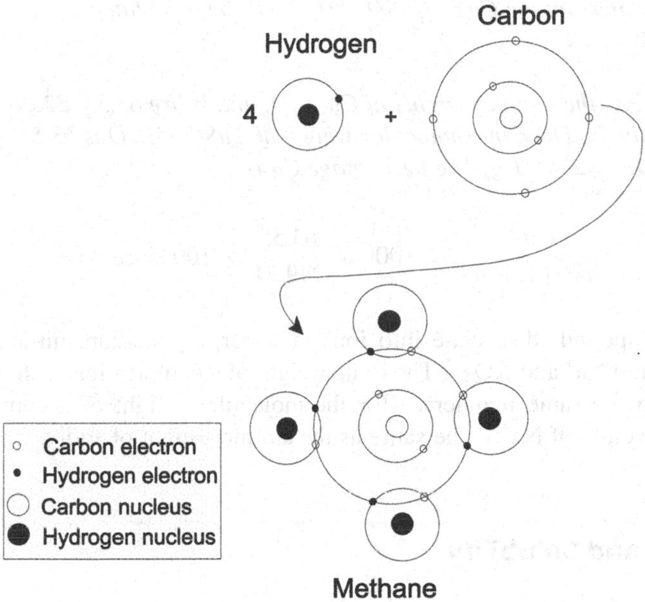

Fig. 3.3 Covalent bonding of hydrogen and carbon atoms to form methane

formulas O_2 and N_2, respectively. Sodium chloride is NaCl and sodium carbonate is Na_2CO_3. The molecular weights of molecules can be determined by summing the atomic weights of all of the constituent elements. If an element has a subscript, its atomic weight must be summed the number of times indicated by the subscript. One molecular weight of a substance in grams is a gram molecular weight or a mole as illustrated in Ex. 3.1. The percentage of an element in a compound is determined by dividing the weight of the element by the formula weight of the compound and multiplying by 100 (Ex. 3.2).

Ex. 3.1: *The molecular weights of O_2 and $CaCO_3$ will be calculated from their formulas.*

Solution:
O_2: The atomic weight of O is 15.9994 (Table 3.1*), but it usually is rounded to 16, thus $16 \times 2 = 32$ g/mol.*
$CaCO_3$: The atomic weights of Ca, C, and O are 40.08, 12.01, and 16, respectively (Table 3.1*), thus $40.08 + 12.01 + 3(16) = 100.09$ g/mol.*

The percentage composition of substances can be estimated from their formulas because of the Law of Definite Proportions.

Ex. 3.2: *The percentage Cu in $CuSO_4{\cdot}5H_2O$ will be calculated.*

Solution:
From Table 3.1, *the atomic weights of Cu, S, O, and H are 63.55, 32.06, 16.00 and 1.01, respectively. Thus, the molecular weight of $CuSO_4{\cdot}5H_2O$ is 63.55 + 32.06 + 9 (16) + 10(1.01) = 249.71 g. The percentage Cu is*

$$\frac{Cu}{CuSO_4 \cdot 5H_2O} \times 100 \ or \ \frac{63.55}{249.71} \times 100 = 25.45\%.$$

Some compounds dissociate into ions in water, e.g. sodium nitrate ($NaNO_3$) dissociates into Na^+ and NO_3^-. The ionic weight of a complex ion such as nitrate is calculated in the same manner as for the molecular weight of a compound. Of course, the weight of Na^+ is the same as the atomic weight of sodium.

Solutions and Solubility

Definition of a Solution

A solution consists of a solvent and a solute. A solvent is defined as the medium in which another substance—the solute—dissolves. For example, in an aqueous sodium chlorine solution, water is the solvent and sodium chloride is the solute. Miscibility refers to a solute and solvent mixing in all proportions to form a homogenous mixture or solution. A true solution is by definition a homogenous mixture of two or more components that cannot be separated into their individual ionic or molecular components. Most solutes are only partially miscible in water. If a soluble compound such as sodium chloride is mixed with water in progressively increasing amounts, the concentration of the solute will reach a constant level. Such a solution is said to be saturated, and the amount of the sodium chloride held in solution is the solubility of sodium chloride in water at the particular temperature. Any more sodium chloride added to the saturated solution will settle to the bottom without dissolving.

Much of this book is devoted to a discussion of factors controlling concentrations of dissolved matter in natural waters. This dissolved matter in water includes inorganic ions and compounds, organic compounds, and atmospheric gases. This chapter is devoted to a discussion of the solubility of solids in water. The principles of gas solubility will be discussed in Chap. 6 on dissolved gases.

Methods of Expressing Solute Strength

The solubility of a solute in a saturated solution or its concentration in an unsaturated solution may be expressed in several ways—the most common of which will be described.

Molarity

In a molar solution, the solute strength is expressed in moles of solute per liter, e.g., a 1 molar (1 M) solution contains 1 mol of solute in 1 L of solution. Calculations for a molar solution are shown in Ex. 3.3.

Ex. 3.3: *How much Na_2CO_3 must be dissolved and diluted to 1 L to give a 0.25 M solution?*

Solution:

The molecular weight of Na_2CO_3 is 106 g.

$$106\ \mathrm{g/mol}\ \times 0.25\ \mathrm{mol/L} = 26.5\ \mathrm{g/L}.$$

Normality

Because reactions occur on an equivalent weight basis, it is often more convenient to express concentration in equivalents per liter instead of moles per liter. A solution containing 1 gram equivalent weight of solute per liter is a 1 normal (1 N) solution. In cases where the equivalent weight and formula weight of a compound are equal, a 1 M solution and a 1 N solution have identical concentrations of solute. This is the case with HCl and NaCl.

The following rules may be used to compute equivalent weights of most reactants: (1) the equivalent weight of acids and bases equal their molecular (formula) weights divided by their number of reactive hydrogen or hydroxyl ions; (2) the equivalent weights of salts equal their molecular weights divided by the product of the number and valence of either cation component (positively charged ion) or anion component (negatively charged ion); (3) the equivalent weights of oxidizing and reducing agents may be determined by dividing their molecular weights by the number of electrons transferred per molecular weight in oxidation-reduction reactions.

Oxidation-reduction reactions usually are more troublesome to students than other types of reactions. An example of an oxidation reduction reaction is provided in which manganese sulfate reacts with molecular oxygen:

$$2MnSO_4 + 4NaOH + O_2 \rightarrow 2MnO_2 + 2Na_2SO_4 + 2H_2O \qquad (3.1)$$

Manganese in $MnSO_4$ has a valence of +2, but in manganese dioxide its valance is +4—manganese was oxidized. Molecular oxygen with valence 0 was reduced to a valence of −2 in manganese dioxide. Each molecule of manganese sulfate (the reducing agent) lost two electrons to oxygen (the oxidizing agent). The equivalent weight of magnesium sulfate in this reaction is its formula weight divided by two.

Additional examples illustrating calculations of equivalent weights and normalities of acids, bases, and salts are provided in Exs. 3.4–3.6.

Ex. 3.4: *What is the equivalent weight of sodium carbonate when it reacts with hydrochloric acid?*

Solution:

The reaction is: $Na_2CO_3 + 2HCl = 2NaCl + CO_2 + H_2O$.

One sodium carbonate molecule reacts with two hydrochloric acid molecules. Thus, the equivalent weight of sodium carbonate is

$$\frac{Na_2CO_3}{2} = \frac{106}{2} = 53 \text{ g.}$$

Ex. 3.5: *What are the equivalent weights of sulfuric acid (H_2SO_4), nitric acid (HNO_3), and aluminum hydroxide [$Al(OH)_3$]?*

Solution:

(i) *Sulfuric acid has two available hydrogen ions*

$$H_2SO_4 \rightarrow 2H^+ + SO_4^{2-}.$$

Thus, the equivalent weight is

$$\frac{H_2SO_4}{2} = \frac{98}{2} = 49 \text{ g.}$$

(ii) *Nitric acid has one available hydrogen ion*

$$HNO_3 \rightarrow H^+ + NO_3^-.$$

The equivalent weight is

$$\frac{HNO_3}{1} = \frac{63}{1} = 63 \text{ g.}$$

(iii) *Aluminum hydroxide has three available hydroxide ions*

$$Al(OH)_3 = Al^{3+} + 3OH^-.$$

The equivalent weight is

$$\frac{Al(OH)_3}{3} = \frac{77.99}{3} = 25.99 \text{ g.}$$

Ex. 3.6: *How much Na_2CO_3 must be dissolved and diluted to 1 L to give a 0.05 N solution?*

Solution:

$$Na_2CO_3 \rightarrow 2Na^+ + CO_3^{2-}.$$

The formula weight must be divided by 2 because $2Na^+ = 2$, or because $CO_3^{2-} = -2$. The sign difference ($2Na^+ = +2$ and $CO_3^{2-} = -2$ with respect to valence) does not matter in calculating equivalent weight.

$$\frac{105.99 \text{ g } Na_2CO_3/\text{mol}}{2} = 52.995 \text{ g/equiv.}$$

$$53 \text{ g/equiv.} \times 0.05 \text{ equiv./L} = 2.65 \text{ g/L.}$$

In expressing solute strength for dilute solutions, it is convenient to use milligrams instead of grams. Thus, we have millimoles (mmoles), millimolar (mM), millimoles/L (mmoles/L that is the same as mM), milliequivalents (meq), and milliequivalents/liter (meq/L). A 0.001 M solution is also a 1 mM solution.

Weight per Unit Volume

In water quality, concentrations often are expressed in weight of a substance per liter. The usual procedure is to report milligrams of a substance per liter (Ex. 3.7).

Ex. 3.7: *What is the concentration of K in a solution that is 0.1 N with respect to KNO_3?*

Solution:

$$KNO_3 = 101.1 \text{ g/equiv.}$$

$$101.1 \text{ g/equiv.} \times 0.1 \text{ equiv./L} = 10.11 \text{g/L.}$$

The amount of K in 10.11 g of KNO_3 is

$$\frac{39.1}{101.1} \times 10.11 \text{ g/L} = 3.9 \text{ g/L.}$$

$$3.9 \text{ g/L} = 3{,}900 \text{ mg/L.}$$

The unit, milligrams per liter, is equivalent to the unit, parts per million (ppm) as shown in Ex. 3.8; it is popular in water quality to use parts per million interchangeably with milligrams per liter.

Ex. 3.8: *It will be demonstrated that 1 mg/L = 1 ppm for aqueous solutions.*

Solution:

$$\frac{1\ \text{mg}}{1\ \text{L}} = \frac{1\ \text{mg}}{1\ \text{kg}} = \frac{1\ \text{mg}}{1{,}000\ \text{g}} = \frac{1\ \text{mg}}{1{,}000{,}000\ \text{mg}} = 1\ \text{ppm}.$$

It also is convenient to express the strength of more concentrated solutions in parts per thousand or ppt (Ex. 3.9). One part per thousand is equal to 1 g/L because there are 1,000 g in 1 L. It also is equal to 1,000 mg/L. An alternative way of indicating concentration in parts per thousand is the symbol ‰.

Ex. 3.9: *Express the concentration of sodium chloride in a 0.1 M solution as parts per thousand and milligrams per liter.*

Solution:

$$0.1\ \text{M} \times 58.45\ \text{g}\ NaCl/\text{mol} = 5.845\ \text{g/L}.$$

The salinity is 5.845 ppt that is equal to 5,845 mg/L.

It is easy to convert water quality data in milligrams per liter to molar or normal concentrations (Ex. 3.10).

Ex. 3.10: *The molarity and normality of a 100 mg/L solution of calcium will be calculated.*

Solution:

(i) $$\frac{100\ \text{mg/L}}{40.08\ \text{mg Ca/mmol}} = 2.495\ \text{mmol/L}\ or\ 0.0025\ \text{M}.$$

(ii) *Calcium is divalent, so*

$$\frac{100\ \text{mg/L}}{20.04\ \text{mg/meq}} = 4.99\ \text{meq/L}\ or\ 0.005\ \text{N}.$$

Concentrations of solutions often are in milligrams per liter with no regard for molarity of normality as shown in Ex. 3.11.

Ex. 3.11: How much magnesium sulfate ($MgSO_4{\cdot}7H_2O$) must be dissolved and made to 1,000 mL final volume to provide a magnesium concentration of 100 mg/L?

Solution:

$$\frac{\overset{x}{MgSO_4 \cdot 7H_2O}}{mw = 246.38\ \text{g}} = \frac{\overset{100\ \text{mg/L}}{Mg}}{mw = 24.31\ \text{g}}$$

rearranging,

$$x = \frac{24,631}{24.31} = 1,013.5\ \text{mg/L}\ \textit{of}\ MgSO_4 \cdot 7H_2O.$$

Soluble Versus Insoluble

Solubilities of chemical compounds often are reported as grams of solute in 100 mL of solution. Chemists generally consider substances that will dissolve to the extent of 0.1 g/100 mL in water as soluble. A general list of classes of soluble compounds follows:

- nitrates
- bromides, chlorides, and iodides—except those of lead, silver, and mercury
- sulfates—except those of calcium, strontium, lead, and barium
- carbonates and phosphates of sodium, potassium, and ammonium
- sulfides of alkali and alkaline earth metals and ammonium.

A substance has a finite solubility, and if enough of it is added to water, saturation will occur. The solubilities at 25 °C for some common inorganic compounds are listed in Table 3.2. The concentrations of many important water

Table 3.2 Solubilities in water at 25 °C by some selected, highly soluble compounds

Compound	Formula	Solubility (g/L)
Aluminum sulfate	$Al_2(SO_4)_3{\cdot}18H_2O$	38.5
Ammonium chloride	NH_4Cl	39.5
Ammonium nitrate	NH_4NO_3	212.5
Boric acid	H_3BO_3	5.7
Calcium nitrate	$Ca(NO_3)_2$	143.9
Copper sulfate	$CuSO_4{\cdot}5H_2O$	21.95
Potassium chloride	KCl	35.5
Potassium hydroxide	KOH	120.8
Potassium sulfate	K_2SO_4	12.0
Sodium bicarbonate	$NaHCO_3$	10.2
Sodium carbonate	Na_2CO_3	30.7
Sodium chloride	$NaCl$	36.0
Sodium sulfate	Na_2SO_4	28.1

quality variables in natural water bodies, however, are regulated by the solubility of compounds of low solubility (insoluble compounds by the usual definition).

Types of Dissolutions

There are three basic types of dissolutions. In a simple dissolution, the solute dissolves into individual molecules—table sugar (sucrose) is an example:

$$C_{12}H_{22}O_{11}(s) \xrightarrow{\text{Water}} C_{12}H_{22}O_{11}(aq) \tag{3.2}$$

In (3.2), the (s) designates a solid and (aq) indicates the molecule is dissolved in water. Later, in the discussion of gas solubility (Chap. 6), a (g) beside a molecule will indicate that it is in gaseous form.

The starting compound consists of sucrose molecules in crystalline form; the result of dissolution is sucrose molecules in water. The sugar crystal is not held together strongly, and it has hydroxyl groups in which the bond between hydrogen and oxygen gives a slight negative charge to oxygen and a slight positive charge to hydrogen. Water is dipolar and attracted to the charges forcing the sucrose molecules from the crystal and holding them in aqueous solution. This type of solution is a purely physical process.

Dissolution may result from dissociation of a substance into its ions. This is the common way in which salts dissolve. Gypsum is an ionically bound salt that is hydrated with the formula $CaSO_4{\cdot}2H_2O$. When put in water, the hydrating water is immediately lost into the surrounding solvent, and the salt dissolves as follows:

$$CaSO_4 \rightleftharpoons Ca^{2+} + SO_4^{2-} \tag{3.3}$$

Dipolar water molecules arrange themselves around the ions to insulate them and keep them from recombining into gypsum. This type of solution also is basically a physical process.

Chemical reactions also can cause dissolution as shown for ferrous carbonate in an acidic solution

$$FeCO_3(s) + H^+(aq) \rightleftharpoons Fe^{2+}(aq) + HCO_3^-(aq). \tag{3.4}$$

Chemical weathering is caused by reactions that dissolve minerals in the earth's crust. For example, carbon dioxide reacts with limestone to dissolve it as illustrated in the following equation in which calcium carbonate represents limestone

$$CaCO_3(s) + CO_2(aq) + H_2O \rightleftharpoons Ca^{2+}(aq) + 2HCO_3^-(aq). \tag{3.5}$$

For simplicity, it will be assumed in most instances in this book that the reader can distinguish between solid and aqueous forms and the (s) and (aq) designation will not be used extensively.

Factors Affecting Solubility

Each compound has a specific solubility in a particular solvent—of course, in this book the solvent will be water. Water molecules are polar, and water is a polar solvent particularly good in dissolving polar solutes—hence, the adage "like dissolves like." Nevertheless, two polar compounds may have drastically different solubilities in water; the water solubility of sodium chloride is 359 g/L while that of silver chloride is only 0.89 mg/L at 25 °C.

Temperature has a pronounced effect on solubility, but the effect varies among compounds. Endothermic reactions absorb heat from the environment. Thus, when mixing a solute with water causes the temperature to decrease, increasing temperature will enhance solubility as illustrated for sodium carbonate (Na_2CO_3) in Fig. 3.4. The opposite is true for compounds that release heat when they dissolve as shown for cesium sulfate [$Ce_2(SO_4)_3{\cdot}9H_2O$] in Fig. 3.4. Such reactions are called exothermic reactions. Of course, temperature has much less effect on substances such as sodium chloride (NaCl) that are not appreciably endothermic or exothermic when they dissolve (Fig. 3.4).

The molecules of substances of high molecular weight tend to be larger than the molecules of substances of low molecular weight. It is more difficult for water molecules to surround larger molecules, and substances comprised of larger molecules tend to dissolve less than substances containing smaller molecules.

Pressure seldom has an influence on the solubility of solids in natural waters, but pressure has a tremendous effect on gas solubility in water as explained in Chap. 6.

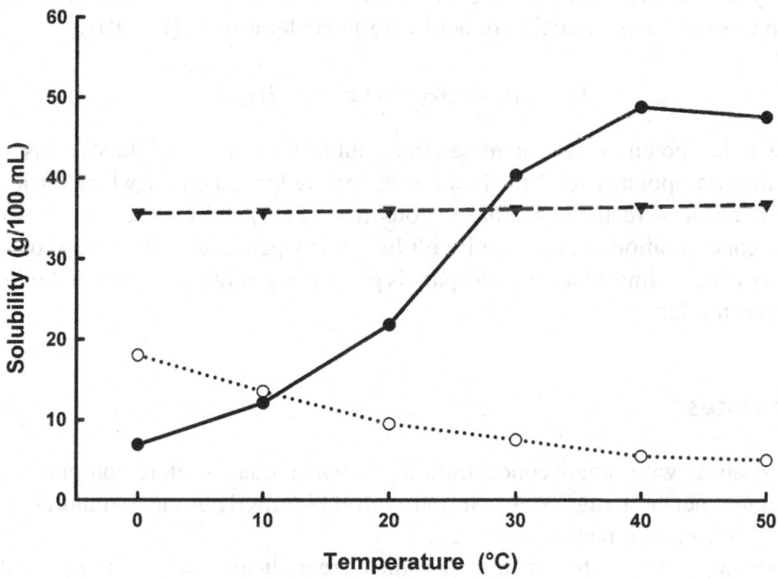

Fig. 3.4 Solubilities of sodium carbonate (*solid dots*), cesium sulfate (*open dots*), and sodium chlorine (*solid triangles*) at water temperatures of 0–50 °C

Dissolution occurs at the surface of minerals and other substances. Thus, dissolution is accelerated by increasing surface area. The more surface area per unit of mass, the quicker a substance will dissolve. Stirring a solution effectively increases the surface area between solute surfaces and solvent thereby accelerating dissolution.

The solubilities of substances often are reported for distilled water, but natural waters already contain various solutes. The solutes already in water can affect the solubility of a substance. For example, if calcium chloride is dissolving in water, the equation is

$$CaCl_2 \rightleftharpoons Ca^{2+} + 2Cl^-. \tag{3.6}$$

Chloride already in the water acts in the same manner as chloride coming from calcium chloride; this lessens the rate of the forward reaction in (3.6) and diminishes the solubility of calcium chloride. This phenomenon is known as the common ion effect.

Ions in a solution that are different from those resulting from the dissolution of a substance can affect the solubility of that substance. This is called the diverse or uncommon ion effect, and it sometimes is known as the salt effect. It occurs because as the total ionic concentration of a water increases, causing interionic attraction to increase in importance. Interaction among ions tends to reduce their effective charges making them less reactive. Because of this, a higher concentration of ions than expected from equilibrium calculations must be present in solution to reach an equilibrium state. This increases the solubility of a substance in water.

Many dissolution reactions are influenced by hydrogen ion concentration or pH. An example is the reaction of acid with ferric hydroxide [$Fe(OH)_3$]

$$Fe(OH)_3 + 3H^+ \rightleftharpoons Fe^{3+} + 3H_2O. \tag{3.7}$$

Low redox potential also increases the solubility of many substances; insoluble ferric iron compounds tend to dissolve at low redox potential when ferric iron (Fe^{3+}) is reduced to more soluble ferrous iron (Fe^{2+}). Of course, low dissolved oxygen concentration is associated with low redox potential. Weathering of some minerals such as limestone and feldspars is greatly accelerated by carbon dioxide as will be seen later.

Some Notes

When dealing with small concentrations in water quality, it is common to use micrograms per liter (μg/L). A solution with 0.006 mg/L of zinc would also be a 6 μg/L solution with respect to zinc.

Some important water quality variables are radicals such as nitrate, nitrite, ammonia, ammonium, phosphate, sulfate, etc. Sometimes, the concentration will be given as the concentration of the radical, and other times, it will be given as the concentration of the element of interest that is contained in the radical. For example, the concentration of ammonia may be given as 1 mg NH_3/L or it may be given as 1 mg NH_3-N/L. It is possible to convert back and forth between the two methods of presenting concentration. In the case of NH_3-N, use of the factor N/NH_3 (14/17 or 0.824): 1 mg NH_3/L × 14/17 = 0.82 mg NH_3-N/L or 1 mg NH_3-N/L ÷ 14/17 = 1.21 mg NH_3/L. Similar reasoning may be used to convert between NO_3^- and NO_3-N, NO_2^- and NO_2^--N, SO_4^{2-} and SO_4^{2-}-S, etc.

It also is useful to note that the dimensions for molarity and normality are moles per liter and equivalents per liter, respectively. Thus, multiplying molarity or normality by volume in liters gives moles and equivalents, respectively. The same logic applies for multiplying millimoles or milliequivalents per milliliter by volume in milliliters.

Equilibrium

When a substance is gradually added to water until saturation is attained, adding more of it will lead to the substance settling to the bottom of the container—no more will dissolve. To illustrate, if a solution is saturated with calcium sulfate, there will be equal molar concentrations of calcium and sulfate in the water and any additional calcium sulfate added to the solution will settle to the bottom.

The equilibrium state that is achieved in a saturated solution conforms to the Law of Mass Action—also frequently known as Le Chatelier's Principle or the Equilibrium Law. This principle basically holds that if a chemical reaction is at equilibrium and conditions of concentration, temperature, volume, or pressure are changed, the reaction will adjust itself to restore the original equilibrium. This idea often is put into equation form as follows:

$$aA + bB \rightleftharpoons cC + dD. \tag{3.8}$$

This equation shows that A and B combine in specific proportions to form C and D also in specific proportions. There is a back reaction, however, in which C and D combine to form A and B. At equilibrium, both the forward and reverse reactions are occurring, but at equal rates, there is no change in concentrations on either side of the equation. At equilibrium, a mathematical relationship exists between the concentrations of the substances on each side of the equation and the equilibrium constant (K)

$$\frac{(C)^c(D)^d}{(A)^a(B)^b} = K \tag{3.9}$$

where the concentrations ideally should be expressed as molar activities (a term to be explained later in this chapter) for solids and in atmospheres of pressure for gases. However, in a general text such as this, calculations are made assuming that molar concentrations and molar activities are equal. Moreover, the equilibrium constant K sometimes is replaced by its negative logarithm, pK; $K = 10^{-9}$ and $pK = 9$. In this book, K will be used in all but one instance where pK is included in a widely used equation.

It is important to realize that until equilibrium is reached in the reaction depicted in (3.8), the quantity $(C)^c(D)^d \div (A)^a(B)^b$ in (3.9) is called the reaction quotient (Q), but when equilibrium is attained $Q = K$. Once equilibrium occurs, addition or removal of any one of the four participants (A, B, C, or D) will disrupt the equilibrium. However, the participants in the reaction will undergo rearrangement in concentrations until equilibrium is attained again and $Q = K$. The participants in mixtures of substances such as illustrated in (3.9) will always react in a manner that reduces stress in the reaction to attain a state of equilibrium.

Relationship of Equilibrium to Thermodynamics

In the simplest terms, the driving force causing a chemical reaction to proceed is the difference in the energy of the products relative to the energy of the reactants. A reaction will progress in the direction of lower energy, and at equilibrium the energy of products equals the energy of reactants. In (3.8), when A and B are brought together, they contain a certain amount of energy. They react to form C and D, and the energy of the "left-hand side" of the equation declines as C and D are formed on the "right-hand side." When the energy on both sides of the equation is equal, a state of equilibrium exists.

In terms of the laws of thermodynamics, the first law states that the amount of energy in the universe is constant, i.e., energy can neither be created nor destroyed. It can, however, be transformed back and forth between heat and work. In a system, heat can be used to do work, work can be changed to heat, almost *et infinitum*. But, at each transformation, some heat will be lost from the system and no longer useful for doing work in the system. Both heat and work usually are reported in the SI unit system as kilojoules (kJ).

In a chemical reaction that has a fixed volume, the output of heat equals the change in internal energy when reactants form products. This output of heat is known as the change in enthalpy (enthalpy is a measure of the internal heat content of a substance), and it is symbolized as ΔH. Thus, $\Delta H = H$ products $-$ H of reactants. In an exothermic reaction, ΔH will have a negative sign because the enthalpy of the reactant is greater than the enthalpy of the products and heat is released to the environment. In an endothermic reaction, heat is absorbed from the environment because the enthalpy of the products is greater than for the reactants (positive ΔH), and heat is required to cause the reactants to form products.

The second law of thermodynamics holds that the universe continuously moves towards greater randomness or disorder. In thermodynamics, disorder or randomness is known as entropy (S). A simple illustration of entropy follows: the degree of entropy of a substance increases from solid, to liquid, to gaseous phases. This results because the complexity of structure of the substance declines from solid to gaseous form. Similarly, compounds have less entropy than their resulting ions upon dissolution, because there are more particles following ionization. The second law is the reason that heat may move only from a warmer substance to a cooler one. Chemical reactions will proceed from lower entropy to higher entropy, and the change in entropy is $\Delta S = S$ products − S reactants.

About 1875, J. W. Gibbs, an American physicist and mathematician, proposed a way to assess chemical reactions based on free energy change in which the energy change was estimated from enthalpy, temperature, and entropy. The free energy change is the amount of the internal energy that can be used to do "work" in the reaction and is not lost as entropy, i.e., free energy $= \Delta H - T\Delta S$ where T is the absolute temperature. The third law of thermodynamics sets absolute zero (−273.15 °C), the temperature at which all molecular movement stops, as the temperature base for thermodynamic considerations. This concept of energy in chemical reactions was referred to as free energy (Garrels and Christ 1965), but it is now called Gibbs free energy (G). Enthalpy and entropy can be measured experimentally and used to estimate the energy of formation of substances. The energy used to form 1 mol of a substance from its basic elements at standard conditions of 1 atm and 25°C is called the standard Gibbs free energy of formation G_f^o. Long lists of G_f^o values are available, and selected values of ΔG_f^o are provided in Table 3.3.

The Gibbs standard-state free energy (ΔG^o) is estimated as follows:

$$\Delta G^o = \Sigma \Delta G_f^o \text{ products} - \Sigma \Delta G_f^o \text{ reactants}. \tag{3.10}$$

The calculation of standard free energy of reaction is illustrated in Ex. 3.12.

Ex. 3.12: *The value of ΔG^o will be calculated for the reaction $HCO_3^- = H^+ + CO_3^{2-}$.*

<u>Solution:</u>
From Table 3.3, ΔG_f^o values are $HCO_3^- = -587.1$ kJ/mol; $H^+ = 0$ kJ/mol; $CO_3^{2-} = -528.1$ kJ/mol. By substitution, we obtain

$$\begin{aligned}\Delta G^o &= \Delta G_f^o CO_3^{2-} + \Delta G_f^o H^+ - \Delta G_f^o HCO_3^- \\ &= (-528.1) + (0) - (-587.1) \\ &= 59 \text{ kJ/mol}.\end{aligned}$$

The following equation may be used to estimate the free energy of reaction during its progress towards the equilibrium state:

$$\Delta G = \Delta G^o + RT \ln Q \tag{3.11}$$

where R = a form of the universal gas law constant (0.008314 kJ/mol/°A), T = absolute temperature (°A), ln = natural logarithm, and Q = the reaction quotient. At equilibrium, $\Delta G = O$ and $Q = K$, and by substitution into (3.11), the standard-state free energy may be expressed as

Table 3.3 Standard Gibbs free energies of formation ΔG_f^o of selected substances[a]

Substance	State[b]	ΔG_f^o (kJ/mol)	Substance	State[b]	ΔG_f^o (kJ/mol)
Al^{3+}	aq	−481.2	Fe_2O_3	c	−741.0
$Al(OH)_3$	am	−1,137.7	I_2	aq	16.43
Ca^{2+}	aq	−553.04	I^-	aq	−51.67
$CaCO_3$	c	−1,128.8	Mn^{2+}	aq	−227.6
$CaSO_4{\cdot}2H_2O$	c	−1,795.9	MnO	c	−363.2
$Ca_3(PO_4)_2$	c	−3,899.5	MnO_2	c	−464.8
$CaHPO_4{\cdot}2H_2O$	c	−2,153.3	MnO_4^{2-}	aq	−503.8
CO_2	g	−394.4	$Mn(OH)_4$	c	−624.8
CO_2	aq	−386.2	Mg^{2+}	aq	−456.01
H_2CO_3	aq	−623.4	$Mg(OH)_2$	c	−833.7
HCO_3^-	aq	−587.1	$MgCO_3$	c	−1,029.3
CO_3^{2-}	aq	−528.1	NO_2^-	aq	−34.5
Cl^-	aq	−131.17	NO_3^-	aq	−110.6
Cl_2	g	0.0	NH_3	aq	−26.6
Cl_2	aq	6.90	NH_4^+	aq	−79.50
HCl	aq	−131.17	OH^-	aq	−157.30
Cu^{2+}	aq	64.98	H_2O	l	−237.19
$CuSO_4{\cdot}5H_2O$	c	−1,879.9	PO_4^{3-}	aq	−1,025.5
H^+	aq	0.0	HPO_4^{2-}	aq	−1,094.1
H_2	g	0.0	$H_2PO_4^-$	aq	−1,135.1
Fe^{2+}	aq	−84.93	H_3PO_4	aq	−1,147.3
Fe^{3+}	aq	−10.6	SO_4^{2-}	aq	−743.0
$Fe(OH)_3$	c	−694.5	H_2S	aq	−27.4
FeS_2	c	−150.6	HS^-	aq	12.61
$FeCO_3$	c	−673.9	S^{2-}	aq	82.74
$Fe(OH)_3$	c	−694.5	H_2SO_4	aq	−742.0

[a]The elemental forms (i.e., Ca, Mg, Al, O_2, H_2, etc.) have $\Delta G_f^o = 0.0$ kJ/mole
[b]aq = aqueous, am = amorphous, c = crystalline, g = gas, l = liquid

$$O = \Delta G^o + RT \ln K$$

$$\Delta G^o = -RT \ln K \tag{3.12}$$

where R = universal gas law constant (0.008314 kJ/°A) and T = absolute temperature (°A; 273.15 + °C).

The term, −RT ln, has the value −5.709 at 25 °C [− (0.008314 kJ/°A) (298.15°A) (2.303) = −5.709], and we may write

$$\Delta G^o = -5.709 \log K. \tag{3.13}$$

Equation 3.13 provides a convenient technique for calculating the equilibrium constant of any reaction for which we know ΔG^o as shown in Exs. 3.13 and 3.14.

Ex. 3.13: *The ΔG^o will be used to estimate K for the reaction $HCO_3^- = CO_3^{2-} + H^+$.*

Solution:
The ΔG^o for the equation was found to be 59 kJ/mol in Ex. 3.12 and

$$\Delta G^o = -5.709 \log K.$$

$$\log K = \frac{\Delta G^o}{-5.709} = \frac{59}{-5.709} = -10.33 \qquad \left(K = 10^{-10.33}\right).$$

This is the K value that is reported in the literature for the dissociation of HCO_3^-.

Ex. 3.14: *The K for the reaction $NH_3 + H_2O = NH_4^+ + OH^-$ will be estimated from ΔG^o.*

Solution:
From Table 3.3, *ΔG^o for the reaction is*

$$\Delta G^o = \left(\Delta G_f^o NH_4^+ + \Delta G_f^o OH^-\right) - \left(\Delta G_f^o NH_3 + \Delta G_f^o H_2O\right)$$

$$\Delta G^o = [-79.50 + (-157.30)] - [(-26.6) + (-237.19)]$$

$$\Delta G^o - (-236.8) - (-263.79) - 26.99.$$

From (3.13)

$$26.99 = -5.709 \log K$$

$$\log K = \frac{26.99}{-5.709} = -4.73$$

$$K = 10^{-4.73}.$$

Temperature affects the equilibrium constant. For temperatures different than 25°C, (3.12) instead of (3.13) must be used to calculate K. The K values for the reaction $HCO_3^- = H^+ + CO_3^{2-}$ at temperatures of 0, 10, 20, 25, 30, and 40°C are: $10^{-11.28}$, $10^{-10.88}$, $10^{-10.51}$, $10^{-10.33}$, $10^{-10.17}$, and $10^{-9.84}$, respectively. The K is almost two orders of magnitude greater at 40 °C than at 0 °C.

Solubility Product

The concept of equilibrium applies to all kinds of chemical reactions, but the solubility product is a special form of Le Chatelier's principle that may be generalized as

$$AB(s) \rightleftharpoons A(aq) + B(aq). \tag{3.14}$$

The concept assumes that there is a large excess of solid phase AB so that the amount that dissolves does not influence its concentration, i.e., the solid phase may be considered unity. This allows the equilibrium constant—also called the solubility product constant (K_{sp})—to be expressed as simply the product of the molar concentrations of A and B in the solution phase

$$(B)(A) = K_{sp}. \tag{3.15}$$

In a case of a more complex molecule we might have

$$A_2B_3(s) \rightleftharpoons 2A(aq) + 3B(aq). \tag{3.16}$$

In this case, the expression for the solubility product constant is

$$(A)^2(B)^3 = K_{sp}. \tag{3.17}$$

The list of selected solubility product constants for compounds of possible interest in water quality in Table 3.4 does not include many common compounds that are highly soluble in water. The reason is that under most conditions, these substances dissolve completely. The solubilities in water of some examples of such compounds are provided (Table 3.2).

Many compounds occur in more than one form. Variation in crystalline structure, and degree of hydration in particular, may influence solubility. Also, an amorphous mineral—one that has not formed the typical crystalline structure of the particular compound—has a different solubility than the crystalline form. Thus, it is not uncommon to find somewhat different solubility product constants and solubility data for the same compound.

The solubility of a compound is not always presented as a simple dissolution in water. Consider the dissolution of gibbsite [$Al(OH)_3$]:

$$Al(OH)_3(s) \rightleftharpoons Al^{3+}(aq) + 3OH^-(aq) \qquad K = 10^{-33}. \tag{3.18}$$

This dissolution also may be presented as

$$Al(OH)_3 + 3H^+ = Al^{3+} + 3H_2O \qquad K = 10^9. \tag{3.19}$$

Either way, the concentration of aluminum ion at equilibrium is the same for a given pH: $(Al^{3+}) = 10^{-6}$ M at pH 5 by both (3.18) and (3.19).

Table 3.4 Selected solubility product constants at 25 °C

Compound	Formula	K_{sp}	Compound	Formula	K_{sp}
Aluminum hydroxide	$Al(OH)_3$	10^{-33}	Lead carbonate	$PbCO_3$	$10^{-13.13}$
Aluminum phosphate	$AlPO_4$	10^{-20}	Lead hydroxide	$Pb(OH)_2$	10^{-20}
Barium carbonate	$BaCO_3$	$10^{-8.59}$	Lead sulfate	$PbSO_4$	$10^{-7.60}$
Barium hydroxide	$Ba(OH)_2$	$10^{-3.59}$	Lead sulfide	PbS	10^{-28}
Barium sulfate	$BaSO_4$	$10^{-9.97}$	Magnesium carbonate	$MgCO_3$	$10^{-5.17}$
Beryllium hydroxide	$Be(OH)_2$	$10^{-21.16}$	Magnesium fluoride	MgF_2	$10^{-10.29}$
Cadmium carbonate	$CdCO_3$	10^{-12}	Manganese carbonate	$MnCO_3$	$10^{-10.65}$
Cadmium sulfate	CdS	10^{-27}	Manganese hydroxide	$Mn(OH)_2$	$10^{-12.7}$
Calcium carbonate	$CaCO_3$	$10^{-8.3}$	Manganese sulfide	MnS	$10^{-10.52}$
Calcium fluoride	CaF_2	$10^{-10.46}$	Mercury carbonate	Hg_2CO_3	$10^{-16.44}$
Calcium hydroxide	$Ca(OH)_2$	$10^{-5.3}$	Mercury sulfate	Hg_2SO_4	$10^{-6.19}$
Calcium magnesium carbonate	$CaCO_3{\cdot}MgCO_3$	$10^{-16.8}$	Mercury sulfide	HgS	$10^{-52.7}$
Calcium phosphate	$Ca_3(PO_4)_2$	$10^{-32.7}$	Nickel carbonate	$NiCO_3$	$10^{-6.85}$
Calcium sulfate	$CaSO_4{\cdot}2H_2O$	$10^{-4.5}$	Nickel hydroxide	$Ni(OH)_2$	$10^{-15.26}$
Cobalt carbonate	$CoCO_3$	10^{-10}	Nickel sulfide	NiS	$10^{-19.4}$
Cobalt sulfide	CoS	$10^{-21.3}$	Silver carbonate	Ag_2CO_3	$10^{-11.08}$
Copper hydroxide	Cu_2O	$10^{-14.7}$	Silver chloride	$AgCl$	$10^{-9.75}$
Copper sulfide	CuS	$10^{-36.1}$	Silver sulfate	Ag_2SO_4	$10^{-4.9}$
Iron carbonate	$FeCO_3$	$10^{-10.5}$	Strontium carbonate	$SrCO_3$	$10^{-9.25}$
Iron fluoride	FeF_2	$10^{-5.63}$	Strontium sulfate	$SrSO_4$	$10^{-6.46}$
Ferrous hydroxide	$Fe(OH)_2$	$10^{-16.31}$	Tin hydroxide	$Sn(OH)_2$	$10^{-26.25}$
Ferrous sulfide	FeS	$10^{-18.1}$	Zinc carbonate	$ZnCO_3$	$10^{-9.84}$
Ferric hydroxide	$Fe(OH)_3$	$10^{-38.5}$	Zinc hydroxide	$Zn(OH)_2$	$10^{-16.5}$
Ferric phosphate	$FePO_4{\cdot}2H_2O$	10^{-16}	Zinc sulfide	ZnS	10^{-25}

Examples of Dissolutions

Some examples showing how to estimate the concentrations of substances in water from chemical reactions and equilibrium constants will be provided in Exs. 3.14–3.16.

Ex. 3.15: *Concentrations of Ca^{2+} and SO_4^{2-} will be calculated for a saturated solution of gypsum ($CaSO_4{\cdot}2H_2O$).*

Solution:
The dissolution equation is

$$CaSO_4 \cdot 2H_2O \rightleftharpoons Ca^{2+} + SO_4^{2-} + 2H_2O$$

and from Table 3.4, K is $10^{-4.5}$.

Because solid phase gypsum and water can be assigned a value of unity, the solubility product expression is

$$(Ca^{2+})(SO_4^{2-}) = 10^{-4.5}.$$

Each gypsum molecule dissolves into one calcium and one sulfate ion; hence, $(Ca^{2+}) = (SO_4^{2-})$ at equilibrium. This allows us to let $x = (Ca^{2+}) = (SO_4^{2-})$, and

$$(x)(x) = 10^{-4.5}$$

$$x = 10^{-2.25}\ \text{M}.$$

This molar concentration is equal to 225 mg/L Ca^{2+} and 540 mg/L sulfate.

Ex. 3.16: *The common ion effect will be illustrated by estimating the calcium concentration expected from the dissolution of gypsum in water already containing 1,000 mg/L ($10^{-1.98}$ M) sulfate, but with 0 mg/L of calcium.*

Solution:
In this situation, we can assume that for calculation purposes, all of the sulfate is from that already in the water, because the water contains more sulfate than expected from equilibrium with gypsum.

$$(Ca^{2+})(SO_4^{2-}) = K_{sp}$$

$$(Ca^{2+}) = \frac{K_{sp}}{(SO_4^{2-})}$$

$$(Ca^{2+}) = \frac{10^{-4.5}}{10^{-1.98}} = 10^{-2.52}\ \text{M}.$$

This is a calcium ion concentration of 121 mg/L as compared to 225 mg/L in Ex. 3.12.

Ex. 3.17: *The solubility of calcium fluoride (CaF_2) will be calculated.*

Solution:
From Table 3.4, the K_{sp} for CaF_2 is $10^{-10.46}$, and the reaction is

$$CaF_2 = Ca^{2+} + 2F^-.$$

The solubility product expression is

$$(Ca^{2+})(F^-)^2 = 10^{-10.46}.$$

Letting $(Ca^{2+}) = X$ and $(F^-) = 2X$,

$$(X)(2X)^2 = 10^{-10.46}$$

$$4X^3 = 10^{-10.46} = 3.47 \times 10^{-11}$$

$$X^3 = 8.67 \times 10^{-12} = 10^{-11.06}$$

$$X = 10^{-3.69}\text{M} = 0.0002\ \text{M}.$$

Thus, $(Ca^{2+}) = 0.0002$ M (8.02 mg/L) and $(F^-) = 0.0002$ M $\times 2 = 0.0004$ M (7.6 mg/L).

Electrostatic Interactions

In natural waters and other solutions, concentrations of ions often are great enough that electrostatic effects among ions—including ion pairing—cause ions to react to a lesser degree than expected from measured molar concentrations, e.g., the reacting concentration is less than the measured concentration. The ratio of the reacting concentration:measured concentration is called the activity coefficient. For dilute solutions, such as most natural waters, the activity coefficient of a single ion may be calculated with the Debye-Hückel equation

$$\log\gamma_i = -\frac{(A)(Z_i)^2(I)^{1/2}}{1 + (B)(a_i)(I)^{1/2}} \qquad (3.20)$$

where γ_i = the activity coefficient for the ion i, A and B are dimensionless constants for standard atmospheric pressure and the temperature of the water (Table 3.5), Z_i = the valence of ion i, a_i = the effective size of ion i (Table 3.6), and I = the ionic strength. In (3.20), the 10^{-8} of a_i in the denominator is cancelled by the 10^{-8} of B in the numerator. Thus, the a_i and B values usually are substituted into the equation without their power of 10 factors.

Table 3.5 Values A and B for substitution in the Debye-Hückel equation at standard atmospheric pressure

Temperature (°C)	A	B ($\times 10^{-8}$)
0	0.4883	0.3241
5	0.4921	0.3249
10	0.4960	0.3258
15	0.5000	0.3262
20	0.5042	0.3273
25	0.5085	0.3281
30	0.5130	0.3290
35	0.5175	0.3297
40	0.5221	0.3305

Table 3.6 Values for ion size (a_i) for use in the Debye-Hückel equation (Hem 1970)

$a_i \times 10^{-8}$	Ion
9	Al^{3+}, Fe^{3+}, H^+
8	Mg^{2+}
6	Ca^{2-}, Cu^{2+}, Zn^{2+}, Mn^{2+}, Fe^{2+}
5	$CO_3{}^{2-}$
4	$PO_4{}^{3-}$, $SO_4{}^{2-}$, $HPO_4{}^{2-}$, Na^+, $HCO_3{}^-$, $H_2PO_4{}^-$
3	OH^-, HS^-, K^+, Cl^-, $NO_2{}^-$, $NO_3{}^-$, $NH_4{}^+$

The ionic strength of a solution may be calculated as

$$I = \sum_{i}^{n} \frac{(M_i)(Z_i)^2}{2} + \cdots + \frac{(M_n)(Z_n)^2}{2} \tag{3.21}$$

where M = the measured concentration of individual ions.

The activity of an ion is

$$(M_i) = \gamma M_i \, [M] \tag{3.22}$$

where (M_i) = the activity and [M] = measured molar concentration. Of course, one can use millimolar concentrations instead of molar concentrations if desired in (3.21) and (3.22) as illustrated in Ex. 3.18.

Ex. 3.18: The ionic strength of a water will be estimated.

Solution:

The measured concentrations (mg/L) must be converted to millimolar concentration:

Ion	*Measured concentration (mg/L)*	*Conversion factor (mg/mM)*	*Concentration (mM)*
$HCO_3{}^-$	*136*	*61*	*2.23*
$SO_4{}^{2-}$	*28*	*96*	*0.29*
Cl^-	*29*	*35.45*	*0.82*

(continued)

Ion	Measured concentration (mg/L)	Conversion factor (mg/mM)	Concentration (mM)
Ca^{2+}	41	40.08	1.02
Mg^{2+}	9.1	24.31	0.37
Na^{+}	2.2	23	0.10
K^{+}	1.2	39.1	0.03

Substituting into (3.21),

$$I = \sum_{i}^{n} \frac{(2.23)(1)^2}{2} + \frac{(0.29)(2)^2}{2} + \frac{(0.82)(1)^2}{2} + \frac{(1.02)(2)^2}{2} + \frac{(0.37)(2)^2}{2} + \frac{(0.1)(1)^2}{2} + \frac{(0.03)(1)^2}{2}$$

$$I = 4.96 \text{ mM } (\textit{also equal to } 0.00496 \text{ M}).$$

The ionic strength of the solution can be used in the Debye-Hückel equation to estimate activity coefficients for individual ions as will be shown in Ex. 3.19.

Ex. 3.19: *The activities will be estimated for Mg^{2+} and Cl^{-} in the water from Ex. 3.18.*

<u>Solution:</u>
Using (3.20) and obtaining the variables A, B, and a_i from Tables 3.5 and 3.6,

$$\log \gamma_{Mg} = -\frac{(0.5085)(2)^2(0.00496)^{1/2}}{1 + (0.3281)(8)(0.00496)^{1/2}}$$

$$\log \gamma_{Mg} = -0.12090$$

$$\gamma_{Mg} = 0.76.$$

$$\log \gamma_{Cl} = -\frac{(0.5085)(1)^2(0.00496)^{1/2}}{1 + (0.3281)(3)(0.00496)^{1/2}}$$

$$\log \gamma_{Cl} = -0.0335$$

$$\gamma_{Cl} = 0.93.$$

Activities will be calculated with (3.22),

$$(Mg^{2+}) = 0.76\ [0.37 \text{ mM}] = 0.28 \text{ mM}$$

$$(Cl^{-}) = 0.93\ [0.82 \text{ mM}] = 0.76 \text{ mM}.$$

Monovalent ion activities deviate less from measured concentrations than do divalent ion activities as can be seen in Ex. 3.19. The discrepancy between measured concentrations and activities also increases with increasing ionic strength.

The hydrogen ion concentration as calculated from pH measured with a glass electrode is an activity term and needs no correction. The activities of solids and water are taken as unity. Under conditions encountered in natural waters, the measured concentration of a gas in atmospheres may be used without correction to activity.

Ion Pairs

Although the Debye-Hückel equation is widely used in calculating activities of single ions, some of the cations and anions in solution are strongly attracted to each other and act as if they are un-ionized or of lesser or different charge than anticipated. Ions that are strongly attracted in this manner are called ion-pairs. To illustrate, Ca^{2+} and SO_4^{2-} form the ion-pair $CaSO_4^0$, Ca^{2+} and HCO_3^- form $CaHCO_3^+$, and K^+ and SO_4^{2-} form KSO_4^-. The degree to which ions from ion-pairs in a solution is a function of the equilibrium constant for the formation of the particular ion pair and may be handled as illustrated for the ion pair $CaHCO_3^+$:

$$CaHCO_3^+ = Ca^{2+} + HCO_3^- \tag{3.23}$$

$$\frac{(Ca^{2+})(HCO_3^-)}{CaHCO_3^+} = K. \tag{3.24}$$

The activity coefficients of ion-pairs are taken as unity. Equilibrium constants for ion-pairs formed by major ions in natural water are given in Table 3.7. These constants can be used to estimate the concentration of an ion-pair from the ion-pair equation. This calculation is illustrated in Ex. 3.20.

Ex. 3.20: *The concentration of the magnesium sulfate ion pair will be estimated for a water with 2.43 mg/L (10^{-4} M) magnesium and 9.6 mg/L (10^{-4} M) sulfate.*

Solution:

The equation from Table 3.5 *is*

$$Mg\,SO_4^0 \rightleftharpoons Mg^{2+} + SO_4^{2-} \qquad K = 10^{-2.23}$$

and $$\frac{(Mg^{2+})(SO_4^{2-})}{(MgSO_4^0)} = 10^{-2.23}$$

Table 3.7 Equilibrium constants (K) at 25 °C and zero ionic strength for ion-pairs in natural waters (Adams 1971)

Reaction	K
$CaSO_4{}^0 = Ca^{2+} + SO_4{}^{2-}$	$10^{-2.28}$
$CaCO_3{}^0 = Ca^{2+} + CO_3{}^{2-}$	$10^{-3.20}$
$CaHCO_3{}^+ = Ca^{2+} + HCO_3{}^-$	$10^{-1.26}$
$MgSO_4{}^0 = Mg^{2+} + SO_4{}^{2-}$	$10^{-2.23}$
$MgCO_3{}^0 = Mg^{2+} + CO_3{}^{2-}$	$10^{-3.40}$
$MgHCO_3{}^+ = Mg^{2+} + HCO_3{}^-$	$10^{-1.16}$
$NaSO_4{}^- = Na^+ + SO_4{}^{2-}$	$10^{-0.62}$
$NaCO_3{}^- = Na^+ + CO_3{}^{2-}$	$10^{-1.27}$
$NaHCO_3{}^0 = Na^+ + HCO_3{}^-$	$10^{-0.25}$
$KSO_4{}^- = K^+ + SO_4{}^{2-}$	$10^{-0.96}$

These constants can be used to estimate the concentration of an ion pair from the ion pair equation. This calculation is illustrated in Ex. 3.20

$$(MgSO_4^0) = \frac{(10^{-4})(10^{-4})}{10^{-2.23}} = 10^{-5.77}\ \mathrm{M}.$$

Note: About 0.041 mg/L of magnesium and 0.16 mg/L of sulfate are tied up in the magnesium sulfate ion-pair, but indistinguishable from free ions of magnesium and sulfate by the usual analytical methods.

Analytical methods (specific ion electrodes excluded) do not distinguish between free ions and ion-pairs. Sulfate in solution might be distributed among $SO_4{}^{2-}$, $CaSO_4{}^0$, $MgSO_4{}^0$, $KSO_4{}^-$, and $NaSO_4{}^-$. The total sulfate concentration may be measured; $SO_4{}^{2-}$ can only be calculated. As actual ionic concentrations are always less than measured ionic concentrations in solutions containing ion-pairs, ionic activities calculated directly from analytical data, as done in Ex. 3.16, are not exact. Adams (1971) developed a method for correcting for ion-pairing in the calculation of ionic activities and demonstrated a considerable effect of ion-pairing on ionic strength, ionic concentrations, and ionic activities in solutions. The method of correcting for ion-pairing involves: (1) using measured ionic concentrations to calculate ionic strength assuming no ion-pairing, (2) calculating ionic activities assuming no ion-pairing, (3) calculating ion-pair concentrations with respective ion-pair equations, equilibrium constants, and initial estimates of ionic activities, (4) revising ionic concentrations and ionic strength by subtracting the calculated ion-pair concentrations, and (5) repeating steps 2, 3, and 4 until all ionic concentrations and activities remain unchanged with succeeding calculations. Adams (1971, 1974) discussed all aspects of the calculations and gave examples. The iterative procedure is tedious and slow unless it is programmed into a computer. The procedure was developed for soil solutions, but it is equally applicable to natural waters.

Boyd (1981) calculated activities of major ions in samples of natural water with analytical data uncorrected for ion-pairing and analytical data corrected for ion-pairing. Ion-pairing had little effect on ionic activity calculations for weakly mineralized water ($I < 0.002$ M). For more strongly mineralized waters, ionic activities corrected for ion-pairing were appreciably smaller than uncorrected ionic activities. Ion pairing by major ions is particularly significant in ocean and other saline waters. As a general rule, if waters contain less than 500 mg/L of total dissolved ions, ion-pairing may be ignored in calculating ionic activities unless highly accurate data are required. For simplicity, equilibrium considerations in this book will assume that activities and measured molar concentrations are the same.

Significance

Most chemical reactions affecting water quality are of the equilibrium type. Thus, an understanding of the concepts in this chapter is essential background for the student of water quality.

From a broader perspective in water quality investigations, concentrations of water quality variables typically are measured rather than calculated. Nevertheless, knowledge of the principles of solubility and use of solubility products and equilibrium constants to calculate anticipated concentrations can be helpful in explaining observed concentrations and predicting changes in water quality. The concept of chemical equilibrium also can be especially useful in predicting changes in concentrations of water quality variables that may result from addition of substances to water by natural or anthropogenic processes.

References

Adams F (1971) Ionic concentrations and activities in soil solutions. Soil Sci Soc Am Proc 35:420–426

Adams F (1974) Soil solution. In: Carson WE (ed) The plant root and its environment. University Press of Virginia, Charlottesville, pp 441–481

Boyd CE (1981) Effects of ion-pairing on calculations of ionic activities of major ions in freshwater. Hydrobio 80:91–93

Garrels RM, Christ CL (1965) Solutions, minerals, and equilibria. Harper & Row, New York

Hem JD (1970) Study and interpretation of the chemical characteristics of natural water. Water-supply paper 1473. United States Geological Survey, United States Government Printing Office, Washington, DC

Dissolved Solids

4

Abstract

The total dissolved solids (TDS) concentration in freshwater is determined by passing water through a 2 μm filter, evaporating the filtrate to dryness, and reporting the weight of the solids remaining after evaporation in milligrams per liter. The TDS concentrations in natural freshwaters typically range from about 20–1,000 mg/L; the solids consist mainly of bicarbonate (and carbonate at pH above 8.3), chloride, sulfate, calcium, magnesium, potassium, sodium, and silicate. The concentration of TDS in inland waters is controlled mainly by geological and climatic factors. The most weakly mineralized waters are found in areas with high-rainfall and heavily leached or poorly developed soils. The most strongly mineralized waters usually occur in arid regions. Examples of TDS concentrations in different regions are presented, and reasons for differences in TDS concentrations among regions and sources of water are discussed. Although the major dissolved inorganic substances in natural water are essential for life, minor dissolved constituents in water often have the greatest effect on aquatic organisms. The main effect of TDS concentration on animals and plants usually is related to osmotic pressure that increases with greater TDS concentration. The average TDS concentration in seawater is about 35,000 mg/L. However, it is more common to use salinity or specific conductance as an indicator of the degree of mineralization of saline and marine waters.

Keywords

Major ions in water • Specific conductance • Ion balance • Dissolved organic matter • Colligative properties

C.E. Boyd, *Water Quality*, DOI 10.1007/978-3-319-17446-4_4

Introduction

Solids in water originate from two main sources: dissolution and suspension of minerals and organic matter from soils and other geological formations, and living aquatic microorganisms and their decaying remains. Dissolved solids are particles so small that they are considered to be in true solution. Chemists usually consider particles 1 μm in diameter or less to be dissolved, but the method for determining dissolved solids in natural waters and waste waters (Eaton et al. 2005) uses a 2.0 μm filter to separate dissolved from suspended solids.

Dissolved solids are largely inorganic, and waters with high dissolved solids concentrations are said to be highly mineralized. Concentrations of dissolved solids tend to vary greatly ranging from near 0 mg/L to over 100,000 mg/L. In regions with highly infertile soils and geological formations of dissolution-resistant rocks, there may be only 10–30 mg/L of dissolved solids. Waters from areas with karst (limestone) geology may have 200–400 mg/L dissolved solids, while in arid regions concentrations above 1,000 mg/L are common. The ocean averages 34,500 mg/L in dissolved solids, but closed-basin lakes like the Dead Sea and Great Salt Lake may contain over 250,000 mg/L.

There is no widely-accepted definition of freshwater, but in most definitions the total dissolved solids limit is 500–1,000 mg/L. Thus, some inland waters are saline waters rather than freshwaters. Drinking water should not exceed about 500 mg/L total dissolved solids.

Raindrops acquire small amounts of dissolved solids from the atmosphere. When rainwater reaches the earth, it contacts vegetation, soil and geological formations and dissolves additional mineral and organic matter. Mineral matter dissolved in water consists largely of calcium, magnesium, sodium, potassium, chloride, sulfate, bicarbonate, and carbonate. Most natural waters also have a significant concentration of silicon as undissociated silicic acid. Concentrations of other dissolved inorganic substances such as hydrogen and hydroxyl ions, nitrate, ammonium, phosphate, borate, iron, manganese, zinc, copper, cobalt, and molybdenum are comparatively small, but they can have a profound influence on water quality.

The degree of mineralization of inland water can be highly variable from place to place and depends upon the solubility of minerals, length of time and conditions of contact of water with minerals, and concentration of substances through evaporation. While ocean waters tend to be similar in mineral content worldwide, estuarine waters vary greatly in degree of mineralization. The most direct way of assessing the mineralization of water is to measure the concentrations of the individual inorganic constituents directly and sum them or to measure them collectively by total dissolved solids analysis. However, both methods are time-consuming and expensive, so indirect methods of assessing mineralization such as analyses for salinity or specific conductance have been developed.

Soluble organic compounds reach water from the decay of dead plant and animal matter and by the excretion of organic matter by aquatic organisms. Thousands of organic compounds occur in water, but dissolved organic compounds usually are measured collectively for convenience. Dissolved organic matter concentrations are highest in waters that have had contact with organic residues or in nutrient rich waters with much biological activity.

Dissolved Solids and Their Measurement

Total Dissolved Solids

A simple method for determining the concentration of total dissolved solids (TDS) is to filter the water through a glass fiber filter with 2 μm or smaller opening, evaporate the filtrate, and weigh the residue to constant weight (Ex. 4.1). During drying, carbon dioxide will be lost from bicarbonate as follows:

$$Ca^{2+} + 2HCO_3^- \underset{\Delta}{\rightarrow} CaCO_3 \downarrow + H_2O \uparrow + CO_2 \uparrow . \quad (4.1)$$

The weight loss of solids during evaporation would be about 38 % of the portion of the TDS concentration resulting from calcium and bicarbonate. The residue left upon evaporation is mainly dissolved inorganic matter, but also will contain organic matter. In spite of these shortcomings, the total dissolved solids concentration is indicative of the degree of mineralization of freshwaters. The procedure, unfortunately, does not work well for saline waters, because the residue left after evaporation will be large and absorb moisture from the air making it impossible to weigh the residue accurately.

Ex. 4.1: *A 100-mL water sample that has passed a 2 μm filter is dried at 102 °C in a 5.2000-g dish. The dish and residue weigh 5.2100 g. The TDS concentration will be estimated.*

Solution:

$$\textit{Weight of residue} = (5.2100 - 5.2000)\text{g} = 0.0100 \text{ g } \textit{or } 10 \text{ mg}$$

$$TDS = 10 \text{ mg} \times \frac{1{,}000 \text{ mL/L}}{100 \text{ mL } \textit{sample}} = 100 \text{ mg/L}.$$

Portable meters are available for measuring dissolved solids. These devices operate on the principle that the ability of water to conduct an electrical current (conductance) increases as a function of ionic concentrations. Measurements can be made rapidly with these meters, and they are especially useful for saline waters.

Salinity

Salinity refers to the total concentration of all ions in water. The salinity concentration in milligrams per liter usually is 95 % or more of the total dissolved solids in a water sample. Salinity can be determined from a complete analysis of water by summing the concentrations of all ions, but it normally is estimated indirectly.

Dissolved ions or salinity increases the density of water (Table 1.4). Hydrometers that have been calibrated versus different salinities are available commercially. These devices often are used to obtain a direct estimate of salinity.

The concentration of dissolved ions in water affects the refractive index, so a refractometer can be calibrated against salinity concentration and used to estimate salinity. This hand-held device (Fig. 4.1) is called a salinometer, and it gives a direct reading of salinity in grams per liter (g/L) that is the same as parts per thousand (ppt).

Hydrometers and salinometers are not highly accurate at salinities below 3 or 4 ppt. Conductivity meters can be calibrated versus salinity concentration and used to measure salinity directly. In seawater, salinity also can be estimated from chloride concentration. The approximate relationship is

$$\text{Salinity (mg/L)} = 30 + (1.805)(\text{Cl}^- \text{ in mg/ L}). \tag{4.2}$$

Specific Conductance

The current (I) flowing in an electrical circuit is related to the voltage (V) applied and the resistance (R) of the circuit as defined by Ohm's law (I = V/R). Thus, it is important to be able to measure the resistance of substances used as conductors of

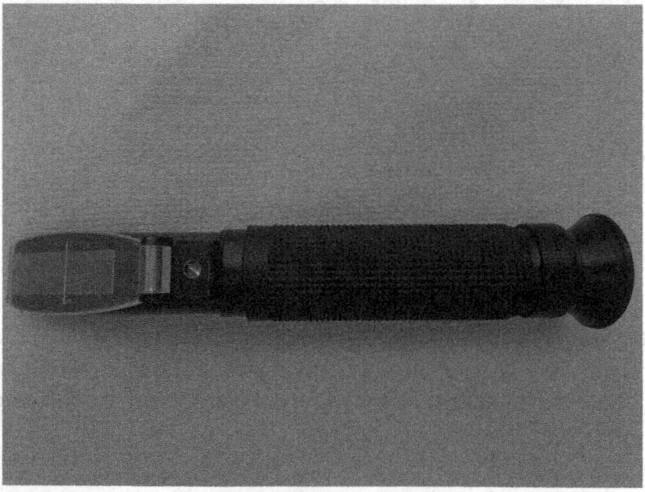

Fig. 4.1 A refractometer or salinometer for measuring salinity

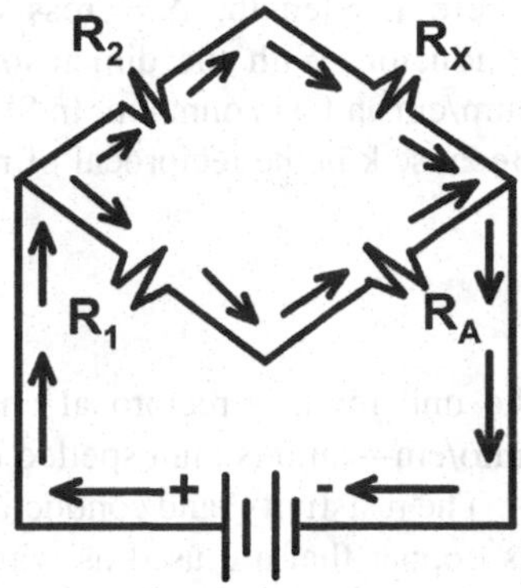

Fig. 4.2 A Wheatstone bridge circuit for determining an unknown resistance (R_X). Resistances R_1 and R_2 are known resistances, and resistance R_A is an adjustable resistance

electricity in circuits. An unknown resistance can be measured with a Wheatstone bridge circuit that consists of four resistors (Fig. 4.2). Resistors R_1 and R_2 have a known resistance, resistor R_A has an adjustable resistance, and the resistance R_X is the material (substance or object) for which the resistance needs to be determined. By adjusting the resistance at R_A, the circuit can be balanced so that the same amount of current flows through the upper part (U) of the circuit containing R_X and R_2 as through the lower part (L) of the circuit containing R_1 and R_A. This condition will exist when there is no current flow through the galvanometer. When this state is achieved, the following relationships exist between current (I) and resistance (R):

$$I_U R_2 = I_L R_1$$

$$I_U R_X = I_L R_A.$$

By dividing the upper expression by the lower and solving for R_X, the resistance can be measured

$$\frac{I_U R_2}{I_U R_X} = \frac{I_L R_1}{I_L R_A}$$

$$R_X = \frac{R_2 R_A}{R_1}. \qquad (4.3)$$

An instrument called a conductivity meter that functions according to the principle of the Wheatstone bridge circuit is used to measure the conductivity of water. Notice, the conductivity rather than the resistance of water usually is determined. Conductance (G) is the reciprocal of the resistance. Resistance and conductance depend upon the dimensions of the conductor, and the terms resistivity and conductivity are used to describe the resistance of a unit length and cross-sectional area of the conductor

$$R = \rho \frac{L}{A} \quad \text{or} \quad \rho = R \frac{A}{L} \qquad (4.4)$$

where L = length, A = cross-sectional area, and ρ = specific resistance of the conductor. With the dimensions of the conductor in centimeters, R will be in ohm/cm and ρ in ohm·cm. In SI units, R will be in siemens/cm and ρ in siemens·cm. Because k is the reciprocal of resistance, that is

$$k = \frac{1}{\rho} \tag{4.5}$$

the unit for k is reciprocal ohm·cm (1/ohm·cm) which usually is referred to as mho/cm—mho is ohm spelled backwards.

The resistivity (and conductivity) of different materials vary greatly. Metals such as copper that are used as wires in electrical circuits conduct electricity via free electrons not attached to cations of the metal than can move around in the metal. In wires, the resistance (R) is greater for an equal length of small diameter wire than for a larger diameter wire, but resistivity (ρ) is equal for both because ρ is a property of the metal and R is the product of ρ multiplied by (L/A). The same reasoning can be applied to conductance (G) and k (conductivity).

Pure water is not a good conductor of electricity, because electrical current in liquids is conducted by dissolved ions. But, as the concentration of ions in water increases, the conductivity increases. The measurement of conductivity or specific conductance is an excellent indicator of the degree of mineralization of a water.

The specific conductance of water is the conductance afforded by a cubic mass of water with edges 1 cm long. A 0.0100 N KCl solution has a specific conductance of 0.001413 mho/cm at 25 °C. To avoid small decimal numbers, it is customary to report specific conductance as micromhos per centimeter, or in the case of 0.0100 N KCl, 1,413 μmho/cm. The unit, μmho/cm, is replaced with its equivalent in the International System (SI) of Units by millisiemens per centimeter (mS/cm); 1 μmho/cm = 1 mS/cm. Some meters calibrated in SI units read in millisiemens per meter (mS/m) and 1 mS/m = 10 μmho/cm.

The conductance of a water sample will decrease with decreasing temperature, and it can be adjusted to 25 °C (k_{25}) with the equation

$$k_{25} = \frac{k_m}{1 + 0.0191(T - 25)} \tag{4.6}$$

where k_m = measured k at any temperature (T in °C). This equation assumes that the specific conductance measurement has been corrected for the cell constant—the value necessary to adjust for variation in the dimensions of the electrode. Specific conductance increases with increasing temperature as illustrated in Ex. 4.2. But, most modern conductivity meters have an automatic temperature compensator built in, and readings are automatically adjusted to 25 °C.

Table 4.1 Equivalent ionic conductance (λ) at infinite dilution and 25 °C for selected ions (Sawyer and McCarty 1967; Laxen 1977)

Cation	λ (mho cm^2/eq.)	Anion	λ (mho cm^2/eq.)
Hydrogen (H^+)	349.8	Hydroxyl (OH^-)	198.0
Sodium (Na^+)	50.1	Bicarbonate (HCO_3^-)	44.5
Potassium (K^+)	73.5	Chloride (Cl^-)	76.3
Ammonium (NH_4^+)	73.4	Nitrate (NO_3^-)	71.4
Calcium (Ca^{2+})	59.5	Sulfate (SO_4^{2-})	79.8
Magnesium (Mg^{2+})	53.1		

Ex. 4.2: *The specific conductance of 1,203 μmho/cm for a sample at 22 °C will be adjusted to 25 °C.*

Solution:
Using (4.6),

$$k_{25} = \frac{1,203}{1 + 0.0191\ (22 - 25)} = 1,276\ \mu\text{mho/cm}.$$

Only ions carry a current, but all ions do not carry the same amount of current (Sawyer and McCarty 1967; Laxen 1977). The equivalent conductances of common ions at infinite dilution are provided in Table 4.1. The equivalent conductance of a particular ion is related to specific conductance as follows:

$$\text{k} = \frac{\lambda \text{N}}{1,000} \tag{4.7}$$

where λ = equivalent conductance and N = the normality of the ion in solution. The specific conductance of a solution will equal the sum of the specific conductances of the individual ions. If specific conductance is calculated with (4.7), it usually will be larger than the measured specific conductance (Ex. 4.3).

Ex. 4.3: *The specific conductance of 0.01 N potassium chloride will be estimated and compared to the measured value of 1,413 μmho/cm.*

Solution:

$$(i)\quad k_{K^+} = \frac{(73.5\ \text{mho}\cdot\text{cm}^2/\text{eq})(0.01\ \text{eq/L})}{1,000\ \text{cm}^3/\text{L}} = 0.000735\ \text{mho/cm}.$$

$$k_{Cl^-} = \frac{(76.3\ \text{mho}\cdot\text{cm}^2/\text{eq})(0.01\ \text{eq/L})}{1,000\ \text{cm}^3/\text{L}} = 0.000763\ \text{mho/cm}.$$

Table 4.2 Relationship between concentration of potassium chloride solutions and measured specific conductance at 25 °C

	Specific conductance (μmho/cm)	
Concentration (*N*)	Measured	Calculated
0	0	
0.0001	14.94	14.98
0.0005	73.90	
0.001	147.0	149.8
0.005	717.8	
0.01	1,413	1,488
0.02	2,767	
0.05	6,668	
0.1	12,900	14,980
0.2	24,820	
0.5	58,640	
1.0	111,900	150,800

Specific conductance values estimated with (4.7) are provided for selected concentrations (Eaton et al. 2005)

$$k_{KCl} = k_{K^+} + \; = k_{Cl^-} = 0.001498\ \text{mho/cm} = 1,498\ \mu\text{mho/cm}.$$

(*ii*) *This is greater than the measured value of 1,413 μmho/cm.*

The reason that measured specific conductance is less than the estimated value lies in electrostatic interactions among anions and cations in solution which neutralize a portion of the charge on the ions and reduces their ability to conduct electricity (see section on "Electrostatic Interactions" in Chap. 3). The effect becomes greater as concentration increases as illustrated in Table 4.2 for standard KCl solutions. The electrostatic effect is even greater in solutions containing divalent and trivalent ions.

Because all ions do not have the same equivalent conductance, the specific conductance of water samples containing the same concentration of total ions may have different specific conductance values. For example, a water containing mainly potassium, chloride, and sulfate would have a higher specific conductance than a water containing the same amount of total ions but with calcium, magnesium, and bicarbonate predominating (refer to equivalent conductances in Table 4.1). Nevertheless, for a particular water body, there usually is a strong, positive correlation between specific conductance and TDS concentration as illustrated in Fig. 4.3. This results because there usually is a fairly constant proportionality among the ions in surface waters from a particular region irrespective of the TDS concentration. For the river in Fig. 4.3, the factor for converting specific conductance to total dissolved solids was 0.61. Silipajarn et al. (2004) reported a factor of 0.67 for surface waters in the Blackland Prairie region of Alabama. The factor is about 0.70 for seawater.

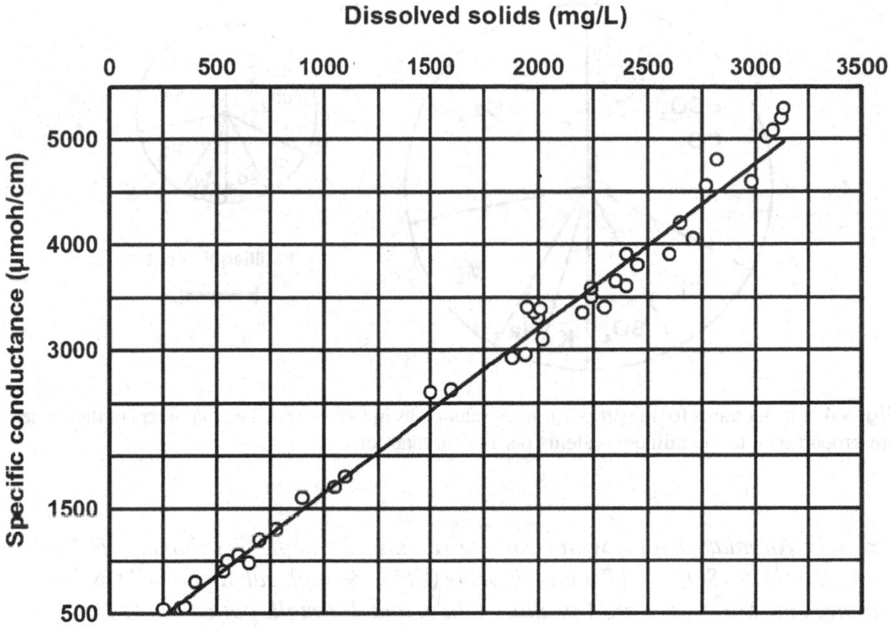

Fig. 4.3 Dissolved solids determined by water analysis compared with specific conductance of daily samples from Gila River at Bylas, Arizona, in a 12-month period (Hem 1970)

Ionic proportions are more constant in ocean or estuarine water. Thus, salinity and TDS meters are more reliable for general use in estuarine or ocean waters than in inland waters.

The most weakly mineralized water is distilled water with a specific conductance usually less than 2 µmho/cm. Rainwater normally has specific conductance less than 50 µmho/cm. In humid areas, inland surface waters seldom have conductances above 500 µmho/cm, but values may exceed 5,000 µmho/cm in arid regions. Potable waters normally have specific conductances of 50–1,500 µmho/cm. The upper limit for specific conductance of freshwater is around 1,500 µmho/cm, while the specific conductance of seawater is around 50,000 µmho/cm.

Cation-Anion Balance

The principle of electrical neutrality requires that the sum of the equivalent weights of positively-charged ions (cations) equal the sum of the equivalent weights of the negatively-charged ions (anions). Because the major ions usually represent most of the dissolved ions, there will be almost equal equivalent amounts of major cations and major anions in water. The cation-anion balance concept is useful for checking the validity of water analyses. In an accurate analysis, the sum of the milliequivalents of major cations and major anions should be nearly equal as illustrated in Ex. 4.4.

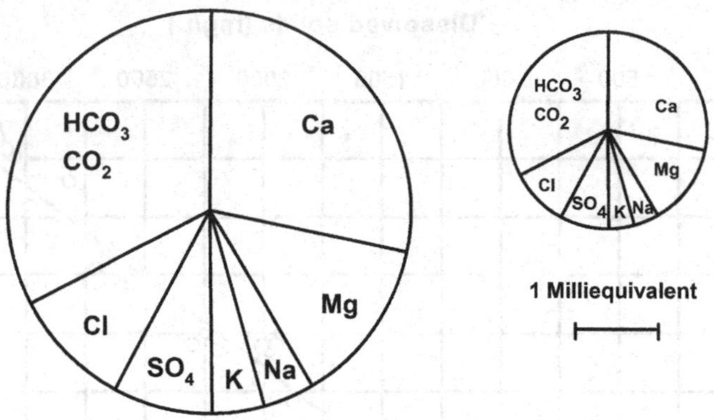

Fig. 4.4 Pie diagrams for the proportions of major ions in two waters. The diameters of the circles are proportional to the milliequivalents per liter of total ions

Ex. 4.4: *An analysis of a water sample reveals 121 mg/L bicarbonate (HCO_3^-), 28 mg/L sulfate (SO_4^{2-}), 17 mg/L chloride (Cl^-), 39 mg/L calcium (Ca^{2+}), 8.7 mg/L magnesium (Mg^{2+}), 8.2 mg/L sodium (Na^+), and 1.4 mg/L potassium (K^+). Is this analysis likely accurate?*

Solution:

Anions

$121\ mg\ HCO_3/L \div 61\ mg\ HCO_3/meq = 1.98\ meq\ HCO_3/L$
$28\ mg\ SO_4/L \div 48\ mg\ SO_4/meq = 0.58\ meq\ SO_4/L$
$17\ mg\ Cl/L \div 35.45\ mg\ Cl/meq = 0.48\ meq\ Cl/L$
Total $= 3.04\ meq\ anions/L$

Cations

$39\ mg\ Ca/L \div 20.04\ mg\ Ca/meq = 1.95\ meq\ Ca/L$
$8.7\ mg\ Mg/L \div 12.16\ mg\ Mg/meq = 0.72\ meq\ Mg/L$
$8.2\ mg\ Na/L \div 23\ mg\ Na/meq = 0.36\ meq\ Na/L$
$1.4\ mg\ K/L \div 39.1\ mg\ K/meq = 0.04\ meq\ K/L$
Total $= 3.07\ meq\ cations/L$

The anions and cations balance closely, so the analysis probably is quite accurate.

The anion-cation balance concept also allows the use of pie diagrams to provide quick comparisons of the equivalent proportions of ions and degrees of mineralization of different waters (Fig. 4.4). One-half of each pie is for anions and the other half for cations. The proportion of a pie given to an ion is proportional to the ion's concentration in milliequivalents per liter. For comparing the degree of mineralization of different waters, the diameters of the pies can be made proportional to the total milliequivalents of ions in a sample.

The anion-cation balance may not always provide irrefutable proof of the accuracy of a water analysis. The author once sent a sample of water from the Pacific Ocean along the coast of New Caledonia, a rather remote island a considerable distance from the northwest coast of Australia, to a laboratory for analyses of major ions. The water had a specific conductance measured at sampling of 49,500 μmho/cm. The laboratory reported a similar specific conductance, and the anion-cation ratio was 0.98 suggesting an accurate analysis of major ions. However, upon adding, the total measured concentrations of major ions provided by the laboratory was 65,000 mg/L. This did not agree with the expected total for the concentrations of major ions based on specific conductance—ocean water should have around 35,000 mg/L of total ions and a specific conductance of about 50,000 μmho/cm. Further investigation revealed that the instrument used for the analysis was incorrectly calibrated and had given erroneously high values for all major ions.

Alkalinity and Hardness

Most readers will have heard about the total alkalinity and total hardness of water. The alkalinity results primarily from bicarbonate and carbonate, while hardness is caused mainly by calcium and magnesium. Hard waters contain a high concentration of calcium, magnesium, or both, but they may or may not have a high alkalinity. These two variables are discussed in Chaps. 8 and 9, respectively.

Concentrations of Major Ions in Natural Waters

As a general rule, TDS concentrations in natural water bodies increase in the following order: rainfall < surface water < groundwater < estuarine water < ocean water and water of closed basin lakes. Some closed-basin lakes have greater TDS concentration than ocean water.

Rainwater

The atmosphere contains salts of marine origin, dust, products of combustion, and nitrate produced by electrical activity in clouds. The particles in the atmosphere are swept out by falling rain, and as a result, rainwater contains dissolved mineral matter. The average concentration of seven mineral constituents in rainwater collected from 18 locations across the conterminous US (Carroll 1962) are provided in Table 4.3. The concentrations of the constituents vary greatly from place to place with the highest concentration for variables being about 6–220 times greater than the lowest. The highest total concentration of several variables—especially sodium and chloride were from locations near the ocean. For example, sodium concentration at Brownsville, Texas and Tacoma, Washington were 22.30 and 14.50 mg/L, respectively, while at

Table 4.3 Concentrations of seven constituents in rainwater from 18 locations across the conterminous US (Carroll 1962)

Constituent	Concentration (mg/L)	
	Mean ± standard deviation	Range
Sodium	2.70 ± 5.95	0.10–22.30
Potassium	0.29 ± 0.31	0.07–1.11
Calcium	1.48 ± 1.71	0.23–6.50
Chloride	3.24 ± 7.10	0.13–22.58
Sulfate	1.48 ± 1.32	0.03–5.34
Nitrate	2.31 ± 1.17	0.81–4.68
Ammonium	0.43 ± 0.52	0.05–2.21

inland sites of Columbia, Missouri and Ely, Nevada, the respective values were 0.33 and 0.69 mg/L. The average specific conductance of US rainfall calculated from ion concentrations reported in Table 4.3 and assuming a pH of 6 was 25.7 μmhos/cm.

Average concentrations of the six major ions in rainwater over the entire United States were lower than concentrations of these ions typically found in inland surface waters. Ammonium and especially nitrate, however, were of higher average concentration then often found in surface waters—including the ocean. Rainfall in highly populated and industrialized regions may have particularly high sulfate concentrations.

At a given location, rainfall varies in composition from storm to storm because of the origin of the air masses and because of the duration of the rainfall events. Initial rainfall tends to be more concentrated in impurities than rainfall occurring later in a storm event, because the initial rainfall tends to sweep the majority of particles from the atmosphere.

Inland Surface Waters

Inland surface waters include overland runoff, streams, lakes, reservoirs, ponds, and swamps. Overland runoff has brief contact with soil, but this provides the first opportunity for increasing its mineral content above that of rainwater. Streams carry runoff from the land surface (overland flow) and water from groundwater seepage (base flow). Groundwater usually is more concentrated in mineral matter than overland flow because it has resided for months or years within the pore space of geological formations. Groundwater often is depleted of dissolved oxygen and charged with carbon dioxide because it infiltrates through the soil. Carbon dioxide lowers the pH of water, and many minerals are more soluble at low pH and low oxygen content. The presence of carbon dioxide will increase the capacity of water to dissolve limestone, calcium, silicate, feldspars, and some other minerals. Moreover, stream water has contact with soil and geological material in the streambed. These factors contribute to higher concentrations of mineral matter in stream water than in overland flow, but stream water usually is less concentrated in minerals than

Table 4.4 Mean concentrations of major constituents in waters of major rivers in different continents (Livingstone 1963)

Continent	HCO_3^- CO_3^{2-}	SO_4^{2-}	Cl^-	Ca^{2+}	Mg^{2+}	Na^+	K^+	SiO_2	Sum[a]
North America (mg/L)	68	20	8	21	5	9	1.4	9	142
South America (mg/L)	31	4.8	4.9	7.2	1.5	4	2	11.9	69
Europe (mg/L)	95	24	6.9	31.1	5.6	5.4	1.7	7.5	182
Asia (mg/L)	79	8.4	8.7	18.4	5.6	(9.3)[b]		11.7	142
Africa (mg/L)	43	13.5	12.1	12.5	3.8	11	–	23.2	121
Australia (mg/L)	32	2.6	10	3.9	2.7	2.9	1.4	3.9	59
World average (mg/L)	58	11.2	7.8	15.0	4.1	6.3	2.3	13.1	120

[a]Sum is approximately equal to total dissolved solids (TDS)
[b]Sum of sodium and potassium

groundwater. Streams often flow into standing water bodies where the residence time is greater than in streams. This provides longer contact of water with bottom soils and greater opportunity for dissolution of minerals.

Mineral matter in inland surface water results primarily from dissolution of minerals in soil and other geological material. Each mineral has a given solubility depending on the temperature, pH, carbon dioxide concentration, and dissolved oxygen concentration (or redox potential) of the water with which it has contact. If the contact is long enough, minerals will dissolve until equilibrium is reached between the concentrations of dissolved ions and the solid phase minerals (Li et al. 2012). In humid climates where there has been extensive weathering, soils may be highly leached and contain only small quantities of soluble minerals. In areas where there are abundant outcrops of rocks and weathering has not produced deep soils, rainwater falling on the surface will have little opportunity to dissolve minerals. Of course, limestone is quite soluble in comparison to many types of rocks, and rainwater contacting limestone has greater opportunity for mineralization. In more arid areas, there is not enough rainfall to leach minerals from the soil profile. Salts may accumulate in the surface layer of soils in arid regions or at intermediate depths in the soil profile in semi-arid regions. This provides a greater opportunity for mineralization of overland flow than is afforded in most humid regions.

Large rivers have extensive drainage basins, and water comes from a variety of geological and climatic zones resulting in an averaging effect upon ionic composition. Livingstone (1963) gave the average composition of the waters for the larger rivers by continent and computed the world average for the concentration of major constituents of river water (Table 4.4). The world average for dissolved solids is 120 mg/L, but average dissolved solids concentrations for continents ranged from 59 mg/L for Australian rivers to 182 mg/L for European rivers. The variation in individual ionic constituents was even greater than for dissolved solids. On a global basis, bicarbonate was the dominant anion and calcium was the dominant cation. Based on the average composition of river water, normal freshwaters are dilute solutions of alkaline earth (calcium and magnesium) and alkali (sodium and potassium) bicarbonate, carbonate, sulfate, and chloride with a quantity of silica (Hutchinson 1957).

Notice that the concentration of silica was greater than the concentration of many of the major ions in the average river water composition for all continents (Table 4.4). Minerals containing silicon are the most common ones in the earth's crust—they include quartz, micas, feldspars, amphiboles, and other groups of rocks. The common form of silicon is silicon dioxide (called silica) of which common sand is composed. Nearly 28 % of the earth's crust is made of silicon, and 40 % of minerals contain this element. Thus, despite the low solubility of silicon minerals, dissolved silica is a major component of most natural waters.

Silicon dioxide dissolves slightly in water to form silicic acid as follows:

$$SiO_2 + 2H_2O = H_4SiO_4. \quad (4.8)$$

Silicic acid dissociates at high pH as shown below:

$$H_4SiO_4 = H^+ + H_3SiO_4^- \qquad K = 10^{-9.46} \quad (4.9)$$

$$H_3SiO_4^- = H^+ + H_2SiO_4^{2-} \qquad K = 10^{-12.56}. \quad (4.10)$$

The ratio of the parent silicic acid to ionized $H_3SiO_4^-$ is

$$\frac{(H_3SiO_4^-)}{(H_4SiO_4)} = \frac{10^{-9.46}}{(H^+)}. \quad (4.11)$$

The pH must be 9.46 before the ratio has a value of 1.0; at pH values below 9.46, the unionized form of silicic acid predominates. In most natural waters, dissolved silicon exists primarily as un-ionized silicic acid.

Small streams have small drainage basins and concentrations of ionic constituents in their waters reflect geological and climatic conditions within a limited area. The ionic composition of water from small streams is much more variable over a large geographic area than is the composition of waters of larger rivers with more diverse conditions on their drainage basins. Concentrations of major mineral constituents in some relatively small rivers (Table 4.5) illustrate this variation. The Moser River of Nova Scotia has dilute water because it flows through an area where the geological formations consist of hard, highly insoluble rocks and soils are not highly developed or fertile. The Etowah River in Georgia has higher concentrations of substances than the Moser River, but it flows through areas where weathering and leaching of soils has been extensive because of high rainfall and temperature, and has fairly dilute water. The Withlacoochee River of Florida also

Table 4.5 Major mineral constituents in some relatively small rivers in the United States and Canada

Continent	HCO_3^- CO_3^{2-}	SO_4^{2-}	Cl^-	Ca^{2+}	Mg^{2+}	Na^+	K^+	SiO_2	TDS
Moser River, Nova Scotia	0.7	4.3	6.1	3.6	2.5	(5.5)[a]		3.0	27
Etowah River, Georgia	20	3.1	1.4	3.8	1.2	2.5	1.0	10	43
Withlacoochee River, Florida	118	23	10	44	3.8	5.0	0.3	7.6	213
Republican River, Kansas	244	64	7.6	53	15	34	10	49	483
Pecos river, New Mexico	139	1,620	755	497	139	488	10	20	3,673

Values are in milligrams per liter. (Livingstone 1963)
[a]Sum of sodium and potassium

flows through an area with weathered and highly leached soils, but there are limestone formations in its drainage basin. Dissolution of limestone results in a higher degree of mineralization of the water than in either the Moser or Etowah Rivers. The Republican River in Kansas is in an area of relatively low rainfall and soils are deep and fertile. The water is more highly mineralized than stream waters in regions of greater rainfall where soils are either poorly developed or highly leached. The Pecos River in New Mexico flows through an arid region, and its waters are highly mineralized as compared to the other four streams. Sulfate, chloride, and sodium are greatly enriched in the Pecos River. This results from the high concentrations of soluble salts that accumulate in soils of arid regions. Limestone, comprised of calcium carbonate or calcium and magnesium carbonates, is not nearly as soluble as the sulfate and chloride salts of alkali and alkaline earth metals found in soils of arid regions.

The composition of stream flow also tends to vary with discharge. Ionic concentrations increase as discharge decreases and *vice versa*. This phenomenon is particularly noticeable for small streams in arid regions (Fig. 4.5) and results from runoff from rainstorms diluting the more highly mineralized base flow.

Open basin lakes have inflow and outflow, and their composition is generally similar to river waters in the same region (Table 4.6). Lakes in Nova Scotia, Alabama, and Florida are quite similar in dissolved solids and most other constituents to river waters in the same regions (see Table 4.5). The lake in west Texas (Table 4.6) is in an arid region and similar in composition to the Pecos River (see Table 4.5) which is in the same arid region. Closed basin lakes have inflow but no outflow and ions concentrate through evaporation. Little Borax Lake in California and the Dead Sea that lies between Israel and Jordan (Table 4.6) are closed basins.

Minerals dissolve in water until equilibrium between solid phase minerals and dissolved ions is achieved. When water concentrates through evaporation, it becomes supersaturated with some ions and they form solid phase minerals and

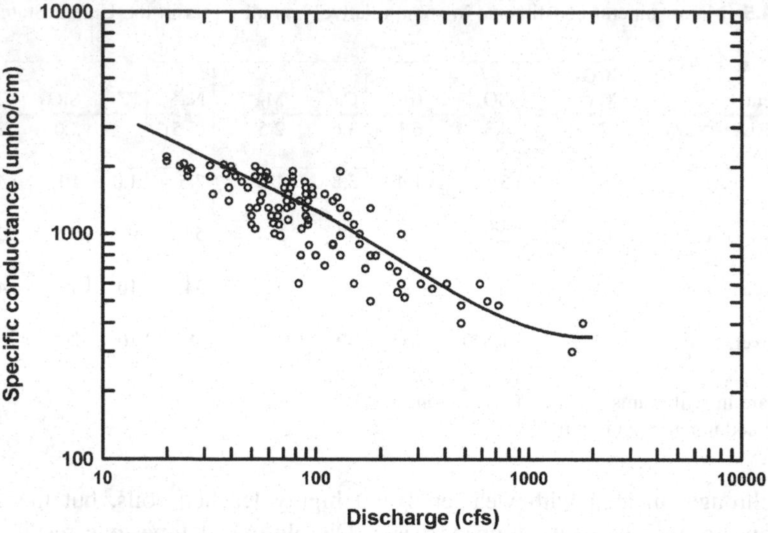

Fig. 4.5 Specific conductance of daily samples and daily mean discharge, San Francisco River at Clifton, Arizona, October 1, 1943 to September 30, 1944 (Hem 1970)

Table 4.6 Major mineral constituents in some selected lakes (Livingstone 1963)

Lake	HCO_3^- CO_3^{2-}	SO_4^{2-}	Cl^-	Ca^{2+}	Mg^{2+}	Na^+	K^+	TDS
Average of nine lakes in Nova Scotia	2.7	5.4	5.2	2.3	0.5	3.2	0.6	20
Lake Ogletree, Auburn, Alabama[a]	25.6	3.2	1.2	3.1	1.7	2.6	1.4	46
Lake Okeechobee, Florida	136	28	29	41	9.1	2.2	1.2	277
Lake Balmorheahak, Texas	159	555	560	38	0.1	(642)[b]		1,970
Little Borax Lake, California	8,166	10	905	8	24	3,390	731	13,600
Dead Sea	240	540	208,020	15,800	41,960	34,940	7,560	315,040

[a]From C. E. Boyd, unpublished data
[b]Sum of sodium and potassium

precipitate. The first substance to precipitate is calcium carbonate and further concentration of ions will lead to calcium sulfate precipitation. In humid regions, calcium and bicarbonate tend to be dominate ions, but as climate becomes drier, magnesium, sodium, potassium, sulfate, and chloride tend to increase relative to calcium and bicarbonate. Finally, in extremely concentrated natural waters, sodium and chloride tend to be the dominant ions. Of course, many possible intermediate mixtures of ions exist among the three extremes of bicarbonate-carbonate, sulfate-chloride, and chloride dominated waters.

Table 4.7 Concentrations of major constituents in waters of manmade ponds in different physiographic regions of Alabama and Mississippi

Physiographic region	HCO_3^- CO_3^{2-}	SO_4^{2-}	Cl^-	Ca^{2+}	Mg^{2+}	Na^+	K^+	TDS
Mississippi Yazoo Basin Alabama	307	–	14.2	61.7	20.8	–	5.1	–
Black Belt Prairie	51.1	4.3	6.8	19.7	1.5	4.3	1.5	94.4
Limestone Valleys and Uplands	42.2	4.2	6.6	11.9	4.7	4.2	3.2	112.0
Appalachian Plateau	18.9	6.6	3.2	5.0	2.8	2.9	1.7	60.2
Coastal Plain	13.2	1.8	5.5	3.4	1.1	2.9	2.8	44.3
Piedmont Plateau	11.6	1.4	2.6	2.7	1.4	2.6	1.4	34.5

Values are in milligrams per liter (Arce and Boyd 1980)

The composition of pond waters reflects the composition of soils in a specific area. Of course, if the climate is dry, concentration of ions through evaporation may overshadow the influence of soil composition. Some data on the average composition of water from several ponds in each of several physiographic regions in Mississippi and Alabama are provided in Table 4.7. Annual rainfall in all regions is between 130 and 150 cm/year, and annual mean air temperature differs by no more than 2 or 3 °C. However, because of differences in soil composition, the concentrations of major constituents in pond waters vary (Arce and Boyd 1980). The Yazoo Basin in Mississippi has fertile, heavy clay soils with a high concentration of exchangeable cations and free calcium carbonate. The soils of the Black Belt Prairie and Limestone Valley and Uplands of Alabama are not as fertile as the Yazoo Basin soils, but they often contain limestone. The other three physiographic regions have less fertile and highly leached soils that do not contain limestone. Notice that the waters in Table 4.7 are primarily dilute calcium and magnesium bicarbonate solutions with variable concentrations of other major ions.

Groundwater can be highly variable in concentrations of major constituents. Several aquifers may occur beneath the ground surface in a particular area and each formation may have different characteristics. Analyses were made of waters from over 100 wells within an 80 km radius in west-central Alabama (Boyd and Brown 1990). There were five distribution patterns among the major ions in the well waters (Fig. 4.6). The most highly mineralized waters (a) (average TDS = 1,134 mg/L) had modest proportions of bicarbonate and calcium, large proportions of chloride and sodium, and relatively small proportions of other major ions. Three patterns were observed in the moderately mineralized water (average TDS of 247–372 mg/L). In some waters (b), bicarbonate was the chief anion and sodium was the primary cation; other ions were relatively scarce. In most moderately mineralized waters (c), bicarbonate was the dominant anion and calcium was the main cation. In other waters of moderate mineralization (d), chloride was proportionally greater than bicarbonate, and calcium and magnesium were roughly proportional to sodium and potassium. In weakly mineralized waters (e) (average TDS = 78 mg/L), bicarbonate was the

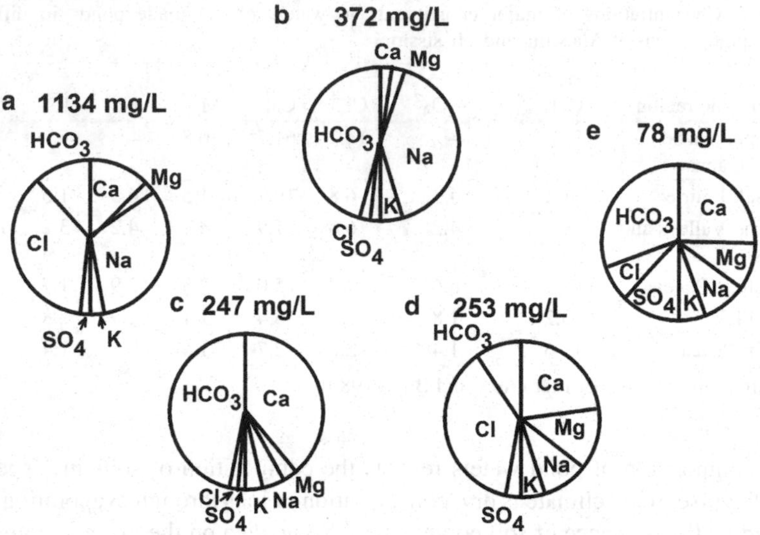

Fig. 4.6 Distribution of major ions in samples of groundwater of different total dissolved solids concentrations from wells in west-central Alabama

dominant anion, but sulfate was proportionally greater than in the other four types. Calcium and magnesium were roughly proportional to sodium and potassium.

The waters containing primarily sodium and chloride and 1,134 mg/L of total dissolved solids (a in Fig. 4.6) came from an aquifer that was influenced by subsurface deposits of sodium chloride that may be found in some areas of Alabama. The water containing primarily bicarbonate and sodium (b) was the result of an interesting phenomenon sometimes found on coastal plains and known as the natural softening of groundwater (Hem 1970). Groundwater high in bicarbonate and sodium concentrations often results in areas where surface formations contain limestone and underlying aquifers have solids which have adsorbed an abundance of sodium. The aquifers contain sodium because they were filled with seawater in an earlier geological time. Seawater was replaced by freshwater as uplifting of the coastal plain occurred. Rainwater infiltrating downward is charged with carbon dioxide and dissolves limestone to increase concentrations of bicarbonate, calcium, and magnesium. Upon reaching the aquifers, calcium in the infiltrating water is exchanged for sodium in the aquifer solids resulting in a dilute sodium bicarbonate solution.

Coastal and Ocean Water

The composition of ocean water has developed over geological time with the import of dissolved mineral matter from erosion of the earth's land masses. Over time, the ocean has become essentially saturated with dissolved inorganic

Table 4.8 The average composition of seawater (Boyd 1990)

Constituent	mg/L	Constituent	mg/L
Cl	19,000	U	0.003
Na	10,500	Mn	0.002
SO_4	2,700 (900 mg S/L)	Ni	0.002
Mg	1,350	V	0.002
Ca	400	Ti	0.001
K	380	Co	0.0005
HCO_3	142 (28 mg C/L)	Cs	0.0005
Br	65	Sb	0.0005
Sr	8	Ce	0.0004
SiO_2	6.4 (3.0 mg Si/L)	Ag	0.0003
B	4.6	La	0.0003
F	1.3	Y	0.0003
N (in NO_3, NO_2, NH_4)	0.5	Cd	0.00011
Li	0.17	W	0.0001
Rb	0.12	Ge	0.00007
P	0.07	Cr	0.00005
I	0.06	Th	0.00005
Ba	0.03	Sc	0.00004
Al	0.01	Ga	0.00003
Fe	0.01	Hg	0.00003
Mo	0.01	Pb	0.00003
Zn	0.01	Bi	0.00002
Se	0.004	Nb	0.00001
As	0.003	Ar	0.000004
Cu	0.003	Be	0.0000006
Sn	0.003		

substances and is more or less at equilibrium with respect to these substances. The average composition of ocean water is provided in Table 4.8. Because they will be discussed later, minor constituents are included in the ocean water data. Notice that the composition of ocean water is dominated by sodium and chloride.

The ocean is quite similar in chemical composition worldwide, but of course, there is some variation. For example, the northern Pacific Ocean has areas where salinity is 31–33 ppt, while some areas in the Atlantic Ocean have salinities of 36–37 ppt. The Mediterranean and Red Seas have areas with salinity above 38 ppt (http://www.nasa.gov/sites/default/files/images/591162main_pia14786-43_full.jpg).

Ocean waters and river waters mix together in estuaries. Because most river waters are much more dilute in ions than ocean waters, ionic proportions in estuarine waters generally reflect those of ocean waters. However, the salinity of estuarine water at a given location in an estuary may vary greatly with water depth, time of day, and freshwater inflow. In coastal reaches of rivers, a density wedge of salt water may occur in the bottoms of rivers causing depth stratification of salinity.

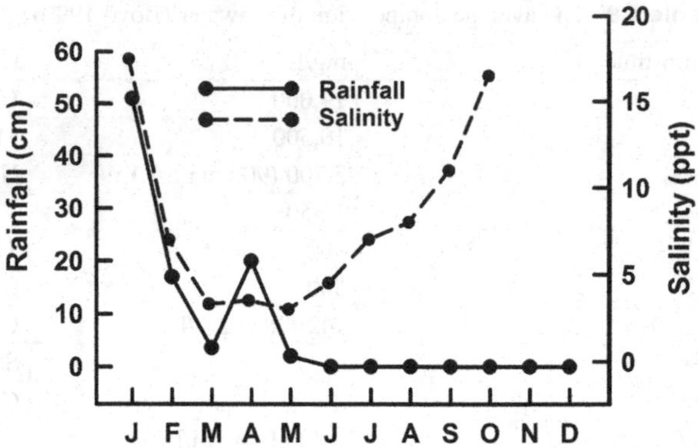

Fig. 4.7 Relationship between rainfall and salinity in coastal water near Guayaquil, Ecuador

Tidal action causes changes in salinity by causing salt water to flow into and out of estuaries. Large inputs of freshwater following rainy weather can dilute the salinity in estuaries as shown in Fig. 4.7. The author once visited a small estuary in Honduras 3 days after a hurricane with much rain had passed. Salinity profiles were taken at several places revealing that only freshwater was present at the time. In estuaries with limited water exchange with the sea, salinity may increase in the dry season in response to less freshwater inflow and concentration of ions by evaporation.

Relative Abundance of Elements

Some readers might find it interesting to compare the relative abundances of the elements in the earth's crust with those of freshwater and ocean water. Thus, the 15 most abundant elements—listed in order of descending abundance—are provided below for the earth's crust (Cox 1989), soil and superficial minerals (Shacklette and Boerngen 1984), freshwater (Livingstone 1963), and oceans (Goldberg 1963):

Earth's crust	Soil	Ocean water	River water
O	O	Cl	O
Si	Si	Na	C
Al	Al	O	Ca
Fe	Fe	Mg	Cl
Ca	C	S	Na
Mg	Ca	Ca	Si
Na	K	K	Mg
K	Na	Br	S
Ti	Mg	C	K
C	H	Sr	H
H	Ti	B	Fe

(continued)

(continued)

Earth's crust	Soil	Ocean water	River water
Mn	S	Si	Mn
P	Ba	H	Zn
F	Mn	F	Cu
S	P	N	B

Water consists of oxygen and hydrogen, but the oxygen dissolved in water is present primarily in bicarbonate, sulfate, silicate, and borate. The main source of dissolved hydrogen is bicarbonate. Ten of the most abundant elements in ocean water are also among the most abundant elements in the earth's crust. There are nine elements in common between soil (the most superficial layer of the earth's crust) and ocean water. The element not in common among the two comparisons is fluoride that is not one of the most abundant elements in soil. Eleven elements among the 15 most abundant elements in both the earth's crust and in soils are among the 15 most abundant elements in river water. Although many of the same elements tend to be abundant in the earth's crust and the soil layer, the relative order of abundance of elements in the earth's minerals and soils is quite different from that of either ocean or river water. Chloride is of high relative abundance in both ocean and river water, but it is not one of the 15 most abundant elements in the earth's crust and soil. Likewise, aluminum and titanium, abundant in the crust and soil, are not abundant in water.

Dissolved Organic Matter

Organic matter consists of remains of plant and animal material, and it is comprised of a wide array of compounds. It can enter water from the dissolution of organic matter or suspension of organic particles from watersheds, leaf fall, etc. Outside sources of organic matter are known as allochthonous sources. Organic matter also is produced within aquatic ecosystems; this material is known as autochthonous organic matter. Organic matter gradually breaks down as a result of physical, chemical, and biological processes, and a fraction of this material becomes small enough that it is soluble. Organisms also can excrete soluble organic compounds directly into the water. Organic matter in soil and water is a mixture of plant and animal remains in various stages of decomposition, compounds synthesized chemically and biologically during decomposition, and microorganisms of decay and their remains.

Generally, organic matter is said to be either nonhumic or humic in nature. The nonhumic substances consist of carbohydrates, proteins, peptides, amino acids, fats, waxes, resins, pigments, and other compounds of relatively low molecular weight. These compounds are readily decomposable by microorganisms and do not occur in water in large concentrations. Humus is a product of synthesis and decomposition by microorganisms. It exists as a series of acidic, yellow to black macromolecules of unknown but high molecular weight. Soil organic matter is 60–80 % humus. Humus chemistry is not well understood, but it consists of a heterogeneous mixture of molecules that form systems of polymers. Molecular weights of polymers in humus range from several hundred to more than 300,000. One hypothesis holds that in humus formation, polyphenols derived from lignin and others synthesized by microorganisms polymerize along or with amino compounds to form polymers of variable molecular weights with functional acidic groups. Humus often is considered to be composed of three classes of compounds: humic acids, fulvic acids, and humin. Fulvic acids have higher oxygen contents, lower carbon contents, lower molecular weights, and higher degrees of acidity than humic acids. Fulvic acids are yellowish, whereas humic acids are dark brown or black. The properties of humin are poorly defined. Humin differs from fulvic and humic acids by being insoluble in alkali. Because humic substances decompose very slowly they can accumulate in some waters and especially in acidic waters.

There are fewer data on concentrations of dissolved organic matter than on concentrations of dissolved minerals. Most relatively clear waters will have less than 10 mg/L organic matter. Nutrient enriched waters with abundant phytoplankton sometimes have much more dissolved organic matter. The highest concentrations will be found in waters that are stained with humic substances such as those of bog lakes and swamps or in polluted waters where dissolved organic matter may exceed 50 mg/L. Organic matter in water causes color and restricts light penetration. Organic compounds chelate metals and increase the solubility of trace metals in water.

Measuring Dissolved Organic Matter

A simple way of determining dissolved organic matter concentration in water is to ignite the residue from the total dissolved solids analysis at 500 °C to burn off the organic matter. The weight loss is the dissolved organic matter.

Ex. 4.5: *The dish and dry residue from the TDS analysis that weighed 5.2100 g (Ex. 4.1) are ignited, and after cooling, the dish and residue weigh 5.2080 g. The dissolved organic matter (DOM) concentration will be estimated.*

<u>Solution:</u>

$$DOM = (5.2100 - 5.2080)\ \text{g} \times 1{,}000\ \text{mg/g} \times \frac{1{,}000\ \text{mL/L}}{100\ \text{mL}\ sample}$$

$DOM = 20\ mg/L.$

The dissolved organic matter concentration may be estimated by a carbon analyzer that combusts the organic matter and measures the amount of carbon dioxide released. This instrument is rather expensive; a cheaper way is by sulfuric acid-potassium dichromate digestion. In this procedure, a sample of water filtered to remove the particulate matter is treated with concentrated sulfuric acid and excess, standard potassium dichromate. It is then held at boiling temperature for 2 h in a reflux apparatus. Certain other reagents are sometimes added as catalysts or inhibitors of chloride of other interfering substances (Eaton et al. 2005). The organic matter is oxidized by the dichromate

$$2Cr_2O_7^{2-} + 3\text{ organic }C^0 + 16H^+ = 4Cr^{3+} + 3CO_2 + 8H_2O. \quad (4.12)$$

The amount of dichromate remaining at the end of the digestion is determined by back-titration with standard ferrous ammonium sulfate. The milliequivalents of dichromate consumed in the reaction are equal to the milliequivalents of dissolved organic carbon in the sample as illustrated in Ex. 4.6.

Ex. 4.6: *Estimation of dissolved organic carbon concentration from dichromate digestion.*

Situation: *A 20 mL water sample that passed a 2-μm filter is treated with concentrated sulfuric acid, 10.00 mL of 0.025 N $K_2Cr_2O_7$, and refluxed at boiling temperature for 2 h. The digestate requires 4.15 mL 0.025 N $FeNH_4SO_4$ (FAS) for reducing the remaining dichromate. The blank titration requires 9.85 mL of the FAS solution.*

Solution:
The unconsumed $K_2Cr_2O_7$ is estimated:

$$\textit{Blank} \quad (0.025\text{ meq/ mL})(9.85\text{ mL }FAS) = 0.246\text{ meq}$$

$$\textit{Sample} \quad (0.025\text{ meq/ mL})(4.15\text{ mL }FAS) = 0.104\text{ meq}$$

The amounts of $K_2Cr_2O_7$ consumed in each are:

$$\textit{Blank} \quad [(0.025\text{ meq/L})(10.00\text{ mL }K_2Cr_2O_7)] - 0.246\text{ meq} = 0.004\text{ meq}$$

$$\textit{Sample} \quad [(0.025\text{ meq/L})(10.00\text{ mL }K_2Cr_2O_7)] - 0.104\text{ meq} = 0.146\text{ meq}$$

Organic matter in the sample equaled the amount of $K_2Cr_2O_7$ consumed or (0.146 – 0.004) meq = 0.142 meq. The equivalent weight of organic carbon in (4.14) can be determined from the observation that three organic carbon atoms with valence of 0 lost four electrons each to obtain a valence of 4+ in carbon dioxide. The equivalent weight of carbon in the reaction is obtained by dividing the atomic weight of carbon (12) by the number of electrons transferred per carbon atom (4).

Thus, the amount of dissolved organic carbon in 20 mL of water was

$$0.142\text{ meq }K_2Cr_2O_7 \times 3\text{ mg }C/\text{meq} = 0.426\text{ mg}.$$

This is equal to 21.3 mg C/L.

Total Dissolved Solids and Colligative Properties

Vapor Pressure

The colligative properties of solutions are physical changes resulting from adding solute to a solvent that depends upon the amount of solute—not the type of solute. These changes include vapor pressure decrease, freezing point depression, boiling point elevation, and osmotic pressure increase that occur as solute concentrations are increased. Thus, the TDS concentration can affect the colligative properties of water.

A French chemist, François Raoult, discovered in the late eighteenth century that at the same temperature the vapor pressure of a solution is less than the vapor pressure of its pure solvent. This phenomenon became known as Raoult's law and can be expressed as

$$P_{sol} = XP^{o} \tag{4.13}$$

where P_{sol} = vapor pressure of the solution, X = the mole fraction of the pure solvent, and P^{o} = the vapor pressure of the pure solvent. The phenomenon is the result of the solute molecules occupying space among the solvent molecules. The surface of the solution does not consist totally of solvent molecules, and the rate of diffusion of the solvent molecules into the air above the solution will be less than if the pure solvent was exposed to the air. Thus, as the concentration of dissolved solids increases in water, the vapor pressure of water decreases.

Raoult's Law provides an explanation of why evaporation rate under the same condition is greater for freshwater than for ocean water (see Chap. 1). The high concentration of salt in ocean water results in less water being exposed at its surface where evaporation occurs than in freshwater.

Freezing and Boiling Points

The boiling point is affected by the vapor pressure of a solution as compared to that of the pure solvent, because the vapor pressure of the solvent molecules escaping the surface of the solution is lower than that of the molecules escaping the surface of the pure solvent. Thus, the solution must be heated to a greater temperature than would the pure solvent to cause boiling.

The freezing point of the solvent of a solution is affected by the vapor pressure, because the freezing point is reached when the vapor pressure of the pure, liquid solvent reaches that of the pure, solid solvent. Thus, the decrease in the vapor pressure caused by dissolved solids in natural water will cause it to freeze at a lower temperature than does pure water. Another way to look at this phenomenon is that for water to freeze, its molecules must arrange themselves into the tetrahedral structure of ice. The presence of dissolved solids interferes with this alignment of molecules resulting in a lower freezing point.

The equations for estimating the lowering of the freezing point and elevation of the boiling point are

$$\Delta T_f = -K_f \times m \quad (4.14)$$

$$\Delta T_b = -K_b \times m \quad (4.15)$$

where ΔT_f and ΔT_b = the changes in freezing and boiling points, respectively; $-K_f$ and $-K_b$ = proportionality constants for freezing and boiling point changes, respectively; and m = molality of the solution (m = g solute/g solution). Typical values for K_f and K_b for water are 1.9 and 5.1, respectively.

Osmotic Pressure

Dissolved solids in water have relatively small effects upon vapor pressure, freezing point, and boiling point. But, they have a major influence upon the other colligative property, osmotic pressure. The osmotic pressure of a solution is the amount of pressure or force needed to prevent the flow of water molecules across a semipermeable membrane from a concentrated to a less concentrated solution (Fig. 4.8).

The semipermeable membrane is permeable to particles of the solvent but not to particles of the solute. Particles of a solvent will pass through a semipermeable membrane from a dilute solution to a more concentrated solution. A simple way to view this phenomenon is to consider that the membrane is being bombarded continuously on both sides by molecules of both solvent and solute. On the side of the membrane facing the dilute solution, more molecules of the solvent will strike the membrane surface than on the side of the membrane facing the more concentrated solution, because there are more solvent molecules per unit volume in the dilute solution than in the concentrated solution. There will be net movement of

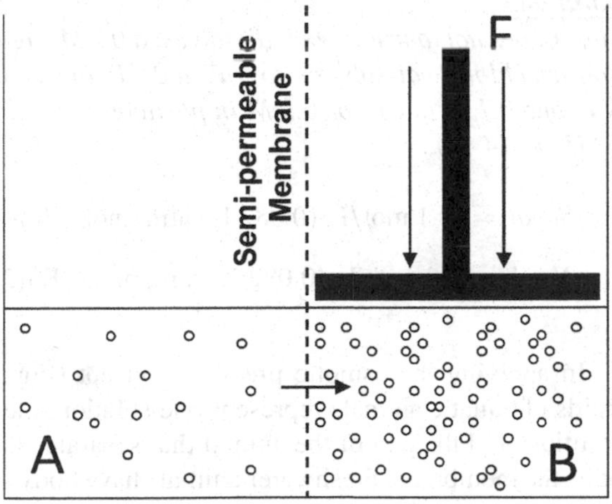

Fig. 4.8 Illustration of the concept of osmotic pressure

solvent molecules from the dilute solution to the concentrated solution until equilibrium (equal concentration) is attained between the two sides. The idea can be expressed in terms of pressure, for pressure can be applied to the concentrated side to stop the net movement of solvent molecules across the membrane (Fig. 4.8). The amount of pressure necessary to accomplish this feat is the osmotic pressure of a solution. The equation for osmotic pressure is

$$\pi = \frac{nRT}{V} \tag{4.16}$$

where π = osmotic pressure (atm), n = number of moles of solute in the solution, R = ideal gas constant (0.082 L·atm/mol·°K), T = absolute temperature (°C + 273.15), and V = volume of solution (L). Because n/V in (4.16) is equal to the molar concentration of the solute, this equation often is written as

$$\pi = CRT \tag{4.17}$$

where C = the molar concentration of the solute. This equation was developed for solutes such as sugar in which the number of particles dissolved in water is exactly equal to the number of dissolved molecules. An inorganic salt like sodium chloride has twice as many ions as the number of sodium chloride molecules, for each molecule dissociates to yield one sodium ion and one chloride ion. Thus, as shown in Ex. 4.7, a 0.1 M NaCl solution has greater osmotic pressure than a 0.1 M sugar solution. In fact, the molarity of the ions in a 0.1 M NaCl solution is 0.2 M. By similar reasoning, the morality of ions in 0.01 M $Al_2(SO_4)_3$ is 0.05 M, because each mole of aluminum sulfate provides two aluminum and three sulfate ions upon dissolution.

Ex. 4.7: *The osmotic pressure of 0.1 M solutions of sugar and sodium chloride will be calculated for 25 °C.*

Solution:

Sugar does not ionize when it dissolves; a 0.1 M sugar solution is 0.1 M in particles. Sodium chloride dissolves into Na^+ and Cl^- ions; a 0.1 M NaCl solution is 0.1 M in Na^+ and 0.1 M in Cl^- or 0.2 M in particles.

Using (4.17),

$$Sugar = (0.1\ \mathrm{mol/L})(0.082\ \mathrm{L \cdot atm/mol \cdot {}^\circ K})(273.15\mathrm{K} + 25) = 2.44\ \mathrm{atm}$$

$$NaCl = (0.2\ \mathrm{mol/L})(0.082\ \mathrm{L \cdot atm/mol \cdot {}^\circ K})(273.15\mathrm{K} + 25) = 4.89\ \mathrm{atm.}$$

In applying the osmotic pressure concept (Fig. 4.8) to aquatic animals, body fluids of aquatic animals represent one solution, the surrounding water is the other solution, and the part of the animal that separates the two solutions can be thought of as the membrane. Freshwater animals have body fluids more concentrated in ions than the surrounding water; they are hypersaline or hypertonic to their environment.

Table 4.9 General standards for total dissolved solids (TDS) and sodium adsorption ratio (SAR) in irrigation water

Salt tolerance of plants	TDS (mg/L)	SAR
All species, no detrimental effects	500	2–7
Sensitive species	500–1,000	8–17
Adverse effects on many common species	1,000–2,000	18–45
Use on tolerant species on permeable soils only	2,000–5,000	46–100

Saltwater species have body fluids more dilute in ions than the surrounding water; they are hyposaline or hypotonic to their environment. Osmoregulation in freshwater fish involves the uptake of ions from the environment and restriction of ion loss. The freshwater fish tends to accumulate water because it is hypertonic to the environment, so it must excrete water and retain ions to maintain its osmotic balance. On the other hand, osmoregulation for saltwater fish requires the constant intake of water and excretion of ions. Because the saltwater fish is hypotonic to the environment, it loses water. To replace this water, the fish takes in saltwater; but to prevent the accumulation of excess salt, it must excrete salt. Each species has an optimum salinity range. Outside of this range the animal must expend considerably more energy than normal for osmoregulation at the expense of other processes such as growth. Of course, if salinity deviates too much from optimum, the animal will die because it cannot maintain homeostasis.

Plants also can have problems with osmoregulation if the water available to them has excessive concentration of ions, because high concentrations of ions increase osmotic pressure. The influence of dissolved solids on the quality of irrigation water is predicted from the sodium adsorption ratio (SAR).

$$\mathrm{SAR} = \frac{(\mathrm{Na})}{0.5(\mathrm{Ca} + \mathrm{Mg})^{0.5}} \tag{4.18}$$

where Na, Ca, and Mg = concentrations in meq/L. Some general standards for use of irrigation water based on TDS and SAR are provided (Table 4.9).

Removal of Dissolved Solids

There is sometimes a necessity to remove dissolved solids from water, e.g., to produce potable water or to make water of low conductivity for use in analytical laboratories or in industrial processes. The natural way of producing relatively pure water is evaporation followed by condensation of water vapor in the atmosphere and rainfall (see Chap. 2). This process is mimicked by distillation in which water is converted to vapor by heating it in a chamber and then condensing the vapor and capturing the condensate. In addition, dissolved solids may be removed from water by ion exchange and reverse osmosis.

Ion Exchange

A deionizer consists of a column packed with an anion exchange resin and a cation exchange resin. These resins typically are high molecular weight polymers; the anion resin exchanges hydroxyl ion with anions in the water, while the cation resins exchanges hydrogen ion with cations in the water. An example of an anion exchange resin is a polymer with many quaternary ammonium sites on it. The nitrogen in the quaternary ammonium group is linked to the polymer and has three methyl groups attached to it and imparting a positive charge to the nitrogen. The resin is treated with hydroxide resulting in the charge on the quaternary ammonium sites being satisfied with hydroxyl ion. When water is passed through the resin, hydroxyl ions are exchanged for anions in the water

$$\text{Polymer} - \text{N(CH)}_3^+\ \text{OH}^- + \text{Cl}^- \rightarrow \text{Polymer} - \text{N(CH)}_3^+\ \text{Cl}^- + \text{OH}^-. \quad (4.19)$$

A cation resin can be illustrated by a polymer with many sulfonic acid groups on it. The sulfur in sulfonic acid groups has two double-bonded oxygens and one single-bonded oxygen with a negative charge. Hydrogen ion neutralizes the charge on the single-bonded oxygen, and when water is passed through the resin, the hydrogen ions can be exchanged for cations in the water

$$\text{Polymer} - \ \text{SO}_3^-\text{H}^+ + \text{K}^+ \rightarrow \text{Polymer} - \ \text{SO}_3^-\text{K}^+ + \text{H}^+. \quad (4.20)$$

In the case of divalent ions, the polymers have many quaternary ammonium or sulfonic acid groups, and the hydroxyl or hydrogen ions on two nearby sites will be exchanged for a divalent anion or divalent cation. Of course, the hydrogen and hydroxyl ions released into the water during the process combine to form water. When ion exchange resins become saturated with anions or cations, they may be backwashed with an alkaline or an acidic solution, respectively, to recharge them for further use.

Dissolved organic matter that is charged—usually it will bear a negative charge—can be removed by ion exchange. Activated carbon filtration can be used to remove dissolved organic carbon that is not charged.

Desalination

Ion exchange is most commonly used for producing relatively small quantities of low conductivity water. For removing dissolved solids from large volumes of water for use in water supply—desalination—distillation or reverse osmosis is used.

High concentrations of dissolved solids make water unsuitable for most beneficial uses, and most of the world's water is ocean water with a high concentration of dissolved solids. Desalination of ocean water has been highly promoted as a means of increasing the supply of freshwater for human use. There are several processes for desalination, but most desalination plants rely on distillation or reverse osmosis. Multi-flash distillation allows the heat recovered from condensing the steam in the first stage vaporization of water to be used to vaporize water in the second stage—and the process is repeated for several stages—making the overall process more energy efficient. In reverse osmosis, pressure is used to force water through membranes that exclude the dissolved solids.

According to the International Desalination Association (idadesal.org), as of June 2011, there were 15,988 desalination plants in 150 countries that produced about 66.5 million m^3/day (24 km^3/year) and supplied water for about 300 million people. This represents only 0.1 % of the world's available, renewable freshwater from natural sources but desalination supplies freshwater to about 4 % of the global population. The future of desalination is uncertain because it requires a large input of energy that is derived mainly from fossil fuels.

When water freezes, dissolved ions are not included in the structure of ice, and they are displaced form the solution. Thus, icebergs and glaciers represent relatively pure water. There have been proposals to move these masses of frozen freshwater to coastal cities for use as a freshwater supply.

Significance of Dissolved Solids

Excessive concentrations of dissolved solids can interfere with various beneficial uses of water. As the TDS concentration rises in inland waters, osmotic pressure increases, and there will be fewer and fewer species that can tolerate the osmotic pressure. There is much variation in dissolved solids concentrations in aquatic ecosystems in different edaphic and climatic regions, and the species composition of aquatic communities is influenced by tolerance to dissolved solids (or tolerance to salinity, specific conductance, or osmotic pressure). Most ecosystems with a good, mixed, freshwater fish fauna will have total dissolved solids concentrations below 1,000 mg/L, but many species of freshwater fish and other organisms may tolerate 5,000 mg/L TDS or even more. To provide protection of aquatic life, dissolved solids concentrations should not be allowed to increase above 1,000 mg/L in freshwater ecosystems. Several states in the United States limit "end of pipe" discharges into natural water bodies to 500 mg/L of total dissolved solids. Of course, in water of higher natural concentrations of total dissolved solids, greater concentrations usually are allowed in discharge. In many estuaries, TDS concentration naturally fluctuates greatly, and species occurring there are adapted to these changes. The TDS concentration usually is not an issue in effluents discharged directly into the ocean—an exception is desalination plants that discharge water much more concentrated in salts than ocean water.

The main physiological influence of high TDS concentration in water to humans is a laxative effect caused by sodium and magnesium sulfates, and sodium has adverse effects on individuals with high blood pressure and cardiac disease and pregnant women with toxemia. Normally, an upper limit for chlorides and sulfates of 250 mg/L each is suggested. Excessive solids also impart a bad taste to water, and this taste is caused primarily by chloride. Problems with encrustations and corrosion of plumbing fixtures also occur when waters have excessive TDS concentration. If possible, public water supplies should not have TDS concentration above 500 mg/L, but some may have up to 1,000 mg/L.

Livestock water can contain up to 3,000 mg/L total dissolved solids without normally causing adverse effects on animals.

References

Arce RG, Boyd CE (1980) Water chemistry of Alabama ponds. Bulletin 522, Alabama Agricultural Experiment Station, Auburn University, Alabama

Boyd CE, Brown SW (1990) Quality of water from wells in the major catfish farming area of Alabama. In: Tave D (ed) Proceedings 50th Anniversary Symposium Department of Fisheries and Allied Aquacultures, Auburn University, Alabama

Carroll D (1962) Rainwater as a chemical agent of geologic processes—a review. Water-supply paper 1535-G. United States Geological Survey, United States Government Printing Office, Washington, DC

Cox PA (1989) The elements: their origin, abundance, and distribution. Oxford University Press, Oxford

Eaton AD, Clesceri LS, Rice EW, Greenburg AE (eds) (2005) Standard methods for the examination of water and wastewater, 21st edn. American Public Health Association, Washington, DC

Goldberg ED (1963) Chemistry—the oceans as a chemical system. In: Hill MN (ed) Composition of sea water, comparative and descriptive oceanography, Vol II. The sea. Interscience, New York, pp 3–25

Hem JD (1970) Study and interpretation of the chemical characteristics of natural water. Water-supply paper 1473. United States Geological Survey, United States Government Printing Office, Washington, DC

Hutchinson GE (1957) A treatise on limnology, Vol 1, geography, physics, and chemistry. John Wiley and Sons, New York

Laxen DPH (1977) A specific conductance method for quality control in water analysis. Water Res 11:91–94

Li L, Dong S, Tian X, Boyd CE (2012) Equilibrium concentrations of major cations and total alkalinity in laboratory soil-water systems. J App Aquacult 25:50–65

Livingstone DA (1963) Chemical composition of rivers and lakes. Professional Paper 440-G. United States Geological Survey, United States Government Printing Office, Washington, DC

Sawyer CN, McCarty PL (1967) Chemistry for sanitary engineers. McGraw-Hill, New York

Shacklette HT, Boerngen JG (1984) Element concentrations in soils and other surficial materials of the conterminous United States. United States Geological Survey Professional Paper 1280. United States Government Printing Office, Washington, DC

Silipajarn K, Boyd CE, Silapajarn O (2004) Physical and chemical characteristics of pond water and bottom soil in channel catfish ponds in west-central Alabama. Bulletin 655, Alabama Agricultural Experiment Station, Auburn University, Alabama

Particulate Matter, Color, Turbidity, and Light

5

Abstract

Natural waters contain particulate matter that increases turbidity, imparts apparent color, and interferes with light penetration. These particles originate from erosion on watersheds and within water bodies, vegetative debris from watersheds, and microorganisms produced in water bodies. Suspended particles vary from tiny colloids (1–100 mμ) that remain suspended indefinitely in still water to larger silt and sand particles held in suspension by turbulence. The settling velocity of particles in still water is estimated by the Stoke's law equation and it depends mainly on particle diameter and density—large, dense particles settle the fastest. Organic particles settle slowly because of their low density, but planktonic organisms also have adaptions that lessen settling velocity. Solar radiation of all wavelengths is quenched quickly—about 50 % is reflected or converted to heat within the first meter. Within the visible spectrum, red and orange light is absorbed most strongly followed by violet, and by yellow, green, and blue. Plants absorb mostly red and orange light and yellow and green light for use in photosynthesis. Turbidity absorbs all wavelengths of light and diminishes photosynthesis. Two common ways of assessing turbidity in natural waters are by measuring the depth of disappearance of a 20-cm diameter disk (the Secchi disk) and by using nephelometry to determine the amount of light reflected at a 90 ° angle from a water sample. The weight of particles retained on a fine filter allows gravimetric assessment of suspended solids.

Keywords

Characteristics of particles • Sedimentation • Sources of color • Causes of turbidity • Light penetration

C.E. Boyd, *Water Quality*, DOI 10.1007/978-3-319-17446-4_5

Introduction

Particles too large to be in true solution but small enough to remain suspended against the force of gravity comprise the particulate matter in water. Colloidal particles will remain suspended even in still water, but larger suspended particles remain suspended because of turbulence in a water body. Suspended particles along with some dissolved substances impart color to water and interfere with the passage of light, making water less transparent—increases its turbidity. Particulate matter, like dissolved matter, may be of both organic and inorganic origin and includes living organisms, detritus, and soil particles. Turbidity restricts light penetration and has a powerful limiting effect upon growth of plants in aquatic ecosystems. Suspended particles also settle creating bottom deposits, and high rates of sedimentation can result in severe bottom habitat degradation. Water for drinking purposes and for many industrial uses is degraded by significant amounts of turbidity and color, and treatment often must be applied to remove these substances.

Suspended Particles

Sources

The primary sources of suspended soil particles in water are erosion of watersheds by rain and overland flow, erosion of stream beds and banks by flowing water, erosion of banks of lakes by wave action, and resuspension of sediment by water currents in lakes and other static water bodies. Organic particles originate from leaves falling into bodies of water, suspension of organic particles from watersheds by overland flow, microscopic organisms (plankton and bacteria) that inhabit natural waters, and the remains of dead aquatic organisms (detritus).

Suspended particles exhibit a wide range in size and density. Larger and denser particles such as sand and coarse silt tend to settle to the bottom quickly. Turbidity results from fine silt and suspended clay particles that do not settle rapidly, and water containing an abundance of small soil particles appears "muddy." Phytoplankton, zooplankton, and detritus also can make waters turbid. In unpolluted, natural waters, the total amount of suspended organic matter will usually be no greater than 5 mg/L, but in nutrient rich waters with abundant plankton, concentrations of organic particles may reach 50 mg/L or more.

Settling Characteristics

There are two schemes for classification of mineral soil particles according to particle diameter as determined by sieve analysis (Table 5.1). Most particles are not truly spherical, and the classification actually is based on the maximum dimensions of particles. Organic particles also cover a wide size range, but they have not been classified according to particle size. The theory of sedimentation is based on an ideal particle that is perfectly spherical, but it applies well to most non-living particles

Table 5.1 United States Department of Agriculture (USDA) and International Society of Soil Science (ISSS) classifications of soil particles based on sieve analysis

Particle fraction name	USDA (mm)	ISSS
Gravel	>2	>2 mm
Very coarse sand	1–2	–
Course sand	0.5–1	0.2–2 mm
Medium sand	0.25–0.5	–
Fine sand	0.1–0.25	0.02–0.2 mm
Very fine sand	0.05–0.1	–
Silt	0.002–0.05	0.002–0.02 mm
Clay	<0.002	<0.002 mm

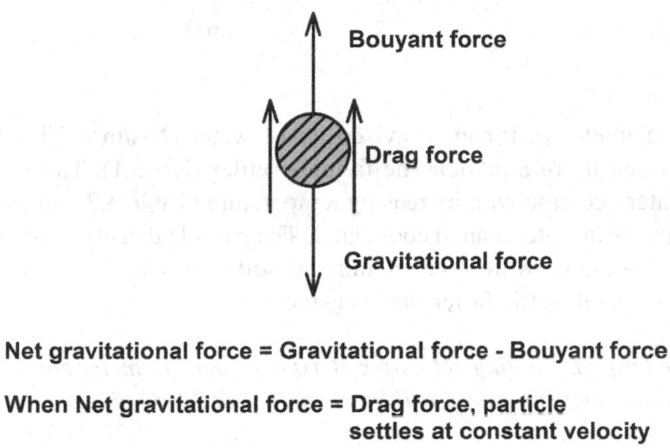
Bouyant force
Drag force
Gravitational force
Net gravitational force = Gravitational force - Bouyant force
When Net gravitational force = Drag force, particle settles at constant velocity

Fig. 5.1 Forces acting on a particle settling in water

found in nature. The settling velocity of a spherical particle is related to its diameter and density. A particle settles in water in response to gravity, but its downward motion is opposed by buoyant and drag forces of water. The buoyant force is equal to the weight of water displaced by the settling particle, and this force increases in direct proportion to particle volume. The net gravitational force is the difference in gravitational and buoyant forces, and it causes the particle to settle. As the particle settles and its velocity increases, the viscous friction force that opposes motion of the particle relative to the water increases. This force is the drag force. When the drag force becomes as great as the net gravitational force, the particle will settle at a constant velocity known as the terminal settling velocity (Fig. 5.1). The variables influencing settling of particles in water have been combined in the Stoke's law equation for estimating the terminal settling velocity of a particle:

$$V_s = \frac{g\left(\rho_p - \rho_w\right)D^2}{18\,\mu} \tag{5.1}$$

where V_s = terminal settling velocity (m/s), g = gravitational acceleration (9.81 m/s^2), ρ_p = particle density (kg/m^3), ρ_w = density of water (kg/m^3),

Table 5.2 Density (ρ) and dynamic viscosity (μ) of water

Temperature (°C)	Density (kg/m^3)	Dynamic viscosity (×10^{-3} N·s/m^2)
0	999.8	1.787
5	999.9	1.519
10	999.7	1.307
15	999.1	1.139
20	998.2	1.022
25	997.0	0.890
30	995.7	0.798
35	994.0	0.719
40	992.2	0.653

D = particle diameter (m), and μ = viscosity of water (N·s/m^2). The greater the diameter and density of a particle the faster it settles (Ex. 5.1). The viscosity and density of water decrease with increasing temperature (Table 5.2), so particles will settle faster in warm water than in cool water. The particle density of organic matter is around 1,050 kg/m^3 while that of mineral soil is around 2,500 kg/m^3. Thus, mineral particles will settle faster than organic ones.

Ex. 5.1: *The terminal settling velocities of a 0.0001-mm diameter clay particle and a 0.02-mm diameter silt particle will be compared at 30 °C.*

Solution:

Using (5.1) and obtaining the density and viscosity of water at 30 °C from Table 5.2,

$$V_s\ silt = \frac{(9.81\ \mathrm{m/s})(2,500 - 995.7)\ \mathrm{kg/m^3}\ \left(2\times 10^{-5}\ \mathrm{m}\right)^2}{18\left(0.798\ \times 10^{-3}\ \mathrm{N\cdot s/m^2}\right)}$$

$$V_s\ silt = 4.1\times 10^{-4}\ \mathrm{m/s}$$

$$V_s\ clay = \frac{(9.81\ \mathrm{m/s})(2,500 - 995.7)\ \mathrm{kg/m^3}\ \left(1\times 10^{-7}\ \mathrm{m}\right)^2}{18\left(0.798\ \times 10^{-3}\ \mathrm{N\cdot s/m^2}\right)}$$

$$V_s\ clay = 1.03\times 10^{-8}\ \mathrm{m/s}.$$

Thus, the silt particle, because of its greater diameter, will sink faster than the clay particle.

Phytoplankton and bacterial cells are quite small, but they still tend to sink. Many species have adaptations to reduce or prevent settling such as gas vacuoles to increase buoyancy or projections to increase drag force. Some species have flagella, cilia, or other methods of motility to avoid sinking. As a result, the Stokes law equation is not applicable to the settling rate of microbial organisms.

In natural waters, turbulence also plays a critical role in settling of particles. Turbulent conditions may keep a particle in suspension much longer than would be expected from calculations made with (5.1). Particles as large as small sand grains or coarse silt may remain suspended as long as sufficient turbulence persists. Thus, the Stokes law equation will not provide a reliable estimate of settling rate in turbulent water.

Colloidal Particles

Colloidal particles range in size from 1 to 100 mμ. Although larger than most molecules, colloidal particles are so small that they remain suspended in water even though they do not dissolve. Colloidal particles exhibit several properties that set them aside from dissolved molecules or from coarse particles that settle. If a narrow beam of light is passed through a true solution, its path cannot be observed. In a colloidal suspension, the path of the light is visible because the colloids are large enough to scatter light while the molecules in true solution are not. This effect of light scattering by colloids is known as the Tyndall effect—named for the British physicist, John Tyndall, who first described it. When a colloidal suspension is viewed under a dark field microscope, the colloidal particles appear as bright points against a dark background, and the bright points move irregularly. This random movement of colloidal particles results from their bombardment by the molecules of the water. Random motion of colloidal particles is called Brownian movement. Colloidal particles cannot be removed from water by filtration unless special filters with very small apertures are used. Colloidal particles have tremendous surface area relative to volume, so they have a large surface adsorptive capacity. Most colloids are electrically charged. The charge may be either positive or negative, but all colloids of the same type have the same charge; thus, colloidal particles repel each other. Most colloids found in natural waters are negatively charged. Neutralization of the charge on colloids causes the dispersed particles to collide, as they no longer repel each other, and they aggregate and precipitate from solution. For example, at municipal water supply plants, suspended, negatively-charged clay particles often are precipitated from water by adding aluminum ion to neutralize the charge on their surfaces.

Color

Suspended particles can impart apparent color of various hues. Phytoplankton blooms color water various shades of green, blue-green, yellow, brown, red, and even black. Suspended mineral particles also can differ greatly in hue—as varied as the color of soils. Tannins and lignins usually impart a yellow-brown, tea-like hue to water, but when concentrations are high the color may be almost black. Well water may contain traces of iron and manganese remaining after iron and manganese oxides precipitated following oxygenation. Iron residues will be rust colored, while those of manganese will be black—when both are present the color may be yellow-brown.

Turbidity and Light Penetration

The reflection of sunlight by water (the albedo) was discussed in Chap. 1. The discussion here will focus on light that passes through the surface and penetrates into the water column.

The electromagnetic spectrum of sunlight includes wavelengths measured in nanometers (nm) that range from <0.01 nm (gamma rays to those >1 billion nm (radio waves) as illustrated in Table 5.3. Water scatters but does not absorb ultraviolet light and it absorbs infrared light quickly—within 1 or 2 m depth. The visible spectrum consists of light with wavelengths between 380 and 750 nm; the positions of different colors of visible light in the electromagnetic spectrum are provided in Table 5.3. Most of the solar radiation reaching the earth is in the form of visible and infrared rays.

In clear water, light undergoes extinction (quenching) quickly—only about 25 % of light passing the surface reaches 10 m depth in the ocean. As a general rule, the rate of light absorption by water is in the order: red and orange > violet > yellow, green, and blue. Thus, the spectrum is rapidly altered with greater depth. Natural waters contain substances that further interfere with light penetration. Phytoplankton absorb light within the visible range with the greatest absorption between 600 and 700 nm in the red and orange range and the lowest adsorption in the green and yellow range of 500–600 nm. Dissolved organic matter such as humic and fulvic acids has a particularly strong preference for blue, violet, and ultraviolet light. Inorganic particles tend to absorb light more uniformly across the entire spectrum. Dissolved salts in water do not interfere with light penetration.

Light is necessary for photosynthesis; the photosynthetically active radiation (PAR) usually is measured as light in wavelength band of 400–700 nm. The peak

Table 5.3 Categories of rays by wavelength in nanometers (nm) and color in the electromagnetic spectrum (spectrum of sunlight)

Types of rays	Wavelength (nm)			
Entire spectrum			Visible spectrum	
Gamma	<0.01		*Color*	(nm)
X	0.01–10		Violet	390–450
Ultraviolet (UV)	10–389		Blue	450–495
Visible	390–750		Green	495–570
Infrared (IR)	$759–10^6$		Yellow	570–590
Micro	$10^6–10^9$		Orange	590–620
Radio	$>10^9$		Red	629–700

absorption of light by plant pigments are: chlorophyll *a*, 430 and 665 nm; chlorophyll *b*, 453 and 642 nm; carotenoids, 449 and 475 nm. These pigments, however, absorb over a wider range and even absorb some green light.

Large bodies of water often appear blue because the surface of the water reflects the color of the sky. But, the true color of natural water results from the unabsorbed light rays remaining from the original light. Because water absorbs light at the red end of the visible spectrum more rapidly than light at the blue end, blue to blue-green light is scattered back and visible—this tends to give waters a blue hue. However, not all waters have a blue or blue-green tint, because substances dissolved or suspended in water also absorb light. True color in water includes color resulting from the natural water and its component of dissolved and colloidal matter. Apparent color is caused by light absorption by suspended matter such as soil particles, phytoplankton, detritus, etc.

The upper, illuminated layer of a water body that supports plant growth is called the photic zone—the bottom of this layer is usually considered to be the depth of 1 % light penetration. The photic zone in the open ocean is about 200 m thick. Freshwater lakes range in nutrient status from oligotrophic (nutrient poor and scant plant growth) to eutrophic (nutrient rich and much plant growth). The thickness of the photic zone based on trophic status is: ultraoligotrophic, 50–100 m; oligotrophic, 5–50 m; mesotrophic, 2–5 m; eutrophic, <2 m (Wetzel 2001). Lakes with turbidity from suspended soil particles or tannins and lignins may be of low productivity yet have a photic zone as thin as a eutrophic water body. The photic zone in small water bodies such as fish ponds may extend to the bottom if waters do not have a dense abundance of plankton (Fig. 5.2).

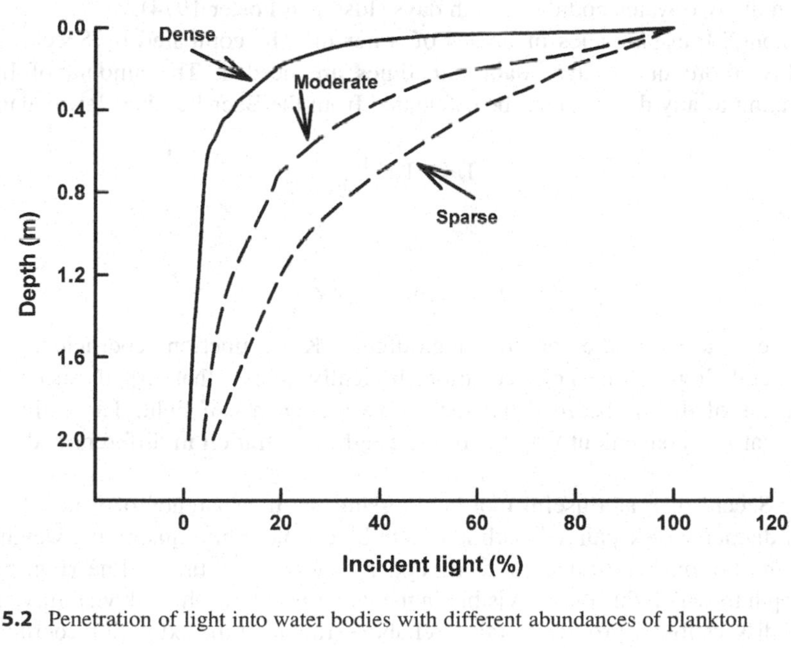

Fig. 5.2 Penetration of light into water bodies with different abundances of plankton

Total energy input in sunlight to the earth's surface can be measured with a pyranometer, and it often is reported in watts per meter square (W/m^2). The amount of solar radiation reaching the earth's cross-section averages about 958 W/m^2. But, the earth is a rotating sphere tilted on its axis of rotation causing the sun's rays to strike at different angles from the vertical at different places and times. The annual mean daily average of instantaneous solar, irradiance values on clear days at 18 locations spread over the conterminous United States varied from 944.2 W/m^2 in Seattle, Washington (47.6097°N; 122.3331°W) to 1,212.8 2 W/m^2 in Albuquerque, New Mexico (35.1107°N; 106.6100°W) with a grand average of 1,071.4 W/m^2 (Hoyt 1978). Cloud cover reduces these values and seasonal variation increases with greater latitude. At Albuquerque, New Mexico, monthly insolation varied from 734 W/m^2 in December to 1,283 W/m^2 in June. Of course, on a daily basis insolation occurs only during daylight—increasing from sunup to solar noon and declining until sundown.

Much of the energy measured by a pyranometer is outside the range of PAR, but in water, the penetration of PAR often is measured. Avogadro's number (6.022×10^{23}) of photons of light is considered 1 mol of light. Underwater light meters usually register the photosynthetic photon flux (PPF) in micromoles per meter square per second ($\mu M/m^2/s$). On a bright day in summer in mid-latitudes, the peak input of PAR might be around 1,400–1,500 $\mu M/m^2/s$.

Temperatures of turbid waters tend to be higher than those of clear waters under similar conditions. This results because turbid waters contain more particulate matter to adsorb heat. For example, afternoon surface water temperature was 31 °C in a small pond at the inception of a phytoplankton bloom and 35 °C at the peak of this algal bloom. Water temperatures at a depth of 60 cm were almost the same in the two water bodies on both days (Idso and Foster 1974).

Although transparencies of bodies of water may be compared by Secchi disk visibility, more quantitative data sometimes are needed. The amount of light penetrating to any depth z may be calculated from the Beer-Lambert law equation

$$I_z = I_o e^{-Kz} \tag{5.2}$$

or

$$\ln I_o - \ln I_z = Kz \tag{5.3}$$

where e = base of the natural logarithms, K = extinction coefficient, and ln = natural logarithm. This equation basically states that equal successive increments of depth absorb equal successive increments of light. The extinction coefficient is a convenient way to compare light penetration in different bodies of water.

The Secchi disk is a useful tool in assessing the transparency of water. It is a 20-cm diameter disk painted with alternate black and white quadrants, weighted under the bottom, and attached at its upper face to a calibrated line (Fig. 5.3). The depth to which this disk is visible in the water is the Secchi disk visibility. The Secchi disk visibility provides a fairly reliable estimate of the extinction coefficient

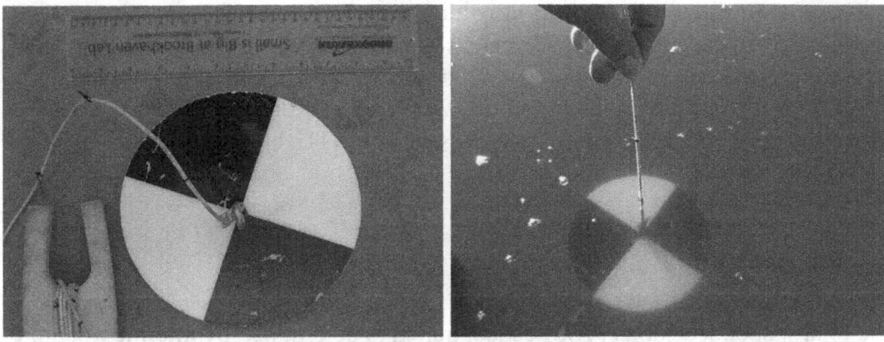

Fig. 5.3 *Left*: a Secchi disk; *Right*: lowering Secchi disk into the water

$$K = \frac{1.7}{Z_{SD}} \tag{5.4}$$

where Z_{SD} = Secchi disk visibility (m).

Assessment of Turbidity and Color

Direct Turbidity Measurement

The condition of reduced clarity in water caused by the scattering or absorption of light by suspended particles in coarse or colloidal suspension or the adsorption of light by colored constituents is known as turbidity. The standard way of measuring turbidity for many years was the Jackson turbidimeter. This instrument consists of a calibrated glass tube, a tube holder, and a candle. The glass tube inside the holder is mounted directly over the candle, and the water sample for turbidity measurement is slowly added to the tube until the candle flame is no longer discernable. The glass tube is calibrated against a silica standard, and 1 mg SiO_2/L equals 1 Jackson turbidimeter unit (JTU). The turbidity in JTU is indicated by the depth of water in the calibrated tube necessary to obscure the candle flame. The Jackson turbidimeter will not read turbidities less than 25 JTU, and it has been largely replaced by turbidity meters that employ the principles of nephelometry.

A nephelometer is an instrument in which a beam of light is passed through a water sample and the amount of light scattered at 90 ° to the light beam is measured. The greater the amount of light that is scattered, the greater the turbidity. The standard for calibrating a nephelometer for turbidity analysis is a suspension of formazin made by combining 5.0 mL volumes of a hydrazine sulfate solution (1 g/100 mL) and a solution of hexamethylenetetramine (10 g/100 mL) and diluting to 100 mL with distilled water. This solution has a turbidity of 400 nephelometer turbidity units (NTU), and it can be further diluted to calibrate the nephelometer (Eaton et al. 2005)

There is no direct relationship in natural waters between the intensity of light scattered 90 ° to the incident light and Jackson candle turbidity. Thus, there is no valid basis for the practice of calibrating a nephelometer in terms of JTU even though a 40 NTU formazin solution may in some cases provide a similar JTU reading.

Most natural waters have turbidities less than 50 NTU, but values can range from <1 NTU to >1,000 NTU.

Suspended Solids

The total suspended solids (TSS) concentration is determined by filtering a sample through a tared glass fiber filter, the filter and residue are dried at 102 °C, and the weight gain of the filter is caused by the suspended solids retained on it (Ex. 5.2). Alternatively, filtered and unfiltered portions of a water sample can be evaporated to dryness in tared dishes, and the difference in weights of residues in the two dishes is the TSS concentration.

Ex. 5.2: *A 100-mL water sample is passed through a glass fiber filter weighing 1.25000 g. Upon drying, the filter weighs 1.25305 g. The TSS concentration will be estimated.*

Solution:

$$TSS = \frac{(1.25305 - 1.25000)\text{g}\ (10^3\ \text{mg/g})(1{,}000\ \text{mL/L})}{100\ \text{mL}} = 30.5\ \text{mg/L}.$$

The organic matter contribution to particulate matter can be determined by igniting the residue from the TSS analysis (Ex. 5.3). The loss of weight upon ignition is equal to the organic component of the residue. This fraction of the TSS concentration is called the suspended volatile solids or the particulate organic matter (POM).

Ex. 5.3: *The glass fiber filter and residue from TSS analysis (Ex. 5.2) weighs 1.25211 g after ignition at 550 °C to constant weight. The POM concentration will be calculated.*

Solution:

$$POM = \frac{(1.25305 - 1.25211)\text{g}\ (10^3\ \text{mg/g})(1{,}000\ \text{mL/L})}{100\ \text{mL}} = 9.4\ \text{mg/L}.$$

The difference between the TSS concentration and the POM concentration represents the suspended inorganic solids. This fraction usually is known as the fixed suspended solids or the particulate inorganic matter. For the illustration in Ex. 5.2 and 5.3, the particulate inorganic matter (PIM) concentration is 21.1 mg/L.

Settleable Solids

Sometimes, it is desired to measure the volume of solids that will rapidly settle from water. This can be done with a 1-L, inverted cone called an Imhoff cone that has graduations to allow the volume of sediment to be estimated. Water is poured into the cone and it is allowed to set undisturbed for 1 h. After 1 h, the volume of settleable solids in milliliters per liter is read from calibration marks in the bottom of the cone. The volume of settleable solids must be 1 mL or greater for accurate measurements, and this technique normally is used only for studies of effluents. Only natural waters with large concentrations of suspended solids exhibit measurable concentrations of settleable solids.

Secchi Disk

The Secchi disk is used to estimate water clarity as mentioned above. However, there is usually a close relationship between Secchi disk visibility and the concentration of suspended particles in water. The source of turbidity associated with a reduction in Secchi disk visibility usually can be determined from the apparent color of the water body and the appearance of suspended solids particles.

Color

The usual method of measuring color in water is relatively primitive in comparison with methods for determining concentrations of other variables. It relies upon comparison of natural water color to a series of standard color solutions in clear tubes (Nessler tubes) of 17.5 or 20 cm in height and with volumes of 50 or 100 mL, respectively. A solution containing 500 color units is prepared by dissolving 1.246 g potassium chloroplatinate and 1.00 g cobaltous chloride with 100 mL nitric acid and diluting to 1,000 mL with distilled water (Eaton et al. 2005). Standards containing from 0 (distilled water) to 70 color units are prepared and placed in Nessler tubes. The water sample is filtered and placed in a Nessler tube and a color unit is assigned to the sample after comparison with the standards. The water sample may be diluted with distilled water if necessary to assign a color reading to a highly colored sample. A spectrophotometric color method is available; it reveals the hue of the color but not the intensity (Eaton et al. 2005).

Significance

Particulate matter in water can result in high rates of turbidity and sedimentation. Sedimentation in water bodies makes affected areas shallower, and it may destroy benthic organisms and fish eggs. Turbidity seldom has direct toxic or mechanical effects on fish and other aquatic organisms, but it restricts light penetration. Water that is turbid with suspended soil particles or high concentrations of dissolved organic matter and organic detritus will not be very productive of phytoplankton or other aquatic plants. Turbidity in water also detracts from its value for recreational purposes and makes it less aesthetically pleasing. Turbidity and color in public water supplies is undesirable because consumers want clear water. Suspended particles also can be troublesome in irrigation water by causing a crust when water evaporates in fields or on plant surfaces. Where sprinklers and other devices are used to distribute irrigation water, clogging of lines and water discharge devices may result.

Water quality criteria usually have some reference to total suspended solids, turbidity, and color. In order to protect aquatic ecosystems, total suspended solids, turbidity, and color should not change by more than 10 % of the seasonal mean concentration, or the compensation point for photosynthetic activity should not be reduced by more than 10 %. Water quality criteria for effluents may contain limits for total suspended solids. A TSS limit of 25 mg/L usually will provide a good level of protection against excessive turbidity and sedimentation in aquatic ecosystems, and fair protection may be achieved by a limit of 50 mg/L. Sometimes, effluent discharge permits may specify that a visible turbidity plume is not permissible. The maximum allowable turbidity concentration for municipal drinking water is 5 NTU and at filter plants, 95 % of samples must be less than 1 NTU.

Although many associate clear water with high purity and turbid water with contamination, clear water could contain harmful microorganisms. Thus, despite the degree of turbidity or color in water sources for human use, the possibility for contamination with pathogens should always be considered. Turbidity and color in water is mainly an aesthetic issue. Most water users want clear drinking water, and do not desire that fabrics be stained during washing or that sinks, tubs, showers, and glassware be stained by exposure to water with a high degree of color. There are no mandated standards for color, the recommended allowable limit is 15 color units.

The removal of turbidity usually is affected by coagulation with aluminum sulfate (alum) or other coagulant, sedimentation to remove the resulting floc, and filtration. This treatment will remove phytoplankton and both coarse and colloidal mineral particles, but it may not clear water of true color from tannins and lignins. High-dose chlorination sometimes will remove tannins and lignins, but in some cases tannins and lignins are removed by activated carbon filters, filtration through anion exchange resins, and treatment with potassium permanganate.

References

Eaton AD, Clesceri LS, Rice EW, Greenburg AE (eds) (2005) Standard methods for the examination of water and wastewater. American Public Health Association, Washington, DC

Hoyt DV (1978) A model for the calculation of solar global insolation. Sol Energy 21:27–35

Idso SB, Foster JM (1974) Light and temperature relations in a small desert pond as influenced by phytoplanktonic density variations. Water Resour Res 10:129–132

Wetzel RG (2001) Limnology, 3rd edn. Academic, San Diego

Dissolved Oxygen and Other Gases

6

Abstract

Concentrations of atmospheric gases in water at saturation vary with the partial pressures and solubilities of the gases, and the temperature and salinity of the water. Dissolved oxygen is used to illustrate the principles of gas solubility. Concentrations of dissolved oxygen at saturation decrease with greater elevation (lower barometric pressure), higher salinity, and rising temperature. The concentration of dissolved oxygen in water may be expressed in milligrams per liter, but it also can be reported in milliliters per liter, percentage saturation, oxygen tension, or other units. The rate of diffusion of dissolved oxygen into water bodies is related to various factors, but the most important are the concentration of the dissolved oxygen already in the water, the area of contact between the air and water, and the amount of wind-driven or other turbulence in the water. Oxygen will diffuse into the air from oxygen-supersaturated water bodies. Dissolved oxygen is of utmost importance in water quality, because it is essential for aerobic respiration. Absorption of oxygen by fish and other aquatic animals is controlled by the pressure of oxygen in the water rather than the dissolved oxygen concentration in milligrams per liter. Low oxygen pressure can stress or even kill aquatic organisms. Excessive dissolved gas (including oxygen) concentrations in water can lead to gas bubble trauma in fish and other aquatic animals.

Keywords

Gas solubility • Percentage saturation • ΔP • Gas transfer • Respiration

Introduction

The atmosphere is comprised of nitrogen, oxygen, argon, carbon dioxide, water vapor, and traces of a few other gases. These gases are soluble in water, and an equilibrium state can be attained between atmospheric gases and dissolved gases.

C.E. Boyd, *Water Quality*, DOI 10.1007/978-3-319-17446-4_6

Surface waters usually are near equilibrium (saturation) with nitrogen, argon, and trace gases, but oxygen and carbon dioxide concentrations vary greatly because of biological process. Groundwater may vary greatly in gas concentrations because of both biological and geologic factors.

Dissolved oxygen is important in water quality because it is essential in aerobic respiration and in maintaining oxidized conditions in the water column and at the sediment-water interface. Carbon dioxide is a source of carbon for photosynthesis, and a major determinant of pH in natural waters. Dissolved nitrogen usually has a higher concentration in water than other gases, but its influence on biological and chemical processes is much less than that of oxygen or carbon dioxide. Other gases usually have little influence on water quality.

The purpose of this chapter is to provide a discussion of gas solubility in water, with emphasis on dissolved oxygen. Additional information on the dynamics of dissolved oxygen, carbon dioxide, and nitrogen and their effects on aquatic life will be provided in other chapters

Gas Solubility

Partial Pressure

The major gases and their approximate percentages by volume in dry, atmospheric air are: nitrogen (N_2), 78.08 %; oxygen (O_2), 20.95 %; argon (Ar), 0.93 %; carbon dioxide (CO_2), 0.039 %. According to Dalton's Law of Partial Pressures, the total pressure of a mixture of gases that do not react with each other is the sum of the partial pressures of the individual gases in the mixture. In the case of atmospheric gases, the total pressure or barometric pressure (BP) can be expressed in terms of the partial pressures of the individual gases as follows:

$$BP = P_{N_2} + P_{O_2} + P_{Ar} + P_{CO_2} + P_{H_2O} + P_{\text{Other gases}} \tag{6.1}$$

where P = partial pressures of the gases indicated by subscripts. Dalton's Law also states that the partial pressure of each gas is directly proportional to its volume percentage in the mixture.

At standard temperature (0 °C) and pressure (760 mmHg) or STP, the approximate partial pressures of the gases are

$$P_{N_2} = (760)(0.7808) = 593.4 \text{ mm}$$

$$P_{O_2} = (760)(0.2095) = 159.2 \text{ mm}$$

$$P_{Ar} = (760)(0.0093) = 7.1 \text{ mm}$$

$$P_{CO_2} = (760)(0.00039) = 0.30 \text{ mm.}$$

Other gases such as neon, helium, and methane are present in the atmosphere at concentrations of a few parts per million, and krypton, xenon, and radon are even rarer.

Gas pressures are reported in atmospheres for some purposes. The atmospheric pressure at STP is 1 atm; thus, the volume percentage of a gas in the atmosphere divided by 100 is its pressure in atmosphere, e.g., nitrogen comprises 78.08 % of the atmosphere by volume, and its pressure is 0.7808 atm. Completely dry air seldom is found in nature. The water vapor pressure varies tremendously in air at different places and saturation vapor pressure increases with increasing air temperature (Table 1.2). Water vapor in air dilutes other gases slightly. The saturation vapor pressure at STP is 4.38 mm, (0.0058 atm). Thus, the concentration of a gas in the atmosphere will be diluted by a factor of 0.994 [(760 mm − 4.38 mm) ÷ 760 mm] in air saturated with water vapor.

Henry's Law Constant and Bunsen Coefficient

The solubilities of gas in water (or other liquid), according to Henry's Law of gas solubility, are directly proportional to the partial pressures of each gas above the water. Gases do not have equal solubilities at equal partial pressure, and the Henry's Law constant is the ratio of the gas concentration in water at equilibrium to its concentration in air above the water. The Henry's Law constants for atmospheric gases are not presented, because it is more common in considerations of water quality to convert Henry's Law constants to Bunsen coefficients. The Bunsen coefficient is the ratio of dissolved gas to water (or other liquid) at equilibrium typically expressed as milliliter gas per milliliter water. The Bunsen coefficients for the major gases in freshwater at STP are: nitrogen, 0.0237 mL/mL; oxygen, 0.04910 mL/mL; argon, 0.05363 mL/mL; carbon dioxide, 1.7272 mL/mL. Thus, carbon dioxide is the most soluble gas while nitrogen is the least soluble.

Weight of Air in Water

The concentrations of gases in water usually are expressed in milligrams of gas per liter of water in water quality endeavors. The estimation of gas solubility on a weight to volume basis from the Bunsen coefficient will be illustrated for oxygen. One mole of a gas—32 g in the case of O_2—occupies 22.4 L at STP according to Avogadro's Law relating volume to mass of a gas. It follows that 32 mg of O_2 would occupy 22.4 mL, and the density of O_2 at STP is 1.429 mg/mL. Thus, multiplying the Bunsen coefficient for O_2 (0.04910 mL/mL) by density gives 0.070164 mg O_2/mL or 70.164 mg/L of oxygen in water at equilibrium. This concentration is for pure O_2 in contact with water. The atmosphere is only 20.95 % O_2, so multiplying the decimal fraction of oxygen in the atmosphere by the oxygen concentration for water in equilibrium with pure oxygen gives a value of 14.70 mg/L. This concentration is for dry air in contact with oxygen, so the dilution factor of atmospheric gases by saturation water vapor of 0.994 (calculated above) must be multiplied by the dissolved oxygen concentration of 14.70 mg/L estimated for dry air in contact with water—the result is 14.61 mg/L.

The same approach can be applied to nitrogen, argon, and carbon dioxide. The resulting solubilities of these three gases at STP are 23.05 mg/L for nitrogen, 1.32 mg/L for carbon dioxide, and 0.89 mg/L for argon. Nitrogen has the greatest concentration in water despite being the least soluble, because it comprises a much greater volume percentage of the atmosphere than do other gases.

The calculations described above can be arranged into an equation for estimating gas solubility for moist air in water from the Bunsen coefficient

$$C_{i,m} = (1,000\, B_i D_i X_i)\frac{760 - vp}{760} \qquad (6.2)$$

where $C_{i,m}$ = concentration of gas i from moist air (mg/L), B_i = Bunsen coefficient for gas i (mL/mL), D_i = density of gas i, X_i = decimal fraction of gas i in atmosphere (% concentration by volume ÷ 100), and vp = saturation vapor pressure of water (mmHg). Provided the appropriate value of the Bunsen coefficient, gas density, and saturation vapor pressure are used, (6.2) may be used for any gas, temperature, or salinity.

Colt (1984) reported the Bunsen coefficients for atmospheric gases in water of different temperatures and salinities. The solubilities of oxygen, nitrogen, and carbon dioxide from moist air in water for the normal temperature and salinity ranges encountered in water quality efforts are provided in Tables 6.1, 6.2, and 6.3, respectively. The solubility of a gas decreases as temperature increases. This results because the movement of molecules of both water and gas increase as a function of temperature, and as temperature increases the more rapidly moving molecules collide more frequently to hinder the entrance of gas molecules into the water. Dissolved salts hydrate in water and a portion of the water molecules in a given volume of water are bound so tightly to salt ions through hydration that they are not free to dissolve gases. In reality, salinity has no effect on the solubility of a gas, but increasing the salinity decreases the amount of free water available to dissolve gases in a given volume of water.

Use of Gas Solubility Tables

Oxygen solubilities in Table 6.1 are for moist air at 760 mm pressure. To adjust values to a different barometric pressure, use the following equation:

$$DO_s = DO_t \frac{BP}{760} \qquad (6.3)$$

where DO_s = concentration of dissolved oxygen at saturation corrected for pressure, and DO_t = dissolved oxygen concentration at saturation at 760 mm from Table 6.1.

Table 6.1 The solubility of oxygen (mg/L) as a function of temperature and salinity (moist air, barometric pressure = 760 mmHg) (Benson and Krause 1984)

Temperature (°C)	Salinity (g/L)								
	0	5	10	15	20	25	30	35	40
0	14.602	14.112	13.638	13.180	12.737	12.309	11.896	11.497	11.111
1	14.198	13.725	13.268	12.825	12.398	11.984	11.585	11.198	10.825
2	13.813	13.356	12.914	12.487	12.073	11.674	11.287	10.913	10.552
3	13.445	13.004	12.576	12.163	11.763	11.376	11.003	10.641	10.291
4	13.094	12.667	12.253	11.853	11.467	11.092	10.730	10.380	10.042
5	12.757	12.344	11.944	11.557	11.183	10.820	10.470	10.131	9.802
6	12.436	12.036	11.648	11.274	10.911	10.560	10.220	9.892	9.573
7	12.127	11.740	11.365	11.002	10.651	10.311	9.981	9.662	9.354
8	11.832	11.457	11.093	10.742	10.401	10.071	9.752	9.443	9.143
9	11.549	11.185	10.833	10.492	10.162	9.842	9.532	9.232	8.941
10	11.277	10.925	10.583	10.252	9.932	9.621	9.321	9.029	8.747
11	11.016	10.674	10.343	10.022	9.711	9.410	9.118	8.835	8.561
12	10.766	10.434	10.113	9.801	9.499	9.207	8.923	8.648	8.381
13	10.525	10.203	9.891	9.589	9.295	9.011	8.735	8.468	8.209
14	10.294	9.981	9.678	9.384	9.099	8.823	8.555	8.295	8.043
15	10.072	9.768	9.473	9.188	8.911	8.642	8.381	8.129	7.883
16	9.858	9.562	9.276	8.998	8.729	8.468	8.214	7.968	7.730
17	9.651	9.364	9.086	8.816	8.554	8.300	8.053	7.814	7.581
18	9.453	9.174	8.903	8.640	8.385	8.138	7.898	7.664	7.438
19	9.261	8.990	8.726	8.471	8.222	7.982	7.748	7.521	7.300
20	9.077	8.812	8.556	8.307	8.065	7.831	7.603	7.382	7.167
21	8.898	8.641	8.392	8.149	7.914	7.685	7.463	7.248	7.038

(continued)

Table 6.1 (continued)

Temperature (°C)	Salinity (g/L)								
	0	5	10	15	20	25	30	35	40
22	8.726	8.476	8.233	7.997	7.767	7.545	7.328	7.118	6.914
23	8.560	8.316	8.080	7.849	7.626	7.409	7.198	6.993	6.794
24	8.400	8.162	7.931	7.707	7.489	7.277	7.072	6.872	6.677
25	8.244	8.013	7.788	7.569	7.357	7.150	6.950	6.754	6.565
26	8.094	7.868	7.649	7.436	7.229	7.027	6.831	6.641	6.456
27	7.949	7.729	7.515	7.307	7.105	6.908	6.717	6.531	6.350
28	7.808	7.593	7.385	7.182	6.984	6.792	6.606	6.424	6.248
29	7.671	7.462	7.259	7.060	6.868	6.680	6.498	6.321	6.148
30	7.539	7.335	7.136	6.943	6.755	6.572	6.394	6.221	6.052
31	7.411	7.212	7.018	6.829	6.645	6.466	6.293	6.123	5.959
32	7.287	7.092	6.903	6.718	6.539	6.364	6.194	6.029	5.868
33	7.166	6.976	6.791	6.611	6.435	6.265	6.099	5.937	5.779
34	7.049	6.863	6.682	6.506	6.335	6.168	6.006	5.848	5.694
35	6.935	6.753	6.577	6.405	6.237	6.074	5.915	5.761	5.610
36	6.824	6.647	6.474	6.306	6.142	5.983	5.828	5.676	5.529
37	6.716	6.543	6.374	6.210	6.050	5.894	5.742	5.594	5.450
38	6.612	6.442	6.277	6.117	5.960	5.807	5.659	5.514	5.373
39	6.509	6.344	6.183	6.025	5.872	5.723	5.577	5.436	5.297
40	6.410	6.248	6.091	5.937	5.787	5.641	5.498	5.360	5.224

Table 6.2 The solubility of nitrogen (mg/L) as a function of temperature and salinity (moist air, barometric pressure = 760 mmHg)

	Salinity (g/L)								
Temperature (°C)	0	5	10	15	20	25	30	35	40
0	23.04	22.19	21.38	20.60	19.85	19.12	18.42	17.75	17.10
1	22.45	21.63	20.85	20.09	19.36	18.66	17.98	17.33	16.70
2	21.88	21.09	20.33	19.60	18.89	18.21	17.56	16.93	16.32
3	21.34	20.58	19.84	19.13	18.45	17.79	17.15	16.54	15.95
4	20.82	20.08	19.37	18.68	18.02	17.38	16.77	16.17	15.60
5	20.33	19.61	18.92	18.26	17.61	16.99	16.40	15.82	15.26
6	19.85	19.16	18.49	17.84	17.22	16.62	16.04	15.48	14.94
7	19.40	18.73	18.08	17.45	16.85	16.26	15.70	15.15	14.63
8	18.96	18.31	17.68	17.07	16.48	15.92	15.37	14.84	14.33
9	18.54	17.91	17.30	16.71	16.14	15.59	15.06	14.54	14.05
10	18.14	17.53	16.93	16.36	15.81	15.27	14.75	14.25	13.77
11	17.76	17.16	16.58	16.02	15.49	14.97	14.46	13.98	13.51
12	17.39	16.81	16.24	15.70	15.18	14.67	14.18	13.71	13.25
13	17.03	16.47	15.92	15.39	14.88	14.39	13.91	13.45	13.01
14	16.69	16.14	15.61	15.09	14.60	14.12	13.65	13.20	12.77
15	16.36	15.82	15.31	14.81	14.32	13.86	13.40	12.97	12.54
16	16.04	15.52	15.02	14.53	14.06	13.60	13.16	12.74	12.32
17	15.73	15.23	14.74	14.26	13.80	13.36	12.93	12.51	12.11
18	15.44	14.94	14.47	14.00	13.56	13.12	12.71	12.30	11.91
19	15.15	14.67	14.21	13.76	13.32	12.90	12.49	12.09	11.71
20	14.88	14.41	13.96	13.52	13.09	12.68	12.28	11.89	11.52

(continued)

Table 6.2 (continued)

Temperature (°C)	Salinity (g/L)								
	0	5	10	15	20	25	30	35	40
21	14.61	14.16	13.71	13.28	12.87	12.47	12.08	11.70	11.33
22	14.36	13.91	13.48	13.06	12.65	12.26	11.88	11.51	11.15
23	14.11	13.67	13.25	12.84	12.45	12.06	11.69	11.33	10.98
24	13.87	13.44	13.03	12.63	12.25	11.87	11.51	11.16	10.81
25	13.64	13.22	12.82	12.43	12.05	11.69	11.33	10.99	10.65
26	13.41	13.01	12.61	12.23	11.86	11.51	11.16	10.82	10.49
27	13.20	12.80	12.42	12.04	11.68	11.33	10.99	10.66	10.34
28	12.99	12.60	12.22	11.86	11.50	11.16	10.83	10.51	10.19
29	12.78	12.40	12.04	11.68	11.33	11.00	10.67	10.36	10.05
30	12.58	12.21	11.85	11.50	11.17	10.84	10.52	10.21	9.91
31	12.39	12.03	11.68	11.34	11.00	10.68	10.37	10.07	9.77
32	12.21	11.85	11.51	11.17	10.85	10.53	10.23	9.93	9.64
33	12.02	11.68	11.34	11.01	10.69	10.39	10.09	9.79	9.51
34	11.85	11.51	11.18	10.86	10.55	10.24	9.95	9.66	9.39
35	11.68	11.34	11.02	10.71	10.40	10.10	9.82	9.54	9.26
36	11.51	11.18	10.87	10.56	10.26	9.97	9.69	9.41	9.14
37	11.35	11.03	10.72	10.42	10.12	9.84	9.56	9.29	9.03
38	11.19	10.88	10.57	10.28	9.99	9.71	9.44	9.17	8.91
39	11.04	10.73	10.43	10.14	9.86	9.58	9.32	9.06	8.80
40	10.89	10.59	10.29	10.01	9.73	9.46	9.20	8.94	8.70

Table 6.3 Solubility of carbon dioxide (mg/L) in water at different temperatures and salinities exposed to moist air containing 0.04 % carbon dioxide at a total air pressure of 760 mmHg

	Salinity (g/L)								
Temperature (°C)	0	5	10	15	20	25	30	35	40
0	1.34	1.31	1.28	1.24	1.21	1.18	1.15	1.12	1.09
5	1.10	1.08	1.06	1.03	1.01	0.98	0.96	0.93	0.89
10	0.93	0.91	0.87	0.85	0.83	0.81	0.79	0.77	0.75
15	0.78	0.77	0.75	0.73	0.70	0.68	0.66	0.65	0.64
20	0.67	0.65	0.63	0.62	0.61	0.60	0.58	0.57	0.56
25	0.57	0.56	0.54	0.53	0.52	0.51	0.50	0.49	0.48
30	0.50	0.49	0.48	0.47	0.46	0.45	0.44	0.43	0.42
35	0.44	0.43	0.42	0.41	0.40	0.39	0.39	0.38	0.37
40	0.39	0.38	0.37	0.36	0.36	0.35	0.35	0.34	0.33

Ex. 6.1: *Estimate the solubility of dissolved oxygen in freshwater at 26 °C when BP is 710 mm.*

Solution:
The solubility of dissolved oxygen in freshwater at 26 °C from Table 6.1 is 8.09 mg/ L. Thus,

$$DO_s = 8.09 \times \frac{710}{760} = 7.56\ \text{mg/L}.$$

Because barometric pressure declines with increasing elevation, the solubility of gases is less for the same temperature at a higher elevation than at sea level.

Ex. 6.2: *How much less is the solubility of oxygen in freshwater (20 °C) at 3,000 m than at sea level?*

Solution:
From Fig. 1.4 or (1.1), BP should be about 536 mm at 3,000 m, and the solubility of oxygen in freshwater at 20 °C and 760 mm is 9.08 mg/L. Thus,

$$DO_s = 9.08 \times \frac{536}{760} = 6.40\ \text{mg/L}.$$

The solubility of oxygen will be 2.68 mg/L less at 3,000 m than at sea level.

Oxygen solubility data in Table 6.1 are for the water surface. The pressure holding a gas in solution at some depth beneath the surface is greater than the barometric pressure by an amount equal to the hydrostatic pressure (see Chap. 1). The hydrostatic pressure expressed in millimeters of mercury per meter of depth for different water temperatures and pressures are provided in Table 6.4. Thus, to

Table 6.4 The specific weight (mmHg/m depth) as a function of temperature and salinity (Colt 1984)

Temperature (°C)	Salinity (g/L)								
	0	5	10	15	20	25	30	35	40
0	73.54	73.84	74.14	74.44	74.73	75.03	75.33	75.62	75.92
5	73.55	73.85	74.14	74.43	74.72	75.01	75.30	75.59	75.88
10	73.53	73.82	74.11	74.39	74.68	74.97	75.25	75.54	75.83
15	73.49	73.77	74.05	74.34	74.62	74.90	75.18	75.47	75.75
20	73.42	73.70	73.98	74.26	74.54	74.82	75.10	75.38	75.66
25	73.34	73.62	73.89	74.17	74.44	74.72	75.00	75.27	75.55
30	73.24	73.51	73.78	74.06	74.33	74.60	74.88	75.15	75.43
35	73.12	73.39	73.66	73.93	74.20	74.48	74.75	75.02	75.30
40	72.98	73.25	73.52	73.79	74.06	74.34	74.61	74.88	75.15

estimate the gas solubility at some depth below the water surface, the total pressure must be used in (6.3) instead of barometric pressure as done in Ex. 6.3.

Ex. 6.3: *Estimate the solubility of dissolved oxygen in water at 25 °C with a salinity of 20 g/L at a depth of 4.5 m when BP is 752 mm.*

Solution:
From Table 6.1, the solubility of oxygen at 760 mm, 20 g/L salinity, and 25 °C is 7.36 mg/L. The specific weight of water at 25 °C and 20 g/L salinity from Table 6.4 is 74.44 mm/m. The computation is as follows:

$$DO_s = 7.36\ \mathrm{mg/L} \times \frac{752\ \mathrm{mm} + (4.5\ \mathrm{m})(74.54\ \mathrm{mm/m})}{760\ \mathrm{mm}} = 10.53\ \mathrm{mg/L}.$$

Dissolved oxygen concentrations may sometimes be expressed in milliliters per liter. To convert milligrams of oxygen to milliliters of oxygen, the density of dissolved oxygen must be determined for existing temperature and pressure. The density of oxygen can be computed if the volume of 1 mol of oxygen is known for existing conditions.

The law that describes the relationships about pressure, temperature, volume, and number of moles of an ideal gas is called the ideal gas law or the universal gas law. The simplest way of expressing the law is by the universal gas law equation.

$$\mathrm{PV} = \mathrm{nRT} \qquad (6.4)$$

where P = pressure (atm), V = volume (L), n = number of moles of gas, R = universal gas law constant (0.082 L·atm/mol·°K), and T = absolute temperature (°A = 273.15 + °C).

The universal gas law equation is a useful tool for solving both theoretical and practical problems related to gases. Thus, the equation will be derived in Ex. 6.4 in order to reinforce the readers' understanding of the gas laws.

Ex. 6.4: *The universal or ideal gas law equation will be derived.*

Solution:
The number of molecules (n) in a mass of a gas depends upon its volume (V), temperature (T), and pressure (P). If P is doubled without a change in V and T, there will be twice as many molecules in the same volume, i.e., n = kP. The number of molecules of a gas also varies directly with V if T and P are constant, i.e., V = kP. But if V and P are constant, an increase in T will increase P and gas molecules would have to be removed to maintain constant V. Thus, the number of molecules varies inversely with temperature, i.e., n = 1/T. We can summarize the relationships mentioned above in the following expression:

$$n = \frac{kPV}{T}$$

where units for n, P, V, and T are the same as already defined in (6.4).

Because it is known that 1 mol of a gas occupies 22.4 L at STP, the previous expression becomes

$$n = k\,\frac{(1\ \text{atm})(22.4\ \text{L/mol})}{273\ \text{°K}} = k(0.082\ \text{L·atm/mole·°K}).$$

If we assign 1 mol to n, k becomes 1/0.082 L·atm/mol·°K, and we may rearrange the equation as follows:

$$\frac{1}{0.082}PV = nT$$

$$\text{and } PV = n\,0.082\,T.$$

The quantity 0.082 L·atm/mol·°K is known as the universal or ideal gas law constant. There are other values for this constant depending upon the unit of pressure used. The final form of the universal gas law equation usually is given as

$$PV = nRT.$$

The gas law equation can be used for many purposes; an example is the conversion of the weight of a dissolved gas in water to the volume of the gas in water (Ex. 6.5).

Ex. 6.5: *A water contains 7.54 mg/L dissolved oxygen. The pressure is 735 mm and the temperature is 30 °C. The dissolved oxygen concentration will be converted to milliliters per liter.*

Solution:

The volume of 1 mol of oxygen is

$$\left(\frac{735\ \text{mm}}{760\ \text{mm}}\right)(V) = (1\ \text{mole})(0.082\ \text{L·atm/mole·°K})(303.15\text{A})$$

$$V = 25.7\ \text{L/mol}.$$

One mole is 32 g of oxygen, so the density is

$$\frac{32\ \text{g/mol}}{25.7\ \text{L/mol}} = 1.245\ \text{g/L } \textit{or } 1.245\ \text{mg/mL}.$$

Thus, the concentration in milliliters per liter is

$$\frac{7.54\ \text{mg/L}}{1.245\ \text{mg/mL}} = 6.06\ \text{mL/L}.$$

Percentage Saturation and Oxygen Tension

Because of biological activity, a water may contain more or less dissolved oxygen than the saturation concentration for existing conditions. The percentage saturation of water with dissolved oxygen is calculated as follows:

$$\text{PS} = \frac{\text{DO}_\text{m}}{\text{DO}_\text{s}} \times 100 \tag{6.5}$$

where PS = percentage saturation (%) and DO_m = the measured concentration of dissolved oxygen (mg/L). A calculation of PS is made in Ex. 6.6.

Ex. 6.6: *Calculate the percentage saturation at the surface of a freshwater body containing 10.08 mg/L dissolved oxygen when the water temperature is 28 °C and BP is 732 mm.*

Solution:

The concentration of dissolved oxygen at saturation is

$$DO_s = 7.81\ \text{mg/L} \times \frac{732}{760} = 7.52\ \text{mg/L}.$$

The percentage saturation is

$$PS = \frac{10.08}{7.52} \times 100 = 134\%.$$

Suppose the following morning, the body of water in Ex. 6.6 contained only 4.5 mg/L dissolved oxygen and the temperature and BP were 24 °C and 740 mm, respectively. The percentage saturation would now be only 55 %.

Organisms respond to the pressure of dissolved oxygen in the water rather than to the absolute concentration. Physiologists often refer to the pressure of dissolved oxygen as the oxygen tension. The tension of dissolved oxygen refers to the pressure of dissolved oxygen in the atmosphere required to hold an observed concentration of dissolved oxygen in the water. The partial pressure of oxygen in air is 159.2 mm at STP. To estimate oxygen tension, the partial pressure of oxygen in the atmosphere is multiplied by the factor $\text{DO}_\text{m}/\text{DO}_\text{s}$—oxygen tension is closely linked to percentage saturation.

Ex. 6.7: *Estimate the oxygen tension in water with 10 g/L salinity and 24 °C where dissolved oxygen is 9.24 mg/L and BP is 760 mm.*

Solution:
The concentration at saturation is 7.93 mg/L. Thus,

$$Oxygen\ tension = \frac{9.24}{7.93} \times 159.2 = 185.5\ \text{mm}.$$

Waters saturated with dissolved oxygen and at different temperatures will have the same oxygen tension. For example, a freshwater at 20 °C with 9.08 mg/L dissolved oxygen has the same oxygen tension as a freshwater at 10 °C with 11.28 mg/L dissolved oxygen. This results because the factor DO_m/DO_s is unity for both. Aquatic organisms would be exposed to the same oxygen pressure (tension) even though the oxygen concentration would be greater for the cooler water. It also is significant to note that when water holding a certain concentration of gas warms without loss of gas to the air, the gas tension and the percentage saturation increase. For example, if a water holding 6.35 mg/L dissolved oxygen at 760 mm and 18 °C warms to 23 °C at the same atmospheric pressure and without loss of oxygen, the value for saturation increases from 67 to 74 % and oxygen tension increases from 107 to 118 mm.

The ΔP

There are various reasons why waters become supersaturated with gases. Rainwater in winter or snow melt may be saturated with gases, and when this water percolates downward to the water table and warms, supersaturation results. In warm weather, water from wells or springs often will be cooler than ambient temperature at ground level, and supersaturated with gas. When water falls over a high dam, air bubbles may be entrained. The water will plunge to a considerable depth beneath the surface in the pool behind the dam into which it falls. The resulting increase in hydrostatic pressure increases the dissolved oxygen concentration at saturation, and when the water ascends and flows into shallower areas, it will be supersaturated with gases (Boyd and Tucker 2014).

Air leaks in or improper submergence of the intake of pumps can result in gas supersaturation. Air bubbles can be sucked into the water, and as a result of the pressure increase caused by the pump, discharge will be supersaturated with gases (air). Photosynthesis also can result in gas supersaturation through production of high dissolved oxygen concentrations.

The difference between the total gas pressure (TGP) and the total pressure at a given depth is called ΔP. At the water surface, ΔP can be expressed as

$$\Delta P = TGP - BP. \quad (6.6)$$

The total gas pressure can be determined from the partial pressures (tensions) of the individual gases in water

$$\Delta P = (P_{O_2} + P_{N_2} + P_{Ar} + P_{CO_2}) - BP. \quad (6.7)$$

For practical purposes, P_{Ar} and P_{CO_2} can be omitted from (6.7), but the analysis for dissolved nitrogen concentration is difficult. Fortunately, a relatively inexpensive instrument known as a saturometer may be used to measure ΔP directly.

At depths below the water surface, some of the ΔP is offset by hydrostatic pressure. The actual ΔP to which aquatic animals are exposed is called the uncompensated ΔP:

$$\Delta P_{uncomp.} = \Delta P - HP. \quad (6.8)$$

Values for the hydrostatic pressure (HP) per meter depth for different temperatures and salinities can be found in Table 6.4.

Gases dissolve in the blood of fish, shrimp, and other aquatic animals and reach equilibrium with external conditions. Thus, if water is supersaturated with gases, the blood of animals living in the water becomes saturated. Also, the blood of an aquatic animal at equilibrium with dissolved gases at a particular temperature will become supersaturated with gases if the animal moves to warmer water. If supersaturating gases do not diffuse rapidly through the gills, gas bubbles can form in the blood.

The occurrence of gas bubbles in the blood causes gas bubble trauma—often called gas bubble disease. Gas bubble trauma leads to stress and mortality. Eggs may float to the surface, larvae and fry may exhibit hyperinflation of the swim bladder, cranial swelling, swollen gill lamellae, and other abnormalities. A common symptom of gas bubble trauma in juvenile and adult fish is gas bubbles in the blood that are visible in surface tissues on the head, in the mouth, and in fin rays. The eyes of affected fish also may protrude.

Aquatic animals exposed to ΔP values of 25–75 mmHg on a continuous basis may present symptoms of gas bubble trauma, and some of the affected animals may die if exposure is prolonged for several days. Acute gas bubble trauma may lead to 50–100 % mortality at greater ΔP.

Gas bubble trauma is most common in shallow water or in organisms that cannot escape the surface water by sounding. At greater depths, the hydrostatic pressure increases the equilibrium concentration of gases in water and thereby lowers the ΔP (Ex. 6.8)

Ex. 6.8: *Calculation of reduction in ΔP with depth.*

Situation:

A freshwater water body has a temperature of 28 °C, and a ΔP of 115 mmHg at the surface. The water is thoroughly mixed and gas concentrations and temperature are the same at all depths. The uncompensated ΔP will be estimated for 1.5 m depth.

Solution:
Using (6.8) and obtaining the hydrostatic pressure per meter of water depth at 28 °C (by extrapolation between 25° and 30 °C) from Table 6.4, the uncompensated ΔP *will be calculated.*

$$\Delta P_{uncomp.} = 115\ \text{mm} - (73.30\ \text{mm/m} \times 1.5\ \text{m}) = 5.05\ \text{mm}.$$

Supersaturation of surface waters with dissolved oxygen during afternoons is a common occurrence. This condition usually does not cause harm to aquatic animals because the period of supersaturation is limited to a few hours in the afternoon and early evening, the dissolved oxygen concentration decreases with depth, and most animals can go to greater depth (sound) where the combination of lower dissolved concentration and greater hydrostatic pressure lessens ΔP.

Gas Transfer

Most of the information on gas transfer between air and water has been developed for oxygen, but principles presented for oxygen are applicable to other gases. In natural waters, dissolved oxygen concentrations are constantly changing because of biological, physical, and chemical processes. The air above water has a constant percentage of oxygen, even though the partial pressure of oxygen in air may vary because of differences in atmospheric pressure. When water is at equilibrium with atmospheric oxygen, there is no net transfer of oxygen between air and water. Transfer of oxygen from air to water will occur when water is undersaturated with dissolved oxygen, and oxygen will diffuse from the water to the air when water is supersaturated with oxygen. The driving force causing net transfer of oxygen between air and water is the difference in oxygen tension. Once equilibrium is reached, oxygen tension in the air and in the water is the same and net transfer ceases. The oxygen deficit or oxygen surplus may be expressed as

$$D = DO_s - DO_m \tag{6.9}$$

$$S = DO_m - DO_s \tag{6.10}$$

where D = oxygen deficit (mg/L), S = oxygen surplus (mg/L), DO_s = solubility of oxygen in water at saturation under existing conditions (mg/L), and DO_m = measured concentration of dissolved oxygen (mg/L).

Oxygen must enter or leave a body of water at the air-water interface, so for the thin film of water in contact with air, the greater D or S, the faster oxygen will enter or leave the film. For undisturbed water, the net transfer of oxygen will depend upon the magnitude of D or S, the area of the air-water interface, the temperature, and time of contact. Even when D or S is great, the rate of net transfer is slow because the surface film quickly reaches equilibrium and further net transfer requires oxygen to diffuse from the film to the greater mass of water below, or from the greater mass of water to the film. Natural waters are never completely quiescent, and oxygen transfer increases with greater turbulence.

The rate of change of dissolved oxygen concentration over time can be expressed as

$$\frac{dc}{dt} = \frac{k}{F}\frac{A}{V}(C_s - C_m) \tag{6.11}$$

where dc/dt = rate of change in concentration, k = diffusion coefficient, F = liquid film thickness, A = area through which the gas is diffusing, V = volume of water into which the gas is diffusing, C_m = saturation concentration of gas in solution, and C_s = concentration of gas in solution.

Gases will be removed from solution (-dc/dt) in those applications in which C_m is greater than C_s. Gas transfer can be accelerated by reducing liquid film thickness (F) and by increasing the surface area (A) through which gas diffuses. Because it is difficult to measure A and F, the ratios A/V and k/F often are combined to establish an overall transfer coefficient (K_La) as follows:

$$\frac{dc}{dt} = K_La(C_s - C_m). \tag{6.12}$$

The overall transfer coefficient reflects conditions present in a specific gas-liquid contact system. Important variables include basin geometry, turbulence, characteristics of the liquid, extent of the gas-liquid interface and temperature. Temperature affects viscosity, which, in turn, influences k, F, and A. Values of K_La can be corrected for the effects of temperature, using the following expression:

$$(K_La)_T = (K_La)_{20}(1.024)^{T-20} \tag{6.13}$$

where $(K_La)_T$ = overall gas transfer coefficient at temperature T, $(K_La)_{20}$ = overall gas transfer coefficient at 20 °C, and T = liquid temperature (°C). Although each gas species in a contact system has a unique value of K_La, relative values for a specific gas pair are inversely proportional to their molecular diameters

$$(K_La)_1/(K_La)_2 = (d)_2/(d)_1 \tag{6.14}$$

where d = diameter of the gas molecule. The K_La determined experimentally for one gas can be used to predict K_La values for other gas species. However, the major atmospheric gases have similar molecular diameters—nitrogen, 0.314 nm, oxygen, 0.29 nm, and carbon dioxide, 0.28 nm.

The air-water interface varies with turbulence and neither variable can be estimated accurately. Nevertheless, it is possible to calculate the gas transfer coefficient empirically between time 1 and time 2 because integration of (6.11) gives

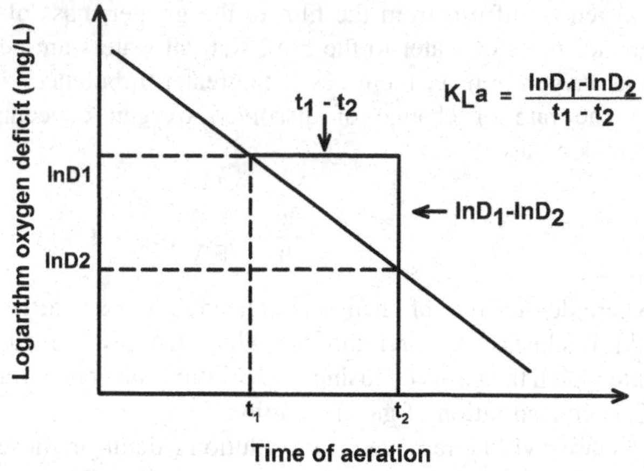

Fig. 6.1 Graphical illustration of the method for determining the transfer coefficient (K_La)

$$K_La = \frac{\ln D_1 - \ln D_2}{t_2 - t_1}. \tag{6.15}$$

where K_La = gas transfer coefficient (h^{-1}), t = time, and subscripts 1 and 2 = at time 1 and time 2. Various units may be used in (6.15), but for convenience K_La often is expressed as 1/h (h^{-1}). A plot of the natural logarithm of D against time for a thoroughly mixed body of water, or the natural logarithm of D as calculated from dissolved oxygen profiles for a less than thoroughly mixed body of water, gives a straight line for which K_La is the slope (Fig. 6.1).

The standard oxygen transfer rate for 20 °C may be estimated as

$$SOTR = (K_La_{20})(C_s)(V) \tag{6.16}$$

where SOTR = standard oxygen transfer rate (g O_2/h), C_s = dissolved oxygen concentration at 20 °C and saturation, and V = water volume in m^3. A sample calculation is provided in Ex. 6.9.

Ex. 6.9: *The K_La_T for 23 °C is 0.0275/h for re-aeration by diffusion on a windy day of 30 m^3 of freshwater in a 20-m^2 tank. The SOTR will be estimated. Assume BP = 760 mm.*

Solution:

From (6.13),

$$K_La_{20} = 0.0275 \div 1.024^{23-20} = 0.0256/\text{h}$$

and

$$SOTR = (0.0256/\text{h})\left(8.56\ \text{g/m}^3\right)\left(30\ \text{m}^3\right) = 6.57\ \text{g O}_2/\text{h}.$$

The oxygen diffused through a 20 m^2 surface, so in terms of surface area,

$$SOTR = \frac{6.57 \text{ g } O_2/\text{h}}{20 \text{ m}^2} = 0.328 \text{ g } O_2/\text{m}^2/\text{h}.$$

The methodology for determining oxygen transfer rate described above often is used to determine the SOTR of mechanical aerators used in wastewater treatment (American Society of Civil Engineers 1992). Dividing the SOTR by the power applied to the aeration device provides an estimate of the standard aeration efficiency of the aerator.

Measurements of SOTR in laboratory tanks range from 0.01 to 0.10 g oxygen/m^2 per hour in still water to as high as 1.0 g oxygen/m^2 per hour in turbulent water. Accurate measurement of SOTR is difficult in natural waters because biological processes add or remove oxygen during the period of measurement, and circulation patterns are difficult to assess. Schroeder (1975) reported SOTR values of 0.01–0.05 g oxygen/m^2 per hour for each 0.2 atm (152 mm) saturation deficit. Welch (1968) measured SOTR values of 0.1 to 0.5 g oxygen/m^2 per hour at 100 % departure from saturation for a pond. The highest SOTR values were for windy days.

Gas exchange with the air is rapid in large bodies of water where waves break to create spray. In small lakes and ponds, water surfaces are calm and oxygen transfer is impeded by lack of disturbance of the water surface. As illustrated in Ex. 6.10, the following equation (Boyd and Teichert-Coddington 1992) can be used to relate wind speed to rate of wind aeration in small ponds:

$$\text{WRR} = (0.153\text{X} - 0.127)\left(\frac{\text{C}_\text{s} - \text{C}_\text{m}}{9.08}\right)\left(1.024^{\text{T}-20}\right) \tag{6.17}$$

where WRR = wind re-aeration rate (g O_2/m^2 per hour) and X = wind speed 3 m above the water surface (m/s).

Ex. 6.10: *The wind re-aeration rate for a wind speed of 3 m/s and the oxygen added to a pond by diffusion during the night (12 h) will be estimated for a 1.5 m deep pond at 25 °C when the barometric pressure is 760 mmHg and dissolved oxygen concentration averages 5 mg/L during the night.*

Solution:

(*i*) $WRR = [(0.153)(3 \text{ m/s}) - 0.127]\left[\frac{8.24 - 5.00}{9.08}\right](1.024)^5$

$$WRR = 0.133 \text{ g } O_2/\text{ m}^2/\text{h}.$$

(ii) There is 1.5 m^3 water beneath 1 m^2 of surface, so

$$\frac{0.133 \text{ g } O_2/m^2/h}{1.5 \text{ m}^3/m^2} = 0.089 \text{ g } O_2/m^3/h$$

$$\textit{or } 0.089 \text{ mg } O_2/L/h.$$

In 12 h, the amount of dissolved oxygen entering the pond by diffusion from the air is

$$0.089 \text{ g } O_2/m^3/h \times 12 \text{ h} = 1.07 \text{ mg/L}.$$

Stream Re-aeration

The re-aeration rate also has been determined for streams by techniques too complex to describe here, and several equations—usually based on stream velocity and average depth—have been proposed. One such equation (O'Connor and Dobbins 1958) was developed for streams with water velocities between 0.15 and 0.5 m/s and average depths between 0.3 and 9 m follows:

$$k_s = 3.93\nu^{0.5}H^{-1.5} \tag{6.18}$$

where k_s = stream re-aeration coefficient (day^{-1}), ν = velocity (m/s), and H = average depth (m). The k_s can be adjusted for temperature and oxygen deficit in the stream by the equation

$$k_s' = k_s\left(\frac{C_s - C_m}{C_s}\right)1.024^{T-20} \tag{6.19}$$

where k_s' = adjusted k_s. The daily oxygen input from diffusion from the atmosphere can be estimated using (6.16) in which V is taken as stream flow past a given point in cubic meters per day, i.e., $Q = V = A\nu$.

Stream re-aeration coefficients are lowest for sluggish rivers—usually being below 0.35 day^{-1}. Large rivers with moderate to normal velocities have stream re-aeration coefficients between 0.35 and 0.7 day^{-1}, while swift flowing rivers may have re-aeration constants of 1.0 or slightly greater. The higher rates of re-aeration are for streams with rapids and waterfalls.

Concentrations of Dissolved Oxygen

Diffusion of oxygen from air to water and *vice versa* causes dissolved oxygen concentrations in natural waters to change towards equilibrium or saturation—but, they seldom are at equilibrium. This results because biological activity changes dissolved oxygen concentrations faster than diffusion can produce equilibrium for existing conditions. The biological processes that influence dissolved oxygen concentrations are photosynthesis by green plants and respiration by all aquatic organisms. This topic will be discussed in Chap. 10, but in short, photosynthesis during daylight usually occurs more rapidly than respiration, and dissolved oxygen concentration increases. Afternoon dissolved oxygen concentration will typically be above saturation in waters with healthy plant communities. At night, photosynthesis stops, but respiration uses oxygen and causes dissolved oxygen concentration to decline to less than saturation. Dissolved oxygen concentration also tends to decline with increasing water depth because there is less illumination in deeper waters.

Oxygen and Carbon Dioxide in Respiration

Aquatic organisms absorb molecular oxygen into their blood from the water and use it in respiration, but they must expel carbon dioxide produced in respiration into the water. The gills of fish and crustaceans provide a surface across which gases in water may enter the blood and *vice versa*. Blood cells of fish contain hemoglobin, and those of crustaceans carry hemocyanin; these pigments combine with oxygen to allow the blood to carry more oxygen than can be carried in solution. Both pigments are metalloproteins. Hemoglobin has a porphyrin ring with iron in its center, while hemocyanin has a porphyrin ring with copper in its center. The hemolymph (blood fluids) and hemoglobin load with oxygen at the gills where oxygen tension is high and unload oxygen to the tissues where oxygen use in respiration results in low oxygen tension as illustrated for hemoglobin (Hb) below:

$$Hb + O_2 = HbO_2 \qquad \text{(in gills)} \tag{6.20}$$

$$HbO_2 = Hb + O_2 \text{ (in tissues)}. \tag{6.21}$$

The effect of oxygen tension in the water on the loading and unloading of oxygen by hemoglobin—the oxyhemoglobin dissociation curve—is illustrated in Fig. 6.2. The curve usually is sigmoid for warmwater species and hyperbolic for coldwater species. As a result, warmwater species have a greater capacity than coldwater species to unload oxygen at the tissue level. This is a major reason why coldwater species require a higher dissolved oxygen concentration than do warmwater species.

Carbon dioxide from respiration dissolves in the hemolymph at the tissues and is transported in venous blood to the gills where it diffuses into the water. A high carbon dioxide concentration in the water inhibits the diffusion of carbon dioxide from the blood. Accumulation of carbon dioxide in the blood depresses blood pH leading to several negative physiological consequences.

As carbon dioxide concentration increases in the water, it also interferes with the loading of the hemoglobin with oxygen in the gills (Fig. 6.3). A high carbon dioxide

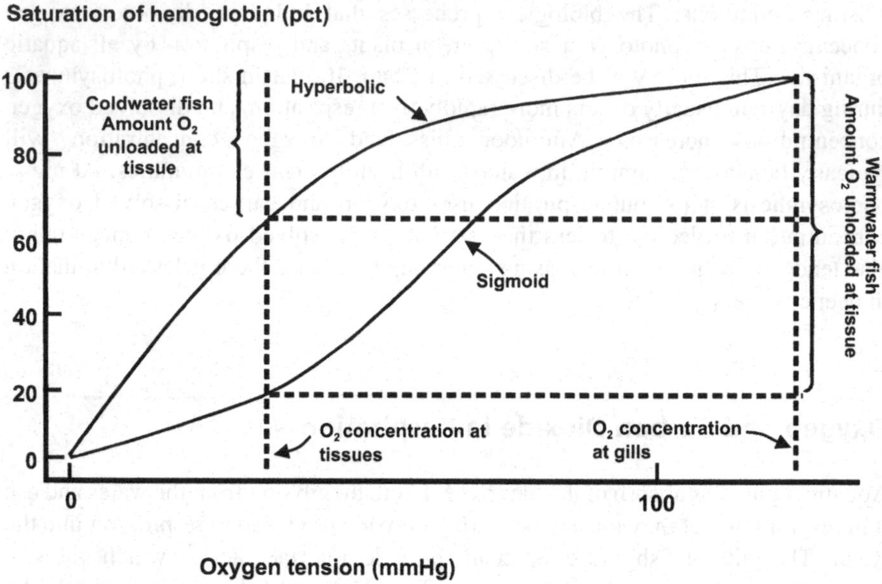

Fig. 6.2 Oxyhemoglobin saturation curves for warmwater and coldwater fish

Fig. 6.3 Effect of carbon dioxide concentration on saturation of hemoglobin with oxygen

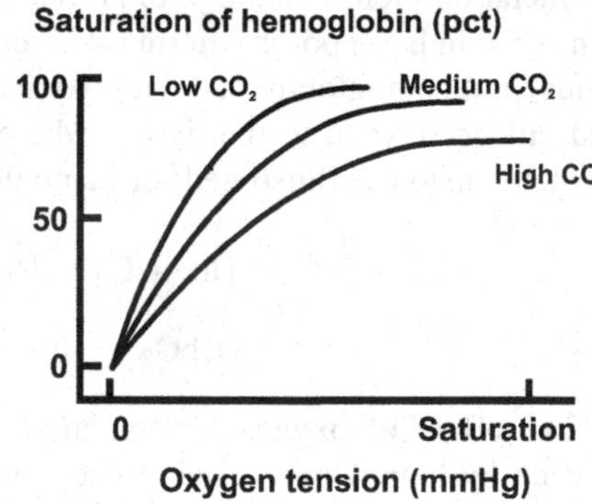

concentration in the water results in organisms requiring a higher dissolved oxygen concentration to avoid oxygen related stress.

All aerobic organisms require dissolved oxygen, and much of the dissolved oxygen in aquatic ecosystems is used by bacteria and other saprophytic microorganisms that oxidize organic matter to carbon dioxide, water, and inorganic minerals. Chemoautotrophic microorganisms such as nitrifying bacteria also require oxygen.

The amount of oxygen used by aquatic animals varies with species, size, temperature, time since feeding, physical activity, and other factors. Small organisms use more oxygen per unit of weight than larger ones of the same species.

For example, 10-g channel catfish were reported to use about twice as much oxygen per unit weight as 500-g channel catfish. A 10 °C increase in water temperature within the range of temperature tolerance of an organism usually will double the oxygen consumption rate. Fish have been shown to use about twice as much oxygen 1 h after feeding than fish fasted overnight. Fish swimming against a current will increase oxygen use as the current velocity increases (Andrews et al. 1973; Andrews and Matsuda 1975).

Average oxygen consumption rates for adult fish usually range from 200 to 500 mg oxygen/kg fish/h (Boyd and Tucker 2014). There is less information on oxygen consumption by crustaceans, but it appears that rates are similar to those for fish. In natural waters where the standing crop of fish populations is typically less than 500 kg/ha, oxygen consumption by fish does not have a great effect on dissolved oxygen concentration (Ex. 6.11)

Ex. 6.11: *The amount of oxygen (in milligrams per liter) removed from a water body of 10,000 m^3 by a fish standing crop of 500 kg will be estimated.*

Solution:

Assume that the fish respire at a rate of 300 mg/kg/h over a 24-h period. The oxygen use will be

$$O_2\ used = 300\ \text{mg/kg/h}\ \times 500\ \text{kg}\ \times 24\ \text{h/day} = 3,600,000\ \text{mg/day}$$
$$= 3,600\ \text{g/day}$$

$$O_2\ concentration = \frac{3,600\ \text{g/day}}{10,000\ \text{m}^3} = 0.360\ \text{g/m}^3/\text{day} = 0.360\ \text{mg/L/day}.$$

Of course, when fish are held at high density in aquaria, holding tanks, or aquaculture production facilities, they cause a much larger demand for dissolved oxygen.

Warmwater fish usually will survive for long periods at dissolved oxygen concentrations as low as 1.0–1.5 mg/L, while coldwater fish may survive at 2.5 3.5 mg/L dissolved oxygen. Nevertheless, fish and other aquatic animals are stressed, susceptible to disease, and grow slowly at low oxygen concentration. As a general rule, aquatic organisms do best when the dissolved oxygen concentration does not fall below 50 % of saturation (Collins 1984; Boyd and Tucker 2014). In freshwater at sea level, 50 % of saturation is around 5 mg/L at 15 °C and 4 mg/L at 26 °C. The effects of dissolved oxygen concentration on warmwater fish are summarized in Table 6.5. The effects on coldwater fish would be similar, but occur at higher dissolved oxygen concentrations. Water quality criteria for aquatic ecosystems usually specify that dissolved oxygen concentrations be above 5 or 6 mg/L, and some may specify that dissolved oxygen concentrations should be at least 80–90 % of saturation. Slightly lower dissolved oxygen standards may be

Table 6.5 Effects of dissolved oxygen concentration on warmwater fish

Dissolved oxygen (mg/L)	Effects
0–0.3	Small fish survive short exposure
0.3–1.5	Lethal if exposure is prolonged for several hours
1.5–5.0	Fish survive, but growth will be slow and fish will be more susceptible to disease
5.0–saturation	Desirable range
Above saturation	Possible gas bubble trauma if exposure prolonged

listed for public water supplies, but few are below 4 mg/L. Even water for irrigation or livestock consumption should have at least 3 or 4 mg/L dissolved oxygen.

Carbon dioxide is not highly toxic to most aquatic animals, but fish have been shown to avoid carbon dioxide concentrations of 10 mg/L or more. Most species tolerate up to 60 mg/L of carbon dioxide provided there is plenty of dissolved oxygen. However, high carbon dioxide concentrations have a narcotic effect on fish.

Significance

Dissolved oxygen is the most important water quality variable in aquatic ecosystems. Molecular oxygen is the electron acceptor in aerobic metabolism, so all aerobic organisms must have an adequate supply of dissolved oxygen. When dissolved oxygen concentrations are low or dissolved oxygen is absent, decomposition of organic matter by anaerobic microorganisms releases many reduced substances, e.g., ammonia, nitrite, ferrous iron, hydrogen sulfide, and dissolved organic compounds, into the water. In the absence of adequate dissolved oxygen, aerobic microorganisms do not function efficiently in oxidizing these reduced substances. The combination of low dissolved oxygen concentration and high concentrations of certain reduce, toxic metabolites cause drastic, negative impacts on the structure and function of aquatic ecosystems. "Clean water" species disappear and only those organisms that can tolerate highly polluted conditions thrive. Such ecosystems are of low biodiversity, unstable, and greatly impaired for most beneficial uses.

References

American Society of Civil Engineers (1992) Measurement of oxygen transfer in clean water, 2nd ed. ANSI/ASCE 2-91, American Society of Civil Engineers, New York

Andrews JW, Matsuda Y (1975) The influence of various culture conditions on the oxygen consumption of channel catfish. Trans Am Fish Soc 104:322–327

Andrews JW, Murai T, Gibbons G (1973) The influence of dissolved oxygen on the growth of channel catfish. Trans Am Fish Soc 102:835–838

Benson BB, Krause D (1984) The concentration and isotopic fractionation of oxygen dissolved in freshwater and seawater in equilibrium with the atmosphere. Lim Oceanog 29:620–632

Boyd CE, Teichert-Coddington D (1992) Relationship between wind speed and reaeration in small aquaculture ponds. Aquacult Eng 11:121–131

Boyd CE, Tucker CS (2014) Handbook for aquaculture water quality. Craftmaster, Auburn

Collins G (1984) Fish growth and lethality versus dissolved oxygen. In: Proceedings specialty conference on environmental engineering, ASCE, Los Angeles, 25–27 June 1984, p 750–755

Colt J (1984) Computation of dissolved gas concentrations in water as functions of temperature, salinity, and pressure. Special Publication 14, American Fisheries Society, Bethesda

O'Connor DJ, Dobbins WE (1958) Mechanism of reaeration in natural streams. Trans Am Soc Civ Eng 123:641–666

Schroeder GL (1975) Nighttime material balance for oxygen in fish ponds receiving organic wastes. Bamidgeh 27:65–74

Welch HE (1968) Use of modified diurnal curves for the measurement of metabolism in standing water. Lim Oceanog 13:679–687

Redox Potential

7

Abstract

Electrons are shed when a substance is oxidized and gained when a substance is reduced. Oxidations and reductions occur in couplets—known as half-cells—in which one substance, the oxidizing agent, accepts electrons from another substance, the reducing agent. The oxidizing agent is reduced and the reducing agent is oxidized. The flow of electrons between two half-cells can be measured as an electromotive force (in volts). The hydrogen half-cell ($H_2 \rightarrow 2H^+ + 2e^-$) has an electrical potential of 0.0 V at 25 °C, 1 atm H_2 and 1 M H^+; it is said to have a standard electrode potential (E^0) of 0.0 V. The flow of electrons to or from the hydrogen half-cell is the reference for determining E^0 values of other half-cells. The greater a positive E^0, the more oxidized a half-cell with respective to the hydrogen electrode; the opposite is true for a negative E^0. The redox potential (E or E_h) for non-standard conditions is measured with a calomel electrode or calculated with the Nernst equation. Water with measureable dissolved oxygen has $E_h = 0.50$ V, and at oxygen saturation, $E_h = 0.56$ V. The redox potential indicates whether given substances may exist in a particular environment and explains the sequence of reactions that occur as dissolved oxygen concentration declines in water bodies and sediments. The redox potential has many practical applications in analytical chemistry and industry, but it is not frequently measured in water quality investigations.

Keywords

Oxidation-reduction reaction • Hydrogen electrode • Standard electrode potential • Redox measurement • Corrosion

C.E. Boyd, *Water Quality*, DOI 10.1007/978-3-319-17446-4_7

Introduction

The oxidation-reduction or redox potential is an important tool in chemistry. This variable is a measurement of both the direction and amount of electron transfer that occurs when some substances react together. The redox potential indicates whether a substance is a reducing agent with a high potential to donate electrons to another substance or an oxidizing agent that has a high potential to accept electrons from another substance.

In a redox reaction, electrons donated (lost) by the reducing agent must be accepted (gained) by the oxidizing agent. Thus, a redox reaction can be thought of as having two parts. In the first part, the electrons are released by the reducing agent. In the second part the electrons released by the reducing agent are accepted by the oxidizing agent. The flow of electrons between the two parts of the reaction can be measured as an electrical current—the redox potential.

Hydrogen (H_2) is the base for comparing redox potentials. Hydrogen has a redox potential of zero at standard conditions (1.0 M; 1 atm pressure; 25 °C). Other substances can be compared to hydrogen, or these redox potentials can be compared to each other. Of course, concentrations of substances, pH, and temperature influence redox potential, but redox can be adjusted for differences in these variables. Redox potential may be used to determine the endpoints of chemical reactions, predict whether or not two substances will react when brought together, ascertain if certain substances are likely to be present in the aquatic environment, and other purposes (Snoeyink and Jenkins 1980; Essington 2004).

Oxidation-Reduction

Principles of the law of mass action, Gibbs free energy change of reaction, and the equilibrium constant can be applied to oxidation-reduction reactions. The driving force for oxidation-reduction reactions can be expressed in terms of a measurable electrical current. Consider the oxidation-reduction reaction

$$I_2 + H_2 = 2H^+ + 2I^-. \tag{7.1}$$

Iodine is reduced to iodide, and hydrogen gas is oxidized to hydrogen ion. The reaction can be divided into two parts, one showing the loss of electrons (e^-) resulting in oxidation and one showing the gain of electrons causing reduction as follows:

$$H_2 = 2H^+ + 2e^- \tag{7.2}$$

$$I_2 + 2e^- = 2I^-. \tag{7.3}$$

Equations 7.2 and 7.3 can be added to give (7.1) with the electrons cancelling during addition. Equation 7.1 is called a cell reaction because it can be separated into two half-cell reactions (7.2 and 7.3).

The standard Gibbs free energy (ΔG^o) of reaction for (7.2) and (7.3) is

$$\Delta G^o = 2\Delta G^0_f H^+ + 2\Delta G^0_f e^- - \Delta G^0_f H_2$$

$$\Delta G^o = 2\Delta G^0_f \mathrm{I}^- - \Delta G^0_f \mathrm{I}_2 - 2\Delta G^0_f e^-.$$

The $\Delta G^0_f e^-$ term cancels during addition of the right-hand sides of these expressions, and the sum is

$$2\Delta G^0_f \mathrm{H}^+ + 2\Delta G^0_f \mathrm{I}^- - \Delta G^0_f \mathrm{H}_2 - \Delta G^0_f \mathrm{I}_2. \quad (7.4)$$

The sum of the two expressions is the same as the expression for estimating ΔG^o from (7.1). Because the $\Delta G^0_f e^-$ term cancels when the two half-cells are added, it is permissible to assign a value of zero to $\Delta G^0_f e^-$. Also, from Table 3.3, $\Delta G^0_f \mathrm{H}^+ = \Delta G^0_f \mathrm{H}_2 = 0$. Thus, $\Delta G^o = 0$ for the reaction in (7.2).

Standard Hydrogen Electrode

The reaction depicted in (7.2) is called the hydrogen half-cell, but it is often written as

$$\tfrac{1}{2}\mathrm{H}_2 = \mathrm{H}^+ + e^-. \quad (7.5)$$

The hydrogen half-cell is a very useful expression. Any oxidation-reduction reaction can be written as two half-cell reactions with the hydrogen half-cell functioning as either the electron-donating or electron-accepting half-cell:

$$\mathrm{Fe}^{3+} + e^- = \mathrm{Fe}^{2+}, \quad \tfrac{1}{2}\mathrm{H}_2 = \mathrm{H}^+ + e^- \quad (7.6)$$

$$\mathrm{Fe(s)} = \mathrm{Fe}^{2+} + 2e^-, \quad 2\mathrm{H}^+ + 2e^- = \mathrm{H}_2\mathrm{(g)}. \quad (7.7)$$

In (7.6), hydrogen is the reductant because it donates electrons that reduce Fe^{3+} to Fe^{2+}. In (7.7), H^+ is the oxidant because it accepts electrons, and thereby oxidizes solid, metallic iron or Fe(s) to Fe^{2+}.

The flow of electrons between half-cells can be measured as an electrical current. The chemical reaction involving reduction of I_2 by H_2 (7.1) can be used to make a cell of two half-cells in which the flow of electrons can be measured (Fig. 7.1). This cell consists of a solution 1 M in hydrogen ion and a solution 1 M in iodine. A platinum electrode coated with platinum black and bathed in hydrogen gas at 1 atm pressure is placed in the solution of 1 M hydrogen ion to form the hydrogen half-cell or hydrogen electrode. A shiny platinum electrode is placed in the iodine solution to form the other electrode. A platinum wire connected between the two electrodes allows free flow of electrons between the two half-cells. A salt bridge connected between the two solutions allows ions to migrate from one side to

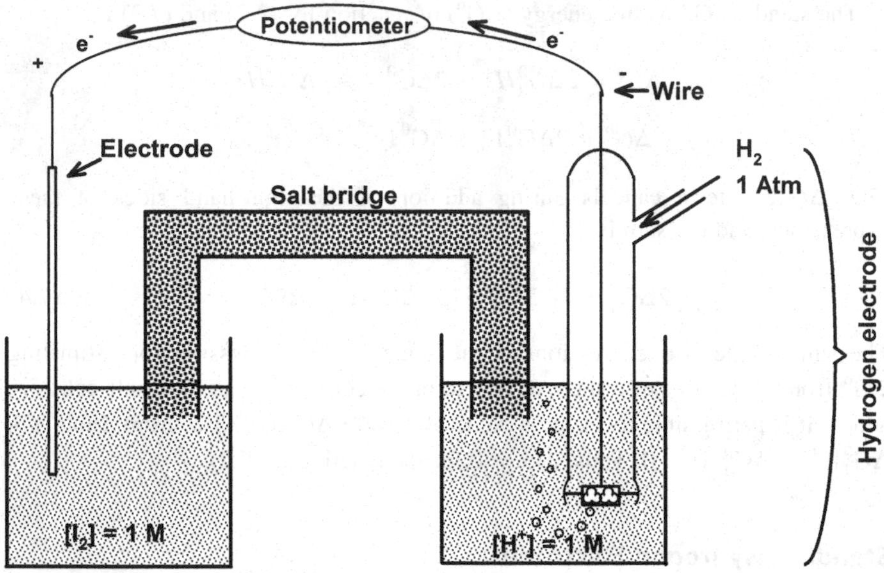

Fig. 7.1 The hydrogen half-cell or electrode connected to an iodine-iodide half-cell

the other and maintain electrical neutrality. The flow of electrons is measured with a potentiometer. Electron flow in the cell shown in Fig. 7.1 is from the hydrogen electrode to the iodine solution. The iodine-iodide half-cell is initially more oxidized than the hydrogen electrode, and electrons move from the hydrogen side of the device to the iodine solution and reduce I_2 to I^-. The oxidation of H_2 to H^+ is the source of the electrons. Electrons continue to flow from the hydrogen electrode to the iodine-iodide half-cell until equilibrium is reached.

The potentiometer in Fig. 7.1 initially will read 0.62 V—the standard electrode potential (E^0) for the iodine half-cell. The value of E^0 declines as the reaction proceeds, and at equilibrium $E^0 = 0$ V. The standard electrode potential refers to the initial voltage that develops between the standard hydrogen electrode (half-cell) and any other half-cell under standard conditions (unit activities and 25 °C). The electrons transferred between the half-cells to drive the oxidation-reduction reaction to equilibrium do not always flow in the direction shown in Fig. 7.1. In some cases the standard hydrogen electrode may be more oxidized than the other half-cell, and electrons will flow toward the hydrogen electrode and the reduction will occur at the hydrogen electrode. It is necessary to give a sign to E^0. The positive sign usually is applied when electrons flow away from the hydrogen electrode and toward the other half-cell as done in Fig. 7.1. The positive sign means that the other half-cell is more oxidized than the hydrogen electrode. When the hydrogen electrode is more oxidized than the other half-cell, electrons flow toward the hydrogen electrode and a negative sign is applied to E^0. The sign is sometimes applied to E^0 in the opposite manner, and the sign of E^0 can cause considerable confusion unless specified.

Table 7.1 Standard electrode potentials at 25 °C

Reaction	E° (V)
$O_3(g) + 2H^+ + 2e^- = O_2(g) + H_2O$	+2.07
$Mn^{4+} + e^- = Mn^{3+}$	+1.65
$2HOCl + 2H^+ + 2e^- = Cl_2(aq) + 2H_2O$	+1.60
$MnO_4^- + 8H^+ + 5e^- = Mn^{2+} + 4H_2O$	+1.51
$Cl_2(aq) + 2e^- = 2Cl^-$	+1.39
$Cl_2(g) + 2e^- = 2Cl^-$	+1.36
$Cr_2O_7^{2-} + 14H^+ + 6e^- = 2Cr^{3+} + 7H_2O$	+1.33
$O_{2(aq)} + 4H^+ + 4e^- = 2H_2O$	+1.27
$2NO_3^- + 12H^+ + 10e^- = N_2(g) + 6H_2O$	+1.24
$MnO_2(s) + 4H^+ + 2e^- = Mn^{2+} + 2H_2O$	+1.23
$O_2(g) + 4H^+ + 4e^- = 2H_2O$	+1.23
$Fe(OH)_3(s) + 3H^+ + e^- = Fe^{2+} + 3H_2O$	+1.06
$NO_2^- + 8H^+ + 6e^- = NH_4^+ + 2H_2O$	+0.89
$NO_3^- + 10H^+ + 8e^- = NH_4^+ + 3H_2O$	+0.88
$NO_3^- + 2H^+ + 2e^- = NO_2^- + H_2O$	+0.84
$Fe^{3+} + e^- = Fe^{2+}$	+0.77
$I_2(aq) + 2e^- = 2I^-$	+0.62
$MnO_4 + 2H_2O + 3e^- = MnO_2(s) + 4OH^-$	+0.59
$O_2 + 2H_2O + 4e^- = 4OH^-$	+0.40
$SO_4^{2-} + 8H^+ + 6e^- = S(s) + 4H_2O$	+0.35
$SO_4^{2-} + 10H^+ + 8e^- = H_2S(g) + 4H_2O$	+0.34
$N_2(g) + 8H^+ + 6e^- = 2NH_4^+$	+0.28
$Hg_2Cl_2(s) + 2e^- = 2Hg(l) + 2Cl^-$	+0.27
$SO_4^{2-} + 9H^+ + 8e^- = HS^- + 4H_2O$	+0.24
$S_4O_6^{2-} + 2e^- = 2S_2O_3^{2-}$	+0.18
$S(s) + 2H^+ + 2e^- = H_2S(g)$	+0.17
$CO_2(g) + 8H^+ + 8e^- = CH_4(g) + 2H_2O$	+0.17
$H^+ + e^- = ½H_2(g)$	0.00
$6CO_2(g) + 24H^+ + 24e^- = C_6H_{12}O_6$ (glucose) $+ 6H_2O$	−0.01
$SO_4^{2-} + 2H^+ + 2e^- = SO_3^{2-} + H_2O$	−0.04
$Fe^{2+} + 2e^- = Fe(s)$	−0.44

In this table, (*g*) gas, (*aq*) aqueous or dissolved, (*s*) solid or mineral, (*l*) liquid form

Values of E^0 have been determined for many half-cells and selected ones are provided (Table 7.1). The E^0 of the hydrogen electrode (0 V) is the reference or standard to which other half-cells are compared in Table 7.1. A half-cell with an E^0 greater than 0 V is more oxidized than the hydrogen electrode, and one with an E^0 less than 0 V is more reduced than the hydrogen electrode. The E^0 of any two half-cells may be compared; it is not necessary to compare only with the hydrogen electrode.

Reference to Table 7.1 can indicate the relative degree of oxidation (or reduction) of different substances. For example, ozone (O_3), hypochlorous acid (HOCl), and permanganate (MnO_4^-) are stronger oxidants than dissolved oxygen [$O_{2(aq)}$], because they have higher E^0 values than dissolved oxygen. But, for the same reason, dissolved oxygen is more oxidized than ferric iron (Fe^{3+}), nitrate (NO_3^-), and sulfate (SO_4^{2-}).

Electrode potentials also can reveal if two or more specific substances can coexist. For example, dissolved oxygen and hydrogen sulfide do not coexist because oxygen is more highly oxidized than sulfide and its sources:

$$O_{2(aq)} + 4H^+ + 4e^- = 2H_2O \quad E^0 = 1.27\,V \tag{7.8}$$

$$SO_4^{2-} + 10H^+ + 8e^- = H_2S(g) + 4H_2O \quad E^0 = 0.34\,V \tag{7.9}$$

$$S(s) + 2H^+ + 2e^- = H_2S(g) \quad E^0 = 0.17\,V. \tag{7.10}$$

If dissolved oxygen is present, hydrogen sulfide will be oxidized to sulfate. There are many other examples pertinent to water quality considerations. One is the occurrence of ferrous iron in water

$$Fe^{3+} + e^- = Fe^{2+} \quad E^0 = 0.77. \tag{7.11}$$

If the water contains dissolved oxygen ($E^0 = 1.27$ V), the redox potential will be too high for Fe^{2+} to exist. Another is nitrite—it seldom exists in appreciable concentration in aerobic water because its redox potential is less than that of dissolved oxygen.

Cell Voltage and Free-Energy Change

The relationship between voltage in cell reactions and the standard Gibbs free energy of reaction is

$$\Delta G^o = -nFE^0 \tag{7.12}$$

where ΔG^o = Gibbs standard free energy of reaction (kJ/mol), n = number of electrons transferred in the cell reaction, F = Faraday constant (96.485 kJ/V-gram equivalent), and E^0 = voltage of the reaction in which all substances are at unit activity. The n and F values are multiplied by the voltage to convert the voltage to an energy equivalent. It follows that

$$\Delta G = -nFE \tag{7.13}$$

where ΔG and E are for non-standard conditions.

Ex. 7.1: *The measured E^0 for the reaction in* (7.1) *is 0.62 V (Table 7.1). It will be shown that (7.12) gives the same value of ΔG^o as calculated by (3.10) using tabular values of ΔG_f^0.*

Solution:
From (7.12), *ΔG^o is*

$$\Delta G^o = \ -(2\ \times 0.62\ \text{V}\ \times 96.48\,\text{kJ}/\textit{volt-gram equivalent}) = \ -119.6\text{kJ}.$$

The ΔG^o may be calculated from ΔG_f^0 values with (3.10) as follows:

$$\Delta G^o = 2\Delta G_f^0 H^+ + 2\Delta G_f^0 I^- - \Delta G_f^0 I_2 - \Delta G_f^0 H_2.$$

Using ΔG_f^0 values from Table 3.3 gives

$$\Delta G^o = 2(0) + 2(-51.56) - (16.43) - 0 = -119.6\text{kJ}.$$

Thus, ΔG^o is the same whether calculated from the standard Gibbs free energy of reaction or from the standard electrode potential (Table 7.1).

It was shown in Chap. 3 that the following equation may be used to estimate the free energy of reaction during a reaction until equilibrium is reached and $\Delta G = 0$ and $Q = K$:

$$\Delta G = G^o + RT\ln Q \tag{7.14}$$

where R = a form of the universal gas law (also called the ideal gas law) equation constant (0.008314 kJ/mol/°A) [°A = absolute temperature (273.15 + °C)], and ln = natural logarithm, and Q = the reaction quotient.

Substituting (7.12) and (7.13) into (7.14) gives

$$-nFe = -nFE^0 + RT\ln Q. \tag{7.15}$$

Dividing (7.15) by –nF gives

$$E = E^0 - \frac{RT}{nF}\ln Q. \tag{7.16}$$

Using values of R = 0.008314 kJ/mol·°A; T = 25 °C + 273.15°; F = Faraday's constant (96.48 kJ/V·gram equivalent), and 2.303 to convert from natural logarithms to common (base 10) logarithms, (7.16) becomes

$$E = E^0 - \frac{0.0592}{n}\log Q. \tag{7.17}$$

Equations 7.16 and 7.17 are both forms of the Nernst equation; (7.17) is for 25 °C and (7.16) is used for other temperatures. Just as with ΔG, when equilibrium is reached $E = 0.0$ and $Q = K$. The electrode potential E is the redox potential; it often is referred to as E_h.

Ex. 7.2: *The redox potential will be calculated using the Nernst equation for water at 25 °C with pH = 7 and a dissolved oxygen concentration of 8 mg/L.*

***Solution*:**

$$\frac{8\,\mathrm{mg/L}O_2}{32,000\,\mathrm{mg/mol}} = 10^{-3.60}\,\mathrm{M}$$

$$(H^+) = 10^{-7}\,\mathrm{M}\ at\ pH\ 7.$$

The appropriate reaction is

$$O_{2(aq)} + 4H^+ + 4e^- = 2H_2O$$

for which $E^0 = 1.27$ *V (Table 7.1)*

$$E_h = E^0 - \frac{0.0592}{n}\log Q$$

where

$$\log Q = \frac{1}{(O_2)(H^+)^4} \quad (H_2O\ is\ taken\ as\ unity).$$

so

$$E_h = 1.27 - \frac{0.0592}{4}\log\frac{1}{(10^{-3.6})(10^{-28})}$$

$$= 1.27 - \frac{0.0592}{4}\log\frac{1}{(10^{-31.6})}$$

$$= 1.27 - (0.0148)(31.6) = 0.802\ \mathrm{V}.$$

The pH has an effect on the redox potential, and E_h is often adjusted to the appropriate value for pH 7. To adjust E_h to pH 7, subtract 0.0592 V for each pH unit below neutrality and add 0.0592 V for each pH unit above neutrality (Ex. 7.3). Many researchers adjust E_h values to pH 7 and report the redox potential using the symbol E_7 instead of E_h.

Ex. 7.3: *What is the* E_h *at pH 5, 6, 7, 8, and 9 for the reaction*

$$O_{2(aq)} + 4H^+ + 4e^- = 2H_2O$$

when dissolved oxygen is 8 mg/L $(10^{-3.60}\,\mathrm{M})$ *and water temperature is 25 °C?*

Solution:
The method of calculating E_h for the reaction was shown already in Ex. 7.2:

$$E_h = 1.27 - \frac{0.0592}{4}\log\frac{1}{(10^{-3.6})(H^+)^4}.$$

$(H^+) = 10^{-5}, 10^{-6}, 10^{-7}, 10^{-8}$, *and* 10^{-9} M *for pH 5, 6, 7, 8, and 9, respectively. Substituting into the preceding equation gives E_h values as follows: pH 5, 0.921 V; pH 6, 0.862 V; pH 7, 0.802 V; pH 8, 0.743 V; pH 9, 0.684 V. Notice that E_h decreases by 0.059 V for each unit increase in pH.*

A change in temperature also causes E_h to change (see 7.17), but the influence of temperature on E_h is not as great as the influence of pH. A temperature increase of 1 °C will cause E_h to decrease by 0.0016 V. This factor can be used to adjust E_h or E_7 to the standard temperature of 25 °C, but modern redox instruments usually have a temperature compensator.

Practical Measurement of Redox Potential

Although the hydrogen electrode is a standard half-cell against which other half-cells are commonly compared, it is not used as the reference electrode in practical redox potential measurement. The most common reference electrode for redox measurement is the calomel (Hg_2Cl_2) electrode

$$Hg_2Cl_2 + 2e^- = 2Hg^+ + 2Cl^-. \tag{7.18}$$

A KCl-saturated calomel electrode has $E^0 = 0.242$ V at 25 °C (Greybowski 1958). Standard electrode potentials for the popular KCl-saturated calomel electrode are 0.242 V less than tabulated E^0 values of standard electrode potentials for the hydrogen electrode as shown in Ex. 7.4.

Ex. 7.4: *E^o values from Table 7.1 will be adjusted to indicate potentials that would be obtained with a KCl-saturated calomel electrode.*

(1) $O_{2(aq)} + 4H^+ + 4e^- = 2H_2O$ $\quad E^0 = 1.27$ V

(2) $Fe^{3+} + e^- = Fe^{2+}$ $\quad E^0 = 0.77$ V

(3) $NO_3^- + 2H^+ + 2e^- = NO_2^- + H_2O$ $\quad E^0 = 0.84$ V.

Solution:

Half-cell	E^0	*Correction factor*	*Calomel potential* (*V*)
(*1*)	*1.27–*	*0.242=*	*1.028*
(*2*)	*0.77–*	*0.242=*	*0.528*
(*3*)	*0.84–*	*0.242=*	*0.598*

As will be seen in other chapters, the redox potential is a useful concept in explaining many chemical and biological phenomena related to water quality. However, the practical use of redox potential in natural ecosystems is fraught with difficulty. In natural waters and sediments, the redox potential is governed by the dissolved oxygen concentration. Although oxygen is used up as it oxidizes reduced substances in the water, oxygen is continually replaced by diffusion of oxygen from the atmosphere or by oxygen produced in photosynthesis, and the redox potential usually remains fairly constant.

Measurement of redox potential is especially difficult in sediment. The redox potential often changes drastically across a few millimeters of sediment depth. But, the redox probe is rather large—about the size of a ballpoint pen—making it impossible to determine the depth in the sediment to which a redox reading applies. The act of inserting the probe in sediment also allows oxygenated water to follow the probe downward in the sediment changing the redox potential.

In the hypolimnion of stratified bodies of water, at the soil-water interface in unstratified water bodies, and in bottom soils and sediments, dissolved oxygen is at low concentration or even absent, and reducing conditions develop. The typical pattern of redox potential in a water body containing dissolved oxygen throughout the water column is illustrated (Fig. 7.2) with data from Mortimer (1942). The redox potential is between 0.5 and 0.6 V in the water column and first centimeter of sediment. The redox potential then drops to less than 0.1 V at a 6- to 10-cm depth in the sediment. The redox increases slightly at greater sediment depths possibly because of less organic matter in the sediment.

The driving force causing a decrease in the redox potential is the consumption of oxygen by microbial respiration. When molecular oxygen is depleted, certain microorganisms utilize relatively oxidized inorganic or organic substances as electron acceptors in metabolism. Redox reactions in reduced environments are very complicated, and they occur spontaneously or microorganisms usually mediate them. There is a wide variety of compounds, concentrations of relatively reduced and relatively oxidized compounds vary greatly both spatially and temporally, and it is impossible to isolate specific components of the system. As a result, the best that can be done is to insert the redox probe into the desired place and obtain a reading. This reading can be compared with the reading obtained for oxygenated water by the same electrode (Masuda and Boyd 1994).

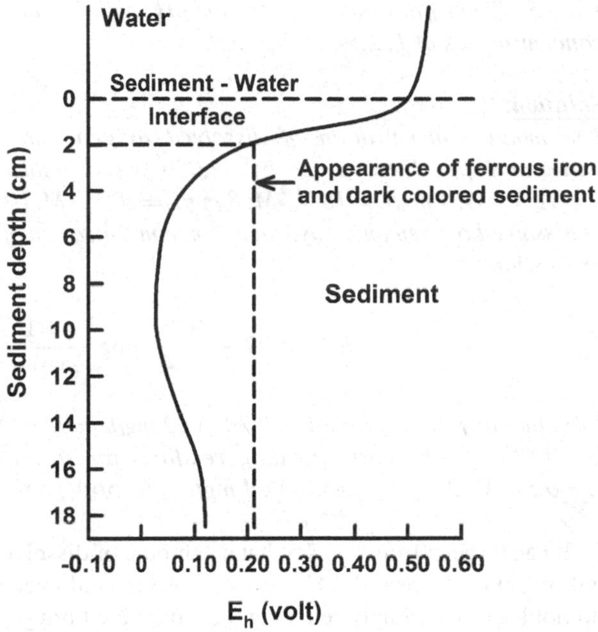

Fig. 7.2 Distribution of redox potential (E_h) in sediment with an oxidized surface

$$O_{2(aq)} + 4H^+ + 4e^- = 2H_2O. \tag{7.19}$$

At 25 °C and pH 7, the oxygen potential in well-oxygenated water should be about 0.802 V (802 mV) against a standard hydrogen electrode (see Ex. 7.4). The oxygen potential of well-oxygenated surface water is about 0.560 V (560 mV) when measured with a KCl-saturated calomel electrode. Redox potential drops in anaerobic waters or soils, and values below −0.250 V have been observed in hypolimnetic waters and in sediments.

Commercially available devices for measuring redox potential do not all provide the same reading for the redox potential of well-oxygenated water or for a sample of reduced water or sediment. This makes it difficult to interpret redox potentials measured in natural waters and sediments (Teasdale et al. 1998). Nevertheless, a decrease in the redox potential below values obtained in oxygen-saturated water with a particular instrument indicates that reducing conditions are developing, and the reducing ability of the environment increases as the redox potential drops. Care must be taken not to introduce air or oxygenated water into the medium where redox potential is to be measured.

The presence of dissolved oxygen in water—even at low concentration—tends to stabilize the redox potential (Ex. 7.5). Thus, the oxygen potential (7.15) is the standard for comparing redox potential among different places in a water body and its sediment.

Ex. 7.5: *The E_h of water at 25 °C and pH 7 will be calculated for dissolved oxygen concentrations of 1, 2, 4, and 8 mg/L.*

***Solution*:**
The molar concentration of dissolved oxygen can be computed by dividing milligrams per liter values by 32,000 mg O_2/mol:1 mg/L = $10^{-4.5}$ M; 2 mg/L = $10^{-4.2}$ M; 4 mg/L = $10^{-3.9}$ M; 8 mg/L = $10^{-3.6}$ M. Use the molar concentrations of dissolved oxygen and a hydrogen ion concentration of 10^{-7} M (pH 7) to solve the expression

$$E_h = 1.27 - \frac{0.0592}{4}\log\frac{1}{(O_2)(H^+)^4}.$$

E_h values as follows: 1 mg/L = 0.789 V; 2 mg/L = 0.793 V; 4 mg/L = 0.798 V; 8 mg/L = 0.802 V. The corresponding readings for a calomel electrode are 1 mg/L = 0.547 V; 2 mg/L = 0.551 V; 4 mg/L = 0.556 V; 8 mg/L = 0.560 V.

It can be seen from Ex. 7.5 that an amount of dissolved oxygen as low as 1 mg/L will maintain E_h near 0.8 V (near 0.56 V for a calomel electrode). Natural systems do not become strongly reducing until dissolved oxygen is entirely depleted.

Corrosion

The standard electrode potential provides a way of predicting metal corrosion. In corrosion, the metal is oxidized to its ionic form with release of electrons (the anode process), and the electrons are transferred to water, oxygen, or other oxidants (the cathode process) as illustrated in a simplified assessment of the corrosion of iron metal by aerobic water (Fig. 7.3). The overall reaction is

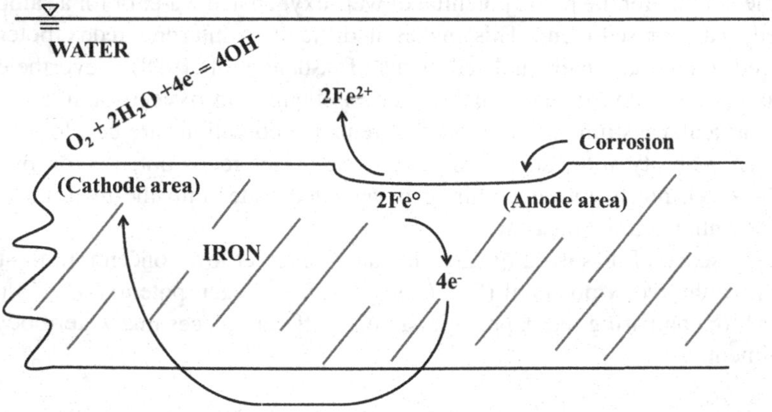

Fig. 7.3 Illustration of the process of metal corrosion

$$2Fe^0 + O_2 + 2H_2O \rightarrow 2Fe^{2+} + 4OH^-. \quad (7.20)$$

The anode process is

$$2Fe^0 \rightarrow 2Fe^{2+} + 4e^- \quad (7.21)$$

while the cathode process is

$$O_2 + 2H_2O + 4e^- \rightarrow 4OH^-. \quad (7.22)$$

In presence of oxygen, Fe^{2+} will precipitate as iron oxide on the surface of the corroding metal.

The potential for corrosion ($E_{corrosion}$) may be assessed by the difference between the E^0 of (7.22) and the E^0 of (7.21):

$$E_{corrosion} = E^0_C - E^0_A$$

where subscripts C and A indicate cathode and anode processes, respectively.

$$E_{corrosion} = +0.40 - (-0.44) = +0.84\ V.$$

The positive value for $E_{corrosion}$ indicates that the reaction is possible and that corrosion will occur under the specific conditions. Of course, this simple approach does not indicate the degree of corrosion that may be expected.

The corrosive potential of water is increased by lower pH, and high concentrations of dissolved oxygen, carbon dioxide, and dissolved solids. Acidic waters have more hydrogen ion to facilitate the cathode reaction, oxygen reacts with hydrogen gas at the cathode speeding up the reaction, carbon dioxide reduces pH, and higher dissolved solids concentration increases electrical conductivity. Bacteria action and presence of oxidants such as nitrate or chlorine also can accelerate corrosion.

A common way of reducing corrosion of metals is to put a corrosion resistant coating on the metal to protect it from contact with the environment, e.g., zinc or tin coating on steel, resins, plastics, paints, or greases. Another way is to reduce the corrosiveness of the environment, e.g., removal of dissolved oxygen from water. There also are electrochemical means of lessening corrosion that will not be discussed.

Significance

The redox potential is directly related to dissolved oxygen concentration. When there is dissolved oxygen at concentrations above 1 or 2 mg/L, the redox potential will be high. The main value of the redox potential in aquatic ecology is for explaining how oxidations and reductions occur in sediment-water systems.

The redox potential is difficult to measure, so it is not generally a useful variable in water quality criteria for aquatic ecosystems. Of course, concepts from electrochemistry discussed above in relation to the redox potential are the basis for understanding and controlling corrosion of metal structures and devices so essential to human society. Redox potential also has far-reaching importance in analytical chemistry and many industrial processes.

References

Essington ME (2004) Soil and water chemistry. CRC Press, Boca Raton

Greybowski AK (1958) The standard potential of the calonel electrode and its application in accurate physiochemical measurement. I. The standard potential. J Phys Chem 62:550–555

Masuda K, Boyd CE (1994) Chemistry of sediment pore water in aquaculture ponds built on clayey Ultisols at Auburn, Alabama. J World Aquacult Soc 25:396–404

Mortimer CH (1942) The exchange of dissolved substances between mud and water in lakes, II. J Ecol 30:147–201

Snoeyink VL, Jenkins D (1980) Water chemistry. John Wiley, New York

Teasdale PR, Minett AI, Dixon K, Lewis TW, Batley GE (1998) Practical improvements for redox potential measurements and the application of a multiple-electrode probe (MERP) for characterizing sediment *in situ*. Anal Chim Acta 367:201–213

pH, Carbon Dioxide, and Alkalinity

8

Abstract

The pH or negative logarithm of the hydrogen ion concentration is a master variable in water quality because the hydrogen ion influences many reactions. Because dissolved carbon dioxide is acidic, rainwater that is saturated with this gas is naturally acidic—usually about pH 5.6. Limestone, calcium silicate, and feldspars in soils and other geological formations dissolve through the action of carbon dioxide to increase the concentration of bicarbonate in water and raise the pH. The total concentration of titratable bases—usually bicarbonate and carbonate—expressed in milligrams per liter of calcium carbonate is the total alkalinity. Total alkalinity typically is less than 50 mg/L in waters of humid areas with highly leached soils, but it is greater where soils are more fertile, limestone formations are present, or the climate is arid. Alkalinity increases the availability of inorganic carbon for photosynthesis, because most phytoplankton species and many higher aquatic plants can obtain carbon from bicarbonate. Water bodies with moderate to high alkalinity are well-buffered against wide daily swings in pH resulting from net removal of carbon dioxide by photosynthesis during daytime and return of carbon dioxide to the water by respiratory process at night when there is no photosynthesis. The optimum pH range for most aquatic organisms is 6.5–8.5, and the acid and alkaline death points are around pH 4 and pH 11, respectively. Fish and other aquatic animals avoid high carbon dioxide concentration, but 20 mg/L or more can be tolerated if there is plenty of dissolved oxygen.

Keywords

pH concept • Reaction of carbon dioxide • Sources of alkalinity • Bicarbonate-carbonate equilibrium • Buffering by alkalinity

C.E. Boyd, *Water Quality*, DOI 10.1007/978-3-319-17446-4_8

Introduction

The pH is a master variable in water quality because many reactions are pH dependent. Normal waters contain both acids and bases and biological processes tend to increase either acidity or basicity. The interactions among these opposing acidic and basic substances and processes determine pH. Carbon dioxide is particularly influential in regulating pH. It is acidic, and its concentration is in continual flux as a result of its utilization by aquatic plants in photosynthesis and release in respiration of aquatic organisms. The alkalinity of water results primarily from bicarbonate and carbonate ions derived from the reaction of carbon dioxide in water with limestone and certain other minerals.

Alkalinity plays an important role in aquatic biology by increasing the availability of inorganic carbon for plants. It also tends to buffer water against excessive pH change. A basic understanding of the relationships among pH, carbon dioxide, and alkalinity are essential to an understanding of water quality.

pH

The Concept

The hydrogen ion is a naked nucleus or proton with a high charge density; it cannot exist in water because it is attracted to the negatively-charged side of water molecules creating an ion pair known as the hydronium ion (H_3O^+). However, for simplicity, it is conventional to write the hydronium ion as H^+ and speak of hydrogen ion concentration rather than hydronium ion concentration. This simplification is followed throughout this book.

The pH is an index of the hydrogen ion concentration (H^+). The concept is based on the dissociation of pure water

$$H_2O = H^+ + OH^-. \tag{8.1}$$

The mass action expression of (8.1) is

$$\frac{(H^+)(OH^-)}{(H_2O)} = K_w. \tag{8.2}$$

But, water dissociates only slightly and can be considered unity in (8.2) giving

$$(H^+)(OH^-) = K_w. \tag{8.3}$$

Water dissociates into equal numbers of hydrogen and hydroxyl (OH^-) ions; hydrogen ion can be substituted into (8.3) to obtain

$$(H^+)(H^+) = K_w.$$

The hydrogen ion concentration is the square root of K_w

$$(H^+) = \sqrt{K_w}. \quad (8.4)$$

Of course, the hydroxyl ion concentration is related to hydrogen ion concentration

$$(OH^-) = \frac{K_w}{(H^+)}. \quad (8.5)$$

In the early 1900s, the Danish chemist, S. P. L. Sørensen recommended taking the negative logarithm of hydrogen ion concentration to avoid the use of small molar concentrations when referring to the concentration of this ion. The negative logarithm of the hydrogen ion concentration was called pH:

$$pH = -\log_{10}(H^+). \quad (8.6)$$

The K_w for water at 25°C is $10^{-14.00}$ (Table 8.1); the pH is

$$pH = -\log_{10}\left(10^{-7.00}\right) = -(-7.00) = 7.00.$$

Pure water is neutral—neither acidic nor basic in reaction—because hydrogen and hydroxyl ion concentrations are equal. The same concept upon which pH is based can be applied to express the hydroxyl ion concentration as pOH—the negative log of the hydroxyl ion concentration. The pOH equals 14 – pH at 25 °C or pK_w – pH at other temperatures. The pOH, however, is not frequently used in discussions of water quality.

It is common practice to use a pH scale of 0–14 so that pH 7 is the midpoint. Acidic reaction intensifies as pH declines below 7.0; basic reaction intensifies as pH rises above 7.0. The pH of pure water is 7.00 only at 25 °C where $K_w = 10^{-14}$. The neutral point increases at lower temperature and decreases at higher temperature (Table 8.1). It is possible to have negative pH values and pH above 14 as well. At 25 °C, a solution that is 2 M ($10^{0.3}$ M) in hydrogen ion has a pH of –0.3; a 2 M hydroxyl solution has a pH of 14.3.

Table 8.1 Ionization constants (Kw), molar concentrations of hydrogen ion (H^+), and pH of pure water at different temperatures. The molar concentrations of hydroxyl ion (OH^-) are the same as (H^+)

Water temperature (°C)	K_w	(H^+)	pH
0	$10^{-14.94}$	$10^{-7.47}$	7.47
5	$10^{-14.73}$	$10^{-7.36}$	7.36
10	$10^{-14.53}$	$10^{-7.26}$	7.26
15	$10^{-14.35}$	$10^{-7.18}$	7.18
20	$10^{-14.17}$	$10^{-7.08}$	7.08
25	$10^{-14.00}$	$10^{-7.00}$	7.00
30	$10^{-13.83}$	$10^{-6.92}$	6.92
35	$10^{-13.68}$	$10^{-6.84}$	6.84
40	$10^{-13.53}$	$10^{-6.76}$	6.76

Calculations

It is easy to estimate pH from solutions with hydrogen ion concentration such as 0.01 M (10^{-2} M) or 0.000001 M (10^{-6} M)—the values are 2 and 6, respectively. But, if the concentration is not an exact multiple of ten, the conversion is slightly more involved (Ex. 8.1).

Ex. 8.1: *The pH of water that is 0.002 M in hydrogen ion concentration will be calculated.*

Solution:

$$0.002\,\mathrm{M} = 2 \times 10^{-3}\,\mathrm{M}$$

$$pH = -log_{10}(H^+) = -\left(log_{10}2 + log_{10}10^{-3}\right) = -[0.301 + (-3)] = 2.70.$$

The pH scale is logarithmic: a solution of pH 6 has a ten times greater hydrogen ion concentration than does one of pH 7. Therefore, the pH of a solution made by mixing 1 L each of solutions of pH 2, 4, and 6 is not 4.0. It is actually 2.47 as illustrated in Ex. 8.2.

Ex. 8.2: *The pH resulting from mixing 1-L each of solutions of pH 2, 4, and 6 will be calculated.*

Solution:

pH	mole H^+ in 1.0 L
2	*0.01*
4	*0.0001*
6	*0.000001*
Sum	*0.010101 in 3.0 L*

$$Average = \frac{0.010101\ \mathrm{mol}\ H^+}{3\,\mathrm{L}} = 3.37 \times 10^{-3}\,\mathrm{mol}\,H^+/\mathrm{L}$$

$$pH = -log_{10}\left(3.37 \times 10^{-3}\right) = -[0.53 + (-3)] = 2.47.$$

Averaging pH directly would give an erroneous pH of 4.0.

Considerable confusion exists over the necessity for transforming pH values to hydrogen ion concentrations for calculating average pH. While obligatory for calculating the average pH of mixtures of different pH as illustrated in Ex. 8.2, hydrogen ion averaging is not necessary—it is actually incorrect—in most practical applications of water quality (Boyd et al. 2011). The averaging of hydrogen ion concentration will not provide the correct average pH, because buffers present in natural waters have a greater effect on final pH than does dilution alone. When data sets are transformed to hydrogen ion to estimate average pH, extreme pH values will distort the average pH. Values of pH conform more closely to a normal distribution than do hydrogen ion concentrations, making pH values more acceptable for use in statistical analyses. Moreover, electrochemical measurements of pH and many biological responses to hydrogen ion concentration are described by the Nernst equation that states that the measured or observed response is linearly related to tenfold changes in hydrogen ion concentration. Of course, in much research, the relationships between pH and other variables generally are reported as the observed effect of measured pH—not H^+ concentration—on the response of the other variables. Thus, the use of pH averaging is correct in most water quality efforts.

pH of Natural Waters

The technique for measuring pH with a glass electrode was commercialized first in 1936. Since then, the measurement of pH in the environment has become common. There have been literally millions of pH measurements in all kinds of natural waters. Reported values range from as low as 1 or 2 to as high as 12 or 13.

Most natural waters have pH within the 6–9 range. Waters in humid areas with highly leached soils tend to have lower pH than waters in areas with limestone formations or those in semi-arid or arid regions.

Waters that have high concentrations of humic substances tend to have a low pH because of acidic groups on the humic molecules as illustrated by the equation below in which R represents the organic (humic) moiety:

$$RCOOH = RCOO^- + H^+. \tag{8.7}$$

Waters from forested areas often have a low pH because of large concentrations of humic acids. For example, waters in rivers flowing from South American jungles often have pH values below 5. Low pH from mineral acids is not as harmful to aquatic organisms as low pH caused by strong, mineral acids. Moreover, species native to rivers from jungles probably are more tolerant to low pH than organisms in other waters.

Carbon Dioxide

The solubility of atmospheric gases, including carbon dioxide, is discussed in Chap. 6. The atmosphere contains an average concentration of roughly 0.040 % carbon dioxide, and raindrops become saturated with this gas while falling to the ground. The equilibrium concentrations of carbon dioxide in water of different temperatures and salinity were provided in Table 6.3.

Dissolved carbon dioxide reacts with water to form carbonic acid (H_2CO_3)

$$CO_2 + H_2O \rightleftharpoons H_2CO_3 \quad K = 10^{-2.75}. \tag{8.8}$$

Carbonic acid is a diprotic acid

$$H_2CO_3 \rightleftharpoons H^+ + HCO_3^- \quad K_1 = 10^{-3.6} \tag{8.9}$$

$$HCO_3^- \rightleftharpoons H^+ + CO_3^{2-} \quad K_2 = 10^{-10.33}. \tag{8.10}$$

The second dissociation, however, can be ignored here, because, as will be explained later, it is not significant at pH less than 8.3.

Only a small fraction of dissolved carbon dioxide reacts with water. Using (8.8), it can be seen that the ratio (H_2CO_3):(CO_2) will be 0.00178:1:

$$\frac{(H_2CO_3)}{(CO_2)} = 10^{-2.75} = 0.00178.$$

Thus, the carbonic acid concentration at equilibrium will be only 0.18 % of the carbon dioxide concentration. Despite carbonic acid being a somewhat strong acid, there is little of it present in a water at equilibrium with atmospheric carbon dioxide. Moreover, common analytical procedures cannot distinguish carbonic acid from carbon dioxide.

A dilemma is avoided by considering the analytical carbon dioxide as a weak acid and deriving an apparent reaction and equilibrium constant. Multiplication of the equilibrium expressions of (8.8) and (8.9) gives

$$\frac{(H_2CO_3)}{(CO_2)} \times \frac{(HCO_3^-)(H^+)}{(H_2CO_3)} = 10^{-2.75} \times 10^{-3.6}$$

$$\frac{(HCO_3^-)(H^+)}{(CO_2)} = 10^{-6.35}.$$

The equilibrium expression derived above typically is used to explain the reaction of carbon dioxide with water.

$$CO_2 + H_2O \rightleftharpoons H^+ + HCO_3^- \quad K = 10^{-6.35}. \tag{8.11}$$

In Chap. 3, it was illustrated that the K for any reaction can be estimated from ΔG^o using (3.13). The K estimated from ΔG^o for (8.11) is $10^{-6.356}$—essentially the same value derived above by combining (8.8) and (8.9) through multiplication.

The pH of rainwater not contaminated with acids stronger than carbon dioxide usually is around 5.6–5.8, and carbon dioxide normally will not lower the pH of water below 4.5. The two points will be illustrated in Exs. 8.3 and 8.4.

Ex. 8.3: The pH of rainwater saturated with carbon dioxide at 25 °C will be estimated.

Solution:
From Table 6.3, the carbon dioxide concentration of rainwater will be 0.57 mg/L ($10^{-4.89}$ M). *Using* (8.11),

$$\frac{(H^+)(HCO_3^-)}{(CO_2)} = 10^{-6.35}$$

but

$$(H^+) = (HCO_3^-)$$

and

$$(H^+)^2 = (10^{-6.35})(CO_2).$$

Thus,

$$(H^+)^2 = (10^{-6.35})(10^{-4.89}) = 10^{-11.24}$$
$$(H^+) = 10^{-5.62}$$
$$pH = 5.62.$$

Rainwater with a lower pH containing an acid stronger than carbon dioxide—usually sulfuric or nitric acids—is discussed in Chap. 13.

Ex. 8.4: Water from a well contains 100 mg/L $(10^{-2.64}\,\mathrm{M})$ *carbon dioxide. The pH will be determined using* (8.11).

Solution:

$$(H^+)^2 = (10^{-6.35})(10^{-2.64}) = 10^{-8.99}$$

$$(H^+) = 10^{-4.50}$$

$$pH = 4.5.$$

Few natural waters contain more than 100 mg/L of carbon dioxide. It follows that a pH below 4.5 suggests that a water contains a stronger acid than carbon dioxide. In surface waters, the most common strong acid is sulfuric acid that originates from deposits of iron pyrite that may be exposed to the air and oxidize resulting in sulfuric acid (see Chap. 13). Rainwater may contain sulfuric and other strong acids as a result of air pollution.

Natural waters often contain more carbon dioxide than suggested by equilibrium concentrations provided in Table 6.3. This occurs when release of carbon dioxide by the biota of a water body exceeds the uptake of carbon dioxide by aquatic plants—usually at night. Most natural waters contain more bicarbonate than that resulting from the reaction of carbon dioxide and water (8.11) as discussed later. There is an equilibrium among H^+, HCO_3^-, and CO_2, and a change in carbon dioxide concentration alters pH. For example, using (8.11), the CO_2 concentrations at equilibrium at 25 °C and pH 7 for 61 mg/L (10^{-3} M) HCO_3^- and 122 mg/L ($10^{-2.55}$ M) HCO_3^- would be 9.85 and 27.8 mg/L, respectively. In pure water, the equilibrium concentration is given as 0.57 mg/L (Table 6.3). Of course, the two waters would be supersaturated with CO_2 with respect to atmospheric CO_2 despite being at equilibrium with respect to (8.11).

Total Alkalinity

The total alkalinity of water is defined as the total concentration of titratable bases expressed as calcium carbonate. Calcium carbonate probably was chosen as the basis for expressing alkalinity because it is commonly added to increase the pH and alkalinity of acidic water. Calcium carbonate also is the substance that precipitates from some waters when they are boiled as will be discussed in Chap. 9.

Examination of (8.11) reveals that bicarbonate results from the reaction of carbon dioxide with water. At equilibrium with atmospheric carbon dioxide, the concentrations of H^+ and HCO_3^- in unpolluted rainwater or other relatively pure water are both $10^{-5.62}$ M (Ex. 8.3). Such a water is not considered to contain alkalinity.

Carbon dioxide is, nevertheless, a major factor in the dissolution of limestone and certain other minerals in soils and other geological formations that imparts bicarbonate (and carbonate) to natural waters. Concentrations of total alkalinity usually range between 10 and 300 mg/L. Although alkalinity often results almost entirely from bicarbonate and carbonate, some waters—especially polluted ones—contain appreciable amounts of hydroxide, ammonia, phosphate, borate, or other bases that contribute alkalinity. The reactions of selected bases in neutralizing acidity are illustrated below:

$$CO_3^{2-} + H^+ \rightleftharpoons HCO_3^-$$

$$HCO_3^- + H^+ \rightleftharpoons CO_2 + H_2O$$

$$OH^- + H^+ \rightleftharpoons H_2O$$

$$NH_3 + H^+ \rightleftharpoons NH_4^+$$

$$PO_4^{3-} + H^+ = HPO_4^-$$

$$HPO_4^- + H^+ \rightleftharpoons H_2PO_4^-$$

$$H_2BO_4^- + H^+ \rightleftharpoons H_3BO_4$$

$$H_3SiO_4^- + H^+ \rightleftharpoons H_4SiO_4$$

$$RCOO^- + H^+ \rightleftharpoons RCOOH.$$

However, bicarbonate and carbonate usually comprise most of the alkalinity in natural waters, and natural waters contain hydroxyl (and hydrogen) ion in accordance with their pH. The alkalinity sometimes is described by the expression

$$\text{Alkalinity} = \left(HCO_3^-\right) + 2\left(CO_3^{2-}\right) + (OH^-) - (H^+)$$

where the ions are entered in moles/L. For example, suppose that a water sample of pH = 9 ($H^+ = 10^{-9}$ M; $OH^- = 10^{-5}$ M) contains 61 mg/L HCO_3^- and 2.81 mg/L CO_3^{2-}. The molar weight of HCO_3^- and CO_3^{2-} are 61 and 60 g/mol, respectively. Thus, (HCO_3^-) = 0.001 M and (CO_3^{2-}) = 0.000047 M. We may write

$$\begin{aligned}\text{Alkalinity} &= 0.001\ \text{M} + 2(0.000047\ \text{M}) + 0.00001\ \text{M} - 0.000000001\ \text{M}\\ &= 0.0011\ \text{M}\end{aligned}$$

Thus, 0.0011 M of acidity would be required to neutralize the alkalinity and total alkalinity would be 55 mg/L as $CaCO_3$. Although the expression above illustrates the concept of alkalinity, it does not provide a practical way of estimating the alkalinity of natural waters.

Measurement of Alkalinity

Total alkalinity is determined by titration with standard sulfuric or hydrochloric acid (Eaton et al. 2005). The titration traditionally was done with a 0.020 N acid, and methyl orange—that changes color from yellow to orange at pH 4.4—was used to signal the endpoint. However, the pH in the water at the end of the titration is influenced by the amount of carbon dioxide produced in the reaction between acid and bicarbonate, and samples of higher alkalinity will be at a lower pH when bicarbonate has been neutralized than will samples of lower alkalinity.

In the situation where all of the alkalinity in a sample is from bicarbonate, the maximum amount of carbon dioxide will be liberated per milligram per liter of alkalinity during titration. Assuming total alkalinity concentrations of 30, 150, and 500 mg/L (36.6, 183, and 610 mg/L alkalinity), and the carbon dioxide liberated would equal 26.4, 132, and 440 mg/L or $10^{-3.22}$, $10^{-2.52}$, and 10^{-2} M, respectively.

With (8.11), it can be shown the potential pH at the endpoint would be 4.78, 4.44, and 4.18, for 30, 150, and 500 mg/L total alkalinity respectively. Of course, some carbon dioxide is lost to the air during titration. Taras et al. (1971) recommended the following endpoint pHs for different alkalinities: <30 mg/L, 5.1; 30–500 mg/L, 4.8; >500 mg/L, 4.5. However, Eaton et al. (2005) revised these recommendations to pHs of 4.9, 4.6, and 4.3 at alkalinities of 30, 150, 500 mg/L, respectively.

Mixed bromocresol green-methyl red indicator can be used to detect pH 5.1 and 4.8, while methyl orange can be used to signal pH 4.5, but many analysts prefer to use a pH electrode to detect the endpoint. Nevertheless, for routine titrations, pH 4.5 often is used as the endpoint for samples of all concentrations of total alkalinity.

The measurement and calculation of total alkalinity will be demonstrated in Ex. 8.5.

Ex. 8.5: *A 100-mL water sample is titrated with 0.0210 N sulfuric acid to an endpoint of pH 4.5. The titration consumes 18.75 mL of sulfuric acid. The total alkalinity will be calculated.*

Solution:

The milliequivalent quantity of alkalinity or $CaCO_3$ equals the milliequivalent amount of acid used in the titration:

$$(18.75\,\text{mL})(0.0210\,\text{N}) = 0.394\,\text{meq}\ CaCO_3.$$

The weight of equivalent $CaCO_3$ is

$$(0.394\,\text{meq})(50.04\,\text{mg}\ CaCO_3/\text{meq}) = 19.72\,\text{mg}\ CaCO_3.$$

To convert to milligrams per liter,

$$19.72\,\text{mg}\ CaCO_3 \times \frac{1,000\,\text{mL/L}}{100\,\text{mL}\ sample} = 197.2\,\text{mg/L}.$$

In water samples with pH $\leq$8.3, the alkalinity is typically considered to be from bicarbonate alone. Samples with a pH >8.3 usually contain both carbonate and bicarbonate, but if the pH is especially high, bicarbonate may be absent. Such samples may contain carbonate only, carbonate and hydroxide, or hydroxide only. Phenolphthalein indicator that is pink above pH 8.3 and colorless at lower pH may be used in samples with pH >8.3 to separate the total alkalinity titration into two steps: determination of the volume of acid needed to decrease pH to 8.3 and continuation of the titration to pH 4.5 (or other selected endpoint) to determine the total volume of acid needed to neutralize all base in the sample. The first step provides an estimate of the phenolphthalein alkalinity, and of course, the result of the second step gives the total alkalinity. The phenolphthalein alkalinity (PA) total alkalinity (TA) may be used to calculate the three possible forms of alkalinity: bicarbonate, carbonate, and hydroxide.

There are five possible cases:

Table 8.2 Relationships of the three kinds of alkalinity to measurements by acid titration of total alkalinity (TA) and phenolphthalein alkalinity (PA)

	Kind of alkalinity		
Titration results	Bicarbonate	Carbonate	Hydroxide
PA = 0	TA	0	0
PA < 0.5TA	TA—2PA	2PA	0
PA = 0.5TA	0	2PA (or TA)	0
PA > 0.5 TA	0	2 (TA—PA)	2PA—TA
PA = TA	0	0	TA (or PA)

1. PA = 0. There is only bicarbonate, and the titration is simply $HCO_3^- + H^+ = CO_2 + H_2O$.
2. PA < 0.5 TA. There are two reactions. In the first step, the reaction is $CO_3^{2-} + H^+ = HCO_3^-$, and HCO_3^- produced from CO_3^{2-} combines with HCO_3^- already in the sample. The second step is the neutralization of all of the bicarbonate. The carbonate alkalinity is 2PA and the bicarbonate alkalinity is TA—2PA.
3. PA = 0.5TA. A sample in which this situation results has only carbonate, and the second part of the titration involves the neutralization of HCO_3^- formed by conversion of carbonate to bicarbonate in the first step. The carbonate alkalinity = 2PA or TA.
4. PA > 0.5 TA. In the first step, the reactions are $OH^- + H^+ = H_2O$ and $CO_3^{2-} + H^+ = HCO_3^-$. The bicarbonate produced in the first step is neutralized in the second step. The hydroxide alkalinity is 2PA—TA, while the carbonate alkalinity is 2(TA—PA). There is no bicarbonate alkalinity.
5. PA = TA. This type of sample contains only hydroxide alkalinity. The pH will fall to 4.5 or less when the hydroxide is neutralized.

The five alkalinity relationships are summarized in Table 8.2.

Sources of Alkalinity

As mentioned above, the major source of alkalinity for many waters is limestone, and calcium carbonate commonly is used as a model for explaining the dissolution of limestone. The solubility of calcium carbonate often is depicted as

$$CaCO_3 = Ca^{2+} + CO_3^{2-}. \qquad (8.12)$$

Several values of K_{sp} for (8.12) have been determined experimentally, but the most appropriate appears to be $10^{-8.3}$ (Akin and Lagerwerff 1965). This K_{sp} agrees well with the K_{sp} of $10^{-8.34}$ estimated from the ΔG^o of (8.12).

The concentration of carbonate in a closed calcium carbonate-water system (free of carbon dioxide) calculated using K_{sp} $CaCO_3 = 10^{-8.3}$ is 4.25 mg/L (7.08 mg/L total alkalinity). This estimate is not exactly correct, because CO_3^{2-} reacts with H^+ forming bicarbonate (reverse reaction in 8.10). This allows more calcium carbonate to dissolve and increases OH^- concentration relative to H^+ concentration increasing the pH above neutral. Nevertheless, the total alkalinity concentration will be far below that of an open calcium carbonate-water system in equilibrium with atmospheric carbon dioxide.

Carbon dioxide reacts with limestone as illustrated below for both calcitic limestone ($CaCO_3$) and dolomitic limestone ($CaCO_3 \cdot MgCO_3$):

$$CaCO_3 + CO_2 + H_2O \rightleftharpoons Ca^{2+} + 2HCO_3^- \quad (8.13)$$

$$CaCO_3 \cdot MgCO_3 + 2CO_2 + 2H_2O \rightleftharpoons Ca^{2+} + Mg^{2+} + 4HCO_3^-. \quad (8.14)$$

Most limestone is neither calcite nor dolomite, but a mixture in which the ratio of $CaCO_3$:$MgCO_3$ usually is greater than unity. Limestones are of low solubility, but their dissolution is increased greatly by the action of carbon dioxide.

The mass action expression for the solubility of calcium carbonate in the presence of carbon dioxide (see 8.13) is

$$\frac{(Ca^{2+})(HCO_3^-)^2}{(CO_2)} = K$$

The same expression and the value of K may be derived by combining the inverse of the mass action form of (8.10) and the mass action forms of (8.11) and (8.12) by multiplication as follows:

$$\frac{(HCO_3^-)}{(H^+)(CO_3^{2-})} \times \frac{(H^+)(HCO_3^-)}{(CO_2)} \times (Ca^{2+})(CO_3^{2-}) = \frac{1}{10^{-10.33}} \times 10^{-6.35} \times 10^{-8.3}.$$

In arranging the expression above, it was necessary to use the inverse of the mass action form of (8.10) in order to have (H^+) and (CO_3^{2-}) in the denominator. Upon multiplication, the terms reduce to

$$\frac{(Ca^{2+})(HCO_3^-)^2}{(CO_2)} = 10^{-4.32}$$

The derived K for (8.13) agrees well with the K of $10^{-4.35}$ estimated from the $\Delta G°$ of the equation. The calcium and bicarbonate concentrations at any carbon dioxide concentration may be estimated using the mass action form of (8.13) as illustrated in Ex. 8.6 for the current atmospheric carbon dioxide level of about 400 ppm.

Ex. 8.6: *The concentration of calcium and bicarbonate at equilibrium between solid phase calcite ($K_{sp} = 10^{-8.3}$) and distilled water in equilibrium with atmospheric carbon dioxide at 25 °C will be calculated.*

Solution:

From Table 6.3, the water contains 0.57 mg/L ($10^{-4.89}$ M) carbon dioxide. From (8.13), each mole of $CaCO_3$ that dissolves results in 1 mol of Ca^{2+} and 2 mol of HCO_3^-.

Thus, we may set $(Ca^{2+}) = X$ *and* $(HCO_3^-) = 2X$ *and substitute these values into the mass action form of* (8.13) *to obtain.*

$$\frac{(X)(2X)^2}{10^{-4.89}} = 10^{-4.32}$$

and

$$4X^3 = 10^{-9.21} = 6.17 \times 10^{-10}$$

$$X^3 = 1.54 \times 10^{-10} = 10^{-9.81}$$

$$X = 10^{-3.27} M = 5.37 \times 10^{-4}\,\mathrm{M}.$$

Thus, $Ca^{2+} = (5.37 \times 10^{-4}\,\mathrm{M})(40.04\,\mathrm{g}\ CaCO_3/\mathrm{mol}) = 0.0215\,\mathrm{g/L} = 21.5\,\mathrm{mg/L}$ *and* $HCO_3^- = (5.37 \times 10^{-4}\,\mathrm{M})(2)(61\,\mathrm{g}\ HCO_3^-/\mathrm{mol}) = 0.0655\,\mathrm{g/L} = 65.5\,\mathrm{mg/L}$ *or* $53.7\,\mathrm{mg/L}$ *total alkalinity.*

The calcium and bicarbonate concentrations calculated in Ex. 8.6 agree quite well with concentrations of 22.4 mg/L calcium and 67.1 mg/L bicarbonate determined experimentally for a carbon dioxide concentration of 0.04 % of total atmospheric gases by Frear and Johnston (1929). The bicarbonate concentration calculated in Ex. 8.6 is equivalent to 53.7 mg/L for both total alkalinity and total hardness (see Chap. 9). Limestone seldom exists as pure calcite, and other forms have different solubilities—some are more soluble than calcite.

Examination of (8.13) reveals that half of the carbon in bicarbonate comes from calcium carbonate and half is from carbon dioxide. Twice as much soluble carbon results from each molecule of calcium carbonate that dissolves as occurs in a closed system (no carbon dioxide). Moreover, the removal of carbon dioxide from the water in the reaction with calcium carbonate allows more atmospheric carbon dioxide to enter the water. Of course, this effect is limited, because when equilibrium is reached, the dissolution of calcium carbonate stops. The upshot is that the action of carbon dioxide on calcium carbonate (or other alkalinity source) provides a means of increasing both the solubility of the calcium carbonate and the capture of atmospheric carbon dioxide.

Equation 8.13 may be used to illustrate the effect of the increasing atmospheric carbon dioxide concentration on the solubility of calcite. In the 1960s, the atmospheric carbon dioxide concentration was about 320 ppm at the Mauna Loa Observatory, Hawaii, but today, the concentration measured at that observatory is about 400 ppm. Using these values, the dissolved carbon dioxide concentration in freshwater at 25 °C has increased from about 0.42 mg/L in the 1960s to 0.57 mg/L today. This has resulted in a corresponding increase in total alkalinity in a calcite-water-air system at 25 °C from 48 mg/L to about 54 mg/L—a 12.5 % increase in alkalinity from a 25 % increase in atmospheric carbon dioxide. Of course, the solubilities of the other sources of alkalinity also increase with greater atmospheric carbon dioxide concentration.

The carbon dioxide concentration in water of meteoric origin tends to increase during infiltration through the soil and deeper formations. This results because hydrostatic pressure increases with depth allowing the water to hold more carbon dioxide at equilibrium. Moreover, there often is abundant carbon dioxide in soil because of respiration by plant roots and soil microorganisms. The solubility of minerals that are sources of alkalinity often will be considerably greater in the soil and underlying geological formations than at the land surface. Groundwater usually is higher in alkalinity than surface water.

In water bodies, decomposition of organic matter also produces carbon dioxide and waters may be supersaturated with carbon dioxide. This phenomenon can result in waters having greater total alkalinity concentrations than expected from calculations based on equilibrium carbon dioxide concentrations.

The dissolution of limestone and other sources of alkalinity will increase at greater concentrations of ions in the aqueous phase, because solubility products are based on activities rather than measured molar concentrations (Garrels and Christ 1965). Activities are estimated by multiplying an activity coefficient by the measured molar concentration. If the ionic strength of a solution increases, the activity coefficients of the ions in the solution decrease. Thus, more of the mineral must dissolve to attain equilibrium at greater ionic strength of the solution, i.e., the measured molar concentration of ions from the dissolving mineral must increase because the activity coefficients of these ions decrease. Thus, calcite—like other minerals—dissolves to an increasing extent as ionic strength of the water increases.

Reactions involving the formation of bicarbonate from carbonate are equilibrium reactions and a certain amount of carbon dioxide must be present to maintain a given amount of bicarbonate in solution. If the amount of carbon dioxide at equilibrium is increased or decreased, there will be a corresponding change in the concentration of bicarbonate ion. Additional carbon dioxide will result in the solution of more calcium carbonate and a greater alkalinity, while removal of carbon dioxide will result in precipitation of calcium carbonate and a lower alkalinity. Thus, when well water or spring water of high alkalinity and enriched with carbon dioxide contact the atmosphere, calcium carbonate often precipitates in response to equilibration with atmospheric carbon dioxide.

Many texts on water quality and limnology tend to leave the reader with the impression that limestone is the only source of alkalinity in water. However, carbon dioxide also reacts with calcium silicate and other silicates such as the feldspars olivine and orthoclase to give bicarbonate:

$$\underset{\text{Calcium silicate}}{CaSiO_3} + 2CO_2 + 3H_2O \rightarrow Ca^{2+} + 2HCO_3^- + H_4SiO_4 \tag{8.15}$$

$$\underset{\text{Olivine}}{Mg_2SiO_4} + 4CO_2 + 4H_2O \rightleftharpoons 2Mg^{2+} + 4HCO_3^- + H_4SiO_4 \tag{8.16}$$

$$\underset{\text{Orthoclase}}{2KAlSi_3O_8} + 2CO_2 + 11H_2O \rightleftharpoons Al_2Si_2O_5(OH)_4 + 4H_4SiO_4 + 2K^+ + 2HCO_3^-. \tag{8.17}$$

Calcium silicate is a major source of alkalinity in natural waters (Ittekkot 2003).

Weathering of feldspars also is an important source of alkalinity in areas with acidic soils that do not contain limestone or calcium silicate.

In the equations above, silicic acid is shown as undissociated, but if pH is elevated to 9 or above, there will be enough $H_3SiO_4^-$ to contribute measurable alkalinity to water. Snoeyink and Jenkins (1980) presented an example from a major water supply in the Bay Area of California where 20 % of alkalinity was derived from silicate. This water had an alkalinity of 20 mg/L, 8 mg/L silica (as SiO_2), and pH of 9.65. It originated from the Sierra Nevada mountains where rock formations are rich in silicate-bearing minerals.

The K for (8.14) can be obtained by the same procedure used above to derive the K for (8.13) or estimated from the ΔG^o. The Ks for (8.15)–(8.17) can be derived from their ΔG^os. These K values will not be estimated because the effect of carbon dioxide on dissolution of minerals that provide alkalinity has been demonstrated adequately. Moreover, the actual alkalinity of a water body must be determined, because the sources of alkalinity differ from one place to another and are often unknown making calculation impossible. Of course, natural waters often are not at equilibrium with atmospheric carbon dioxide or minerals that contribute alkalinity. Nevertheless, exercises in the calculation of alkalinity concentration from the solubility of minerals such as done above for calcite are useful in understanding the processes that impart alkalinity to natural waters.

Comments on Alkalinity Equilibria

The aqueous dissolution products of calcium carbonate are calcium and bicarbonate (8.13), but a very small amount of carbon dioxide and carbonate will be present. Carbon dioxide is present because it is in equilibrium with atmospheric CO_2 and with bicarbonate (8.11) while carbonate results from dissociation of bicarbonate (8.10). Therefore, a bicarbonate solution at equilibrium contains carbon dioxide, bicarbonate, and carbonate. Carbon dioxide is acidic, and carbonate is basic because it hydrolyzes:

$$CO_3^{2-} + H_2O = HCO_3^- + OH^-. \tag{8.18}$$

The hydrolysis reaction also can be written as

$$CO_3^{2-} + H^+ = HCO_3^-. \tag{8.19}$$

Because the hydrogen ion comes from the dissociation of water, hydroxyl ion concentration rises to maintain the equilibrium constant. Equation 8.19 is the reverse reaction of (8.10). In (8.10), bicarbonate is an acid and the equilibrium constant for the forward reaction can be referred to as K_a. In the reverse reaction, carbonate acts as a base and the equilibrium constant for the basic reaction is called K_b. In such reactions, $K_aK_b = K_w$ or 10^{-14}. Thus, the equilibrium constant for the reverse reaction of (8.10) (or the forward reaction of 8.19) is

$$K_b = 10^{-14} - 10^{-10.33} = 10^{-3.67}.$$

Thus, carbonate is stronger as a base ($K_b = 10^{-3.67}$) than either carbon dioxide ($K = 10^{-6.35}$) or bicarbonate ($K_a = 10^{-10.33}$) are as acids. As a result, a bicarbonate solution at equilibrium will be basic.

We can determine the pH of a weak bicarbonate solution at equilibrium with calcium carbonate and atmospheric carbon dioxide by combining the mass action expressions for (8.11) and (8.10) through multiplication.

$$\frac{(H^+)(HCO_3^-)}{(CO_2)} \times \frac{(H^+)(CO_3^{2-})}{(HCO_3^-)} = 10^{-6.35} \times 10^{-10.33}.$$

Because bicarbonate concentration is the same in both expressions and carbon dioxide and carbonate are of negligible concentration when bicarbonate is at maximum concentration, the overall expression reduces to

$$(H^+)^2 = 10^{-16.68}$$

$$(H^+) = 10^{-8.34}.$$

The calculation above shows that for practical purposes water above pH 8.3 does not contain free carbon dioxide, and that carbonate begins to appear at pH 8.3.

In order to see what happens when carbon dioxide is withdrawn from a bicarbonate solution at equilibrium, we can combine the reactions of bicarbonate acing as a base and as an acid:

$$HCO_3^- + H^+ = H_2O + CO_2$$

$$(+)\ HCO_3^- = H^+ + CO_3^{2-}$$

to obtain

$$2HCO_3^- = CO_2 + CO_3^{2-} + H_2O. \tag{8.20}$$

When carbon dioxide is removed, carbonate will increase. The hydrolysis of carbonate will cause pH to increase (8.18 and 8.19). Most natural waters contain calcium, and the pH rise will be moderated because of precipitation of calcium carbonate in response to the increasing carbonate concentration.

During photosynthesis, aquatic plants remove carbon dioxide to cause the pH to rise. Free carbon dioxide is depleted once pH rises above 8.3, but many aquatic macrophytes and most phytoplankton species can use bicarbonate as a carbon source (Korb et al. 1997; Cavalli et al. 2012; Raven et al. 2012). The mechanism by which the carbon dioxide is removed from bicarbonate by aquatic plants involves the enzyme carbonic anhydrase that catalyzes the conversion of HCO_3^- to CO_2. Assuming that the overall affect is that illustrated in (8.20), bicarbonate concentration in the water will diminish while carbonate concentration will increase. Carbonate accumulating in the water will hydrolyze to produce bicarbonate and hydroxyl ion (8.18), but two bicarbonates are removed for each carbonate formed (8.20). Moreover, the hydrolysis reaction only converts a portion of the carbonate to bicarbonate. Thus, carbonate and hydroxyl ion concentrations increase in water as photosynthesis proceeds, and the pH rises. Of course, depending upon the calcium concentration in the water, the rise in pH will be limited by the precipitation of calcium carbonate.

Aquatic plants that tolerate very high pH continue using bicarbonate—including that formed by the hydrolysis of carbonate—resulting in accumulation of hydroxyl ion in the water and extremely high pH (Ruttner 1963). This effect is particularly common in waters of low calcium concentration where the anions are balanced mainly by potassium and magnesium (Mandal and Boyd 1980).

Some aquatic plants accumulate calcium carbonate on their surfaces as a result of using bicarbonate as a carbon source in photosynthesis. An excellent example is species of the macroalgal genus *Chara*—called stoneworts because of their tendency to be coated by calcium carbonate. Samples of *Chara* have been reported to contain as much as 20 % of dry weight as calcium (Boyd and Lawrence 1966). Most macroalgae contain 1 or 2 % of dry weight as calcium, and the *Chara* samples likely were encrusted with an amount of calcium carbonate equal to about half of the dry weight.

Precipitation of $CaCO_3$ also may result when removal of carbon dioxide and bicarbonate by plants raises the pH. This can lead to the phenomenon known as "whiting" or "whitening" in productive water bodies of moderate or high alkalinity (Thompson et al. 1997). The minute $CaCO_3$ particles remain suspended in the water in the afternoon producing a milky appearance. This milkiness typically disappears at night when carbon dioxide increases as a result of respiration in absence of photosynthesis and redissolves the suspended $CaCO_3$. Of course, $CaCO_3$ also may precipitate from the water column—a process known as marl formation. Marl is sediment with a high percentage of calcium and magnesium carbonate mixed with clay and silt.

The interdependence of pH, carbon dioxide, bicarbonate, and carbonate is illustrated in Fig. 8.1. The graph shows that below about pH 5, carbon dioxide is the only significant species of inorganic carbon. Above pH 5, the proportion of bicarbonate increases relative to carbon dioxide until bicarbonate becomes the only significant species at about pH 8.3. Above pH 8.3, carbonate appears and it increases in importance relative to bicarbonate if pH continues to rise.

The amount of inorganic carbon available to aquatic plants depends upon pH and alkalinity. At the same alkalinity, as pH rises, the amount of carbon dioxide decreases, but at the same pH, as alkalinity rises, carbon dioxide concentration increases. Of course, aquatic plants use bicarbonate as a carbon source, and the amount of inorganic carbon available to aquatic macrophytes, and phytoplankton in particular, increases with greater alkalinity. Thus, it is difficult to estimate the actual amount of inorganic carbon available to plants in a particular water body. Saunders et al. (1962) provided a nomograph of converting total alkalinity (mg/L) to available carbon (mg/L) a portion of which is provided in Table 8.3. This nomograph probably provides a reasonable estimate of available inorganic carbon.

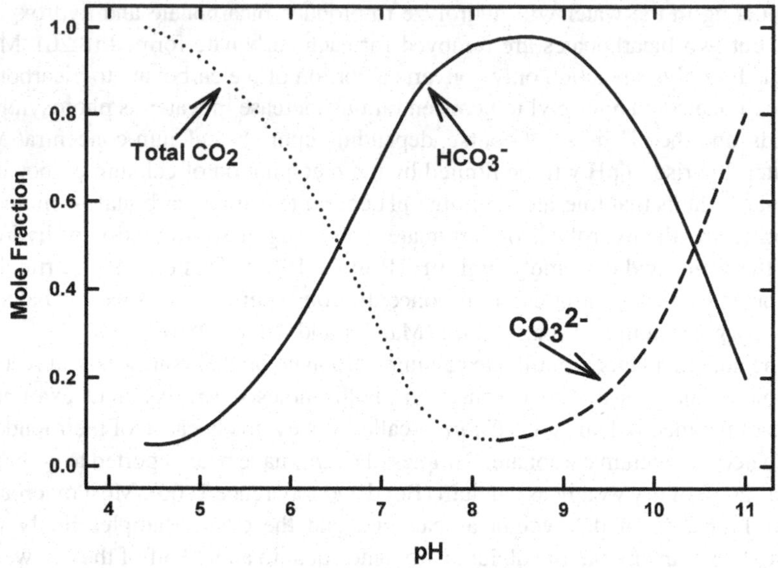

Fig. 8.1 Effects of pH on the relative proportions of total CO_2, HCO_3^-, and CO_3^{2-}. The mole fraction of a component is its decimal fraction of all the moles present

Table 8.3 Factors for converting total alkalinity (mg/L as $CaCO_3$) to carbon available to aquatic plants (mg/L as C)

	Temperature (°C)					
pH	5	10	15	20	25	30[a]
5.0	8.19	7.16	6.55	6.00	5.61	5.20
5.5	2.75	2.43	2.24	2.06	1.94	1.84
6.0	1.03	0.93	0.87	0.82	0.78	0.73
6.5	0.49	0.46	0.44	0.42	0.41	0.40
7.0	0.32	0.31	0.30	0.30	0.29	0.29
7.5	0.26	0.26	0.26	0.26	0.26	0.26
8.0	0.25	0.25	0.25	0.24	0.24	0.24
8.5	0.24	0.24	0.24	0.24	0.24	0.24
9.0	0.23	0.23	0.23	0.23	0.23	0.23

Multiply factors by total alkalinity
[a]Estimated by extrapolation

Buffering

If carbon dioxide is added to a solution containing carbonate or bicarbonate, the pH of the solution will decrease. The drop in pH results from carbon dioxide being a source of hydrogen ion to react with carbonate or bicarbonate. In natural waters, carbon dioxide is released by respiratory processes and added by diffusion from the atmosphere. Removal of carbon dioxide from water causes the pH to climb.

Buffers are substances that allow water to resist pH change. A buffer consists of a mixture of a weak acid and its conjugate base (salt) or a weak base and its conjugate acid. For example, an acidic buffer can be made from acetic acid and its conjugate base sodium acetate, while an alkaline buffer can be made from ammonium hydroxide and its conjugate acid ammonium chloride.

The acidic buffer—acetic acid (CH_3COOH) plus sodium acetate (CH_3COONa)—will be used for illustration. The initial pH is determined by the dissociation of acetic acid

$$CH_3COOH = CH_3COO^- + H^+ \quad K = 10^{-4.74}. \tag{8.21}$$

If more H^+ is added, it combines with CH_3COO^- to form CH_3COOH. If OH^- is added, it reacts with H^+ to form H_2O, CH_3COOH dissociates to provide CH_3COO^-, and pH remains fairly constant. The sodium acetate is a source of CH_3COO^- in excess of that possible from the dissociation of CH_3COOH alone to increase buffering capacity.

The pH of a buffer can be calculated by aid of an equation derived from the equilibrium expression of the weak acid or base as illustrated below for a weak acid:

$$HA = H^+ + A^- \tag{8.22}$$

where HA = a weak acid such as acetic acid and A^- = the conjugate base of the weak acid such as sodium acetate. The equilibrium constant (K_a) for the weak acid is

$$K_a = \frac{(H^+)(A^-)}{(HA)}. \tag{8.23}$$

Taking the negative $\log_{10}$ of both sides of (8.23) gives

$$-\log K_a = -\log\left[\frac{(H^+)(A^-)}{(HA)}\right]$$

which may be rearranged as

$$-\log K_a = -\log(H^+) - \log\left[\frac{(A^-)}{(HA)}\right]. \tag{8.24}$$

By substituting the equivalents pK for $-\log K_a$ and pH for $-\log$ pH, (8.24) becomes

$$pK_a = pH - \log\left[\frac{(A^-)}{(HA)}\right]$$

$$\text{or } pK_a + \log_{10}\left[\frac{(A^-)}{(HA)}\right] = pH$$

which may be rearranged to give

$$pH = pK_a + \log_{10}\left[\frac{(A^-)}{(HA)}\right]. \tag{8.25}$$

Equation 8.25 is known as the Henderson–Hasselbalch equation. It was named after the American biochemist, L. J. Henderson, who first described the relationship between the equilibrium constants of buffers to their pH values, and the Danish biochemist, K. A. Hasselbalch, who first put Henderson's equation into logarithmic form.

This equation reveals that the weak acid and its conjugate base buffer the solution at a pH near the pK_a of the weak acid. Moreover, the pH of the solution will be determined by the ratio of the concentration of the conjugate base:the concentration of the weak acid. When $(A^-) = (HA)$ in (8.25), the ratio $(A^-)/(HA) = 1.0$, $\log 1.0 = 0$, and $pH = pK_a$. When H^+ is added HA is formed, but more A^- is available from the conjugate base to minimize pH change. Adding OH^- results in removal of H^+, i.e., $OH^- + H^+ = H_2O$, but more H^+ is available from the weak acid to minimize pH change.

The pH of a buffer changes as H^+ or OH^- is added, but the changes are relatively small as shown in Ex. 8.7 for addition of OH^- to a buffer.

Ex. 8.7: *The pH change in a buffer ($pK_a = 7.0$) that is 0.1 M in HA and A^- will be calculated for the addition of 0.01 mol/L and 0.05 mol/L of OH^-.*

Solution:

For simplicity, it will be assumed that there is no change in buffer volume as a result of the OH^- addition. Thus, the addition of 0.01 mol/L OH^- will reduce AH to 0.09 M and increase A^- to 0.11 M, while the 0.05 mol/L increase in OH^- will result in concentrations of 0.15 M A^- and 0.05 M AH.

The pH values will be

0.01 mol/L OH^- added:

$$pH = 7.0 + \log\left[\frac{0.11}{0.09}\right]$$

$$pH = 7.0 + \log 1.222 = 7.0 + 0.087 = 7.09.$$

0.05 mol/L OH^- added:

$$pH = 7.0 + \log\left[\frac{0.15}{0.05}\right]$$

$$pH = 7.0 + \log 3 = 7.0 + 0.48 = 7.48.$$

It can be seen from Ex. 8.7 that the addition of hydroxyl ion changed the pH of the buffer. However, the increase in pH was much less than would have occurred had the same quantity of OH^- been added to pure water at pH 7 as illustrated in Ex. 8.8.

Ex. 8.8: *The effect of OH^- additions of 0.01 mol/L and 0.05 mol/L on the pH of pure water initially at pH 7 will be calculated.*

Solution:
0.01 mol/L OH^- added:

$$10^{-7}\,OH^- + 10^{-2}\,\text{mol/L}\,OH^- = 10^{-1.999}\,\text{M}\,OH^-$$

$$pOH = 2$$

$$pH = 12$$

0.05 mol/L OH^- added:

$$10^{-7}OH^- + 0.05\,\text{mol/L}\,OH^- = 10^{-1.30}\,\text{M}\,OH^-$$

$$pOH = 1.30$$

$$pH = 12.7.$$

Carbon dioxide, bicarbonate, and carbonate buffer waters against sudden changes in pH. Water with low alkalinity will exhibit a greater pH fluctuation during a 24-h period as a result of fluctuations in carbon dioxide concentration caused by photosynthesis and respiration than will water of greater alkalinity. At pH below 8.3, if hydrogen ion is added, it reacts with bicarbonate to form carbon dioxide and water so that the pH changes only slightly. A small addition of hydroxyl ion will reduce the hydrogen ion concentration, but carbon dioxide and water react to form more hydrogen ion, thereby minimizing change in pH. The buffer system in natural water for pH below 8.3 may be expressed in terms of the Henderson–Hasselbalch equation (8.25) as follows:

$$\text{pH} = 6.35 + \log_{10}\frac{(\text{HCO}_3^-)}{(\text{CO}_2)} \tag{8.26}$$

where 6.35 is the negative logarithm—or pK—of the reaction constant for (8.11). Notice that in terms of buffers, carbon dioxide is the acid and the bicarbonate ion is the salt or conjugate base.

At pH 8.34 and above, when hydrogen ion is added, it will react with carbonate ion to form bicarbonate. Adding of hydroxide results in a reaction with bicarbonate to form carbonate and water. Putting (8.10) into the Henderson–Hasselbalch equation form gives

$$\text{pH} = 10.33 + \log_{10}\frac{(\text{CO}_3^{2-})}{(\text{HCO}_3^-)} \tag{8.27}$$

Alkalinity and pH of Natural Waters

Total alkalinity concentrations for natural waters may range from 0 mg/L to more than 500 mg/L. The total alkalinity of water may be categorized by concentration as follows:

Less than 10 mg/L	Very low
10–50 mg/L	Low
50–150 mg/L	Moderate
150–300 mg/L	High
More than 300 mg/L	Very high

Waters of moderate to high total alkalinity often are associated with limestone deposits in watershed soils. For example, total alkalinities of waters from different physiographic regions of the Coastal Plain region of Alabama and Mississippi ranged from 2 to 200 mg/L. The lower values consistently were associated with sandy soils and the higher values were for areas where soils contained free calcium carbonate (Boyd and Walley 1975). Livingstone (1963) did not present alkalinity data in his survey of the chemical composition of the world's rivers and lakes, but his bicarbonate data allow inference about alkalinity. Livingstone's data suggest that areas with weakly-developed or highly-leached soils have very low or low alkalinity, humid regions with abundant limestone in watershed soils have moderate alkalinity, and areas with semi-arid or arid climate have moderate to high alkalinity waters. Seawater has an average total alkalinity of 116 mg/L. Alkalinity is very high in closed basin lakes. Waters from places with organic soils tend to have very low or low alkalinity.

The alkalinity of water is closely related to pH. In general, pH tends to increase as alkalinity increases. Waters that do not contain measurable alkalinity have pH values below 5, those with very low or low alkalinity tend to have pH values between 5 and 7, pH usually is between 7 and 8.5 in moderate and high alkalinity waters, and waters of very high alkalinity may have pH of 9 or more. The influence of photosynthesis on pH is greater in low alkalinity waters because of their low buffering capacity. In waters of very low to moderate alkalinity, a dense phytoplankton bloom may cause afternoon pH to rise above 9. At night, pH will usually decline to a much lower value in such waters.

Along coastal plains in several countries including the United States, there are areas where water percolates through calcium carbonates deposit and accumulates moderate to high concentrations of calcium and bicarbonate. This water then enters an aquifer that contains a lot of sodium from having been saturated with seawater in the geological past. Calcium in the water recharging the aquifer is exchanged for sodium in the geological matrix of the formation. The water removed from the aquifer in wells will be high in alkalinity and sodium, but low in calcium. Hem (1970) referred to this process as natural softening of groundwater. When such water is used to fill fish ponds, an extremely high pH that may be toxic to fish can result from the effects of photosynthesis (Mandal and Boyd 1980).

Acidity

The pH is an index of the intensity of hydrogen ions. The acidity of water refers to its reserve capacity to generate additional hydrogen ions through various processes. Thus, pH is an intensity factor and acidity is a capacity factor. The relationship between pH and acidity is similar to the relationship of temperature to heat content. Temperature is a measure of the energy or thermal activity of a substance while heat content represents the amount of heat stored in the mass of the substance. In order to reduce the temperature of the mass of the substance, the internal heat content must be lowered. In a similar way, the acidity of water must be neutralized to raise the pH.

It is convenient to think of water as having both alkalinity and acidity. Carbon dioxide is an acid; waters with pH below 8.3 contain free carbon dioxide and can be said to have acidity. The total acidity is the total titratable acids in water, and it is measured by titrating the water to pH 8.3 with a standard base. As with alkalinity, the results are reported in terms of equivalent calcium carbonate.

Where the pH of water is below 4.5, the source of acidity is an acid stronger than carbon dioxide, because carbon dioxide usually does not depress pH below 4.5 (Fig. 8.1). The exception of course is carbonated beverages in which a large amount of carbon dioxide is added under pressure. This is the reason that these beverages "fizz" when opened—especially if they are warm or shaken. Waters with pH below 4.5 are said to contain mineral acidity. Mineral acidity is measured by titrating the water to pH 4.5 with standard base and reporting results in terms of equivalent calcium carbonate.

Ex. 8.9: *A 100-mL water sample of pH 3.1 is titrated with 0.022 N NaOH to pH 4.5 and then the titration is continued to pH 8.3. The volume of acid to titrate to pH 4.5 is 3.11 mL, and the total acid used in the titration is 3.89 mL. The total acidity and mineral acidity will be estimated.*

Solution:

The same approach used for alkalinity calculation (Ex. 8.5) can be used here because results will again be reported as equivalent $CaCO_3$. The total acidity as meq $CaCO_3$ equals the meq base used in the entire titration:

$$(3.89\,\mathrm{mL})(0.022\,\mathrm{N}) = 0.0856\,\mathrm{meq}\ CaCO_3.$$

The weight equivalent $CaCO_3$ is

$$(0.0856\,\mathrm{meq}\ CaCO_3)(50.04\,\mathrm{mg}\ CaCO_3/\mathrm{meq}) = 4.28\,\mathrm{mg}\ CaCO_3.$$

To convert to milligrams per liter of $CaCO_3$ in sample,

$$(4.28\,\mathrm{mg}\ CaCO_3)\left(\frac{1,000\,\mathrm{mL/L}}{100\,\mathrm{mL}}\right) = 42.8\,\mathrm{mg/L}\ CaCO_3.$$

The same procedure is repeated for the volume of base used for titration to pH 4.5 to obtain the mineral acidity. The result is 34.2 mg/L $CaCO_3$.

The sample had 42.8 mg/L total acidity and 34.2 mg/L mineral acidity—the difference 8.6 mg/L—represents acidity of carbon dioxide in the sample.

Mineral acidity can be visualized as the amount of calcium carbonate needed to raise the pH of the water to 4.5, so it can be thought of as negative alkalinity. The difference in total acidity and mineral acidity is the acidity resulting from free carbon dioxide.

Significance

pH

The pH of water is important in aquatic ecosystems, because it affects aquatic life. Gill tissue is the primary target organ in fish affected by excessively low pH. When fish are exposed to low pH, the amount of mucus on gill surfaces increases. Excess mucus interferes with the exchange of respiratory gases and ions across the gill. Therefore, failure in blood acid–base balance resulting in respiratory stress and decreasing blood concentration of sodium chloride which cause osmotic disturbance are the dominant physiologic symptoms of acid stress. Of course, at low pH, aluminum ion concentration increases in water, and sometimes, toxic effects of aluminum may occur in addition to pH effects.

Gill damage in fish also can occur in alkaline solutions (high pH). Mucus cells at the base of the gill filaments become hypertrophic and the gill epithelium separates from the pilaster cells. Gill damage contributes to problems with respiration and blood acid–base balance. Damage to the lens and cornea of the eye also occurs in fish at high pH (Boyd and Tucker 2014).

The relationship of pH to aquatic animals is summarized in Table 8.4. The acid and alkaline death points are approximately pH 4 and pH 11. Waters with pH values ranging from about 6.5 to 9 are most suitable for aquatic life. In fish and some other aquatic animals, reproduction diminishes as pH declines below 6.5. Most authorities suggest that aquatic ecosystems should be protected from acidic or basic pollutants so that pH remains between 6.5 and 8.5 or 9.0. Of course, many waters may naturally be lower in pH than 6.5 and a few may be higher than 9.0 in pH.

The pH also influences the use of water by humans. Waters with excessively low or high pH will be corrosive. Where water supplies are acidic, liming with

Table 8.4 Effects of pH on fish and other aquatic life

pH	Effects
4	Acid death point
4–5	No reproduction[a]
4–6.5	Slow growth of many species[a]
6.5–9	Optimum range
9–11	Slow growth and adverse reproductive effects
11	Alkaline death point

[a]Some fish in rivers flowing from jungles do very well at low pH

agricultural limestone (crushed limestone), calcium oxide, or calcium hydroxide often is practiced to increase the pH and minimize pipe corrosion. Drinking waters should have pH of 6.5–8.5.

Carbon Dioxide

High concentrations of carbon dioxide have a narcotic effect on fish and even higher concentrations may cause death. Environmental concentrations are seldom high enough to elicit narcotic effects or death; the usual effect is a respiratory one. Carbon dioxide must leave a fish or invertebrate by diffusion from gills, and high external concentration in the surrounding water decrease the rate of loss of carbon dioxide. Thus, carbon dioxide accumulates in the blood and depresses blood pH causing detrimental effects. More importantly, high carbon dioxide concentrations interfere with the loading of hemoglobin in fish blood with oxygen. This results in an elevation of the minimum tolerable dissolved oxygen concentration. In this regard, when dissolved oxygen concentrations in natural waters are low, carbon dioxide concentrations almost always are high.

Fish can sense small differences in free carbon dioxide concentrations and apparently attempt to avoid areas with high concentrations. Nevertheless, 10 mg/L or more of carbon dioxide may be tolerated provided dissolved oxygen concentration is high. Most species will survive in waters containing up to 60 mg/L of free carbon dioxide. Water supporting good fish populations normally contained less than 5 mg/L of free carbon dioxide (Ellis 1937). In low alkalinity, eutrophic waters, free carbon dioxide typically fluctuates from 0 mg/L in the afternoon to 5 or 10 mg/L at daybreak with no obvious ill effect on aquatic organisms.

Carbon dioxide is necessary in photosynthesis, but its concentration is closely related to pH and total alkalinity. At the same pH, a water with a greater alkalinity can contain more available carbon for photosynthesis than a water with lesser alkalinity. At the same alkalinity, the availability of free carbon dioxide (gaseous form) will decrease as pH increases, but when the pH is too high for carbon dioxide to be present, most aquatic plants can use bicarbonate as a carbon source.

Alkalinity

The alkalinity of water is an important variable as productivity is related to alkalinity because of the relationship between alkalinity, pH, and carbon availability. Waters with total alkalinity values of 0–50 mg/L usually are less productive than those with total alkalinity concentrations of 50–200 mg/L (Moyle 1946). Liming sometimes may be used to increase the alkalinity of water. As a general rule, the availability of inorganic carbon for photosynthesis increases as alkalinity rises, because aquatic plants can use bicarbonate as a carbon source (Raven et al. 2012).

Alkalinity has a great influence on the use of water. Total alkalinities up to 400–500 mg/L are sometimes used for public water supplies, but waters with a high carbonate hardness (high alkalinity and high hardness) will be problematic in causing deposits when heated.

Acidity

Water containing mineral acidity—pH below 4.5—is unsuitable for most purposes. The main value of the acidity determination is to provide an index of the degree of acidity. Acidity also allows an estimate of the amount of base needed to neutralize an acidic water.

References

Akin GW, Lagerwerff JV (1965) Calcium carbonate equilibria in aqueous solution open to the air. I. The solubility of calcite in relation to ionic strength. Geochim Cosmochim Acta 29:343–352

Boyd CE, Lawrence JM (1966) The mineral composition of several freshwater algae. Proc Annu Conf Southeast Assoc Game Fish Comm 20:413–424

Boyd CE, Tucker CS (2014) Handbook for aquaculture water quality. Craftmaster Printers, Auburn

Boyd CE, Walley WW (1975) Total alkalinity and hardness of surface waters in Alabama and Mississippi. Bulletin 465, Alabama Agricultural Experiment Station, Auburn University, Auburn

Boyd CE, Tucker CS, Viriyatum R (2011) Interpretation of pH, acidity and alkalinity in aquaculture and fisheries. N Am J Aquac 73:403–408

Cavalli G, Riis T, Baattrup-Pedersen A (2012) Bicarbonate use in three aquatic plants. Aquat Bot 98:57–60

Eaton AD, Clesceri LS, Rice EW, Greenburg AE (eds) (2005) Standard methods for the examination of water and wastewater. American Public Health Association, Washington, DC

Ellis MM (1937) Detection and measurement of stream pollution. US Bur Fish Bull 22:367–437

Frear CL, Johnston J (1929) The solubility of calcium carbonate (calcite) in certain aqueous solutions at 25°C. J Am Chem Soc 51:2082–2093

Garrels RM, Christ CL (1965) Solutions, minerals, and equilibria. Freeman, Cooper, and Company, San Francisco

Hem JD (1970) Study and interpretation of the chemical characteristics of natural water. Water-supply paper 1473, United States Geological Survey, United States Government Printing Office, Washington, DC

Ittekkot V (2003) Geochemistry. A new story from the Ol' Man River. Science 301:56–58

Korb RE, Saville PJ, Johnston AM, Raven JA (1997) Sources of inorganic carbon for photosynthesis by three species of marine diatom. J Phycol 33:433–440

Livingstone DA (1963) Chemical composition of rivers and lakes. Professional paper 440-G, United States Government Printing Office, Washington, DC

Mandal BK, Boyd CE (1980) The reduction of pH in water of high total alkalinity and low total hardness. Prog Fish Cult 42:183–185

Moyle JB (1946) Some indices of lake productivity. Trans Am Fish Soc 76:322–334

Raven JA, Giordano M, Beardall J, Maberly SC (2012) Algal evolution in relation to atmospheric CO_2: carboxylases, carbon-concentrating mechanisms and carbon oxidation cycles. Philos Trans R Soc Lond B Biol Sci 367:493–507

Ruttner F (1963) Fundamentals of limnology. University of Toronto Press, Toronto

Saunders GW, Trama FB, Bachmann RW (1962) Evaluation of a modified C-14 technique for shipboard estimation of photosynthesis in large lakes. Publication No. 8, Great Lakes Research Division, University of Michigan, Ann Arbor

Snoeyink VL, Jenkins D (1980) Water chemistry. John Wiley, New York

Taras MJ, Greenberg AE, Hoak RD, Rand MC (1971) Standard methods for the examination of water and wastewater, 13th edn. American Public Health Association, Washington, DC

Thompson JB, Schultzepam S, Beveridge TJ, Des Marais DJ (1997) Whiting events: biogenic origin due to the photosynthetic activity of cyanobacterial picoplankton. Limnol Oceanogr 42:133–141

Total Hardness

9

Abstract

The total hardness of water results from divalent cations—mainly from calcium and magnesium—expressed as equivalent calcium carbonate. The total hardness equivalence of 1 mg/L calcium is 2.5 mg/L, while 1 mg/L magnesium equates to 4.12 mg/L. Hardness and alkalinity often are similar in concentration in waters of humid regions, but hardness frequently exceeds alkalinity in waters of arid regions. Hardness generally is less important than alkalinity as a biological factor, but it is quite important in water supply and use. High concentrations of calcium and magnesium in water containing appreciable alkalinity lead to scale formation when the water is heated or its pH increases. This leads to clogging of water pipes and scale accumulation in boilers and on heat exchangers. The Langelier Saturation Index often is used to determine if water has potential to cause scaling. Divalent ions also precipitate soap increasing soap use for domestic purposes and in commercial laundries. The traditional method for softening water is to precipitate calcium as calcium carbonate and magnesium as magnesium hydroxide by the lime-soda ash process. Water also may be softened by passing it through a cation exchange medium such as zeolite.

Keywords

Sources of hardness • Types of hardness • Hardness concentration categories • Water softening • Calcium carbonate saturation

Introduction

Some waters do not produce appreciable lather with soap. When such waters are boiled, a scaly residue will be left in the container. These waters are called hard waters, and their failure to produce copious lather results from precipitation of soap by minerals in the water. The effect is illustrated by the precipitation of a common soap, sodium stearate, by calcium in hard water. Sodium stearate dissolves into

C.E. Boyd, *Water Quality*, DOI 10.1007/978-3-319-17446-4_9

stearate ions and sodium ions, but in hard water, stearate ions are precipitated as calcium stearate.

$$2C_{17}H_{35}COO^{-} + Ca^{2+} \rightarrow (C_{17}H_{35}COO)_2Ca \downarrow. \tag{9.1}$$

The scale in the container in which hard water has been boiled results from loss of carbon dioxide to the air by boiling and the precipitation of calcium carbonate

$$Ca^{2+} + 2HCO_3^{-} \underset{\Delta}{\rightarrow} CaCO_3 \downarrow + CO_2 \uparrow + H_2O. \tag{9.2}$$

Scales also can form in water pipes—especially hot water pipes—partially clogging them and lessening flow. Heat exchangers can be rendered less effective by calcium carbonate forming on their surfaces.

Because hardness affects the use of water for domestic and industrial purposes, there has been a tremendous interest in this water quality property for a long time. It is important for those involved in water quality and water supply to have a clear understanding of total hardness.

Sources

The divalent cations—calcium, magnesium, strontium, ferrous iron, and manganous manganese—cause hardness in water. Surface waters usually are oxygenated and will not contain divalent iron or manganese, and few inland waters contain more than 1 or 2 mg/L of strontium. Thus, calcium and magnesium are the major sources of hardness in nearly all surface waters. Groundwater usually contains reduced iron and manganese (Fe^{2+} and Mn^{2+}) because it is anaerobic, and some may have appreciable iron and manganese hardness.

A primary source of calcium and magnesium in water is the dissolution of limestone and calcium silicate. Surface water acquires hardness from contacting these minerals in rock outcrops or soils. As discussed in Chap. 8, carbon dioxide in rainwater accelerates the dissolution of limestone and calcium silicate. As water infiltrates downward into the soil, it dissolves more carbon dioxide from root and microbial respiration that enhances its capacity to dissolve limestone and calcium silicate in subsurface deposits. Some aquifers are actually contained in underground limestone formations, and groundwater from such aquifers has a high hardness. The acquisition of hardness (calcium and magnesium) by the dissolution of dolomitic limestone and calcium silicate are illustrated below:

$$CaCO_3 \cdot MgCO_3 + 2CO_2 + 2H_2O = Ca^{2+} + Mg^{2+} + 4HCO_3^{-} \tag{9.3}$$

$$CaSiO_3 + 2CO_2 + 3H_2O = Ca^{2+} + 2HCO_3^{-} + H_4SiO_4. \tag{9.4}$$

In addition, when gypsum ($CaSO_4{\cdot}2H_2O$) and certain magnesium salts in some soils—particularly in arid regions—dissolve, they impart calcium and magnesium to the water. In arid regions, ions in water are concentrated by evaporation increasing hardness.

Because hardness often is derived from the dissolution of limestone or calcium silicate, concentrations of calcium and magnesium are nearly equal in chemical equivalence to the concentrations of bicarbonate and carbonate in many natural waters—but, this is not always the case. In arid regions, concentration of ions by evaporation tends to cause precipitation of alkalinity as carbonates causing hardness and alkalinity to decline. In areas with acidic soils neutralization of alkalinity may also result in greater hardness than alkalinity. Alkalinity sometimes exceeds hardness in groundwaters that have been naturally softened (see Chap. 8).

Measurement of Hardness

Calculation

If the ionic composition of a water is known, the total hardness may be calculated from the concentrations of hardness cations. As with total alkalinity, total hardness is reported in milligrams per liter of equivalent calcium carbonate as illustrated in thc following example.

Ex. 9.1: *A water sample contains 20 mg/L of calcium and 3 mg/L of magnesium. The total hardness will be calculated.*

Solution:

It is desired to express hardness as equivalent $CaCO_3$. For calcium, $1Ca^{2+} = 1CaCO_3$, 1 mol of Ca weighs 40.08 g, and 1 mol of $CaCO_3$ weighs 100.08 g. Thus, the hardness equivalence of Ca^{2+} is

$$Ca^{2+} \times \frac{100.08}{40.08} = 2.50$$

and 20 mg Ca/L × 2.50 = 50 mg/L of $CaCO_3$.

For magnesium 1 $Mg^{2+} = 1CaCO_3$ and 1 mol Mg^{2+} weighs 24.31 g. Thus, the hardness equivalence of Mg^{2+} is

$$Mg^{2+} \times \frac{100.08}{24.31} = 4.12$$

and 3 mg Mg^{2+}/L × 4.12 = 12.4 mg/L of $CaCO_3$.

The sum of the calcium carbonate equivalence of calcium and magnesium, or the total hardness, is 62.4 mg/L.

The factors for converting other divalent ions to calcium carbonate equivalence are: Fe^{2+}, 1.79; Mn^{2+}, 1.82; Sr^{2+}, 1.14.

Titration

It is more common to measure the concentration of total hardness directly by titration with the chelating agent ethylenediaminetetraacetic acid (EDTA) rather than to calculate it (Eaton et al. 2005). This titrating agent forms stable complexes with divalent cations as illustrated for calcium:

$$Ca^{2+} + EDTA = CaEDTA. \tag{9.5}$$

Each molecule of EDTA complexes one divalent metal ion, and the endpoint is detected with the indicator eriochrome black T. The indicator combines with a small amount of calcium in the sample and holds it rather tightly creating a wine-red color. When all Ca^{2+} in the water sample has been chelated, the EDTA strips the Ca^{2+} from the indicator and the solution turns blue. The endpoint is very sharp and distinct. Each mole of EDTA equals 1 mol of $CaCO_3$ equivalence. Example 9.2 shows the calculation of hardness from EDTA titration.

Ex. 9.2: *Titration of a 100-mL water sample requires 12.55 mL of 0.0095 M EDTA solution. The total hardness will be calculated.*

Solution:
Each mole of EDTA equals a mole of $CaCO_3$, and

$$(12.55\,\text{mL})(0.0095\,\text{M}) = 0.119\,\text{mM}\ CaCO_3$$

$$0.119\,\text{mM} \times 100.08\,\text{mg}\ CaCO_3/\text{mM} = 11.91\,\text{mg}\ CaCO_3$$

$$11.91\,\text{mg}\ CaCO_3 \times \frac{1{,}000\,\text{mL/L}}{100\,\text{mL}} = 119\,\text{mg/L}\ \textit{total hardness as equivalent}\ CaCO_3.$$

The steps in the solution shown in Ex. 9.2 can be combined into an equation for estimating hardness from EDTA titration:

$$\text{Total hardness (mg/L CaCO}_3) = \frac{(\text{M})(\text{V})(100{,}080)}{\text{S}} \tag{9.6}$$

where M = molarity of EDTA, V = volume of EDTA (mL), and S = sample volume (mL).

Hardness of Natural Waters

Concentration

The hardness of inland waters may be around 5–75 mg/L in humid areas with highly-leached, acidic soil, 150–300 mg/L in humid areas with calcareous soil, and 1,000 mg/L or more in arid regions. Average ocean water contains about 1,350 mg/

L magnesium and 400 mg/L calcium (Table 4.8); its calculated hardness is about 6,562 mg/L. Hardness of water for water supply purposes often is classified as follows:

Less than 50 mg/L	Soft
50–150 mg/L	Moderately hard
150–300 mg/L	Hard
More than 300 mg/L	Very hard

This is similar to the classification given in Chap. 8 for total alkalinity.

Types of Hardness

The hardness of water sometimes is separated into calcium hardness and magnesium hardness. This usually can be done by subtracting calcium hardness from total hardness to give magnesium hardness (magnesium hardness is expressed as $CaCO_3$). Calcium hardness can be estimated from calcium concentration or measured directly by EDTA titration using murexide indicator in a sample from which the magnesium has been removed by precipitation at high pH (Eaton et al. 2005). The murexide indicator is stable at the high pH necessary to precipitate magnesium, but it functions in the same manner as eriochrome black T.

Hardness also can be separated into carbonate and noncarbonate hardness. In waters where total alkalinity is less than total hardness, carbonate hardness is equal to total alkalinity. This results because there are more milliequivalents of hardness cations than of the alkalinity anions, carbonate and bicarbonate. In waters where the total alkalinity is equal to or greater than the total hardness, the carbonate hardness equals the total hardness. In such a water, the milliequivalents of alkalinity anions equal or exceed the milliequivalents of hardness cations, calcium and magnesium. Examples of waters with total alkalinity less than total hardness, and waters with total alkalinity equal or greater than total hardness are provided in Table 9.1. Where total alkalinity is less than total hardness, there is a large amount of sulfate and chloride relative to carbonate and bicarbonate. In the water where total alkalinity is

Table 9.1 Hardness fractions and ionic concentrations in waters with and without non-carbonate hardness

Variable	Sample A	Sample B
Total alkalinity (mg $CaCO_3$/L)	52.2	162.5
Total hardness (mg $CaCO_3$/L)	142.5	43.1
Carbonate hardness (mg $CaCO_3$/L)	52.5	43.1
Non-carbonate hardness (mg $CaCO_3$/L)	90.0	0.0
$HCO_3^- + CO_3^{2-}$ (meq/L)	1.04	3.25
$SO_4^{2-} + Cl^-$ (meq/L)	2.64	0.66
$Ca^{2+} + Mg^{2+}$ (meq/L)	2.85	0.86
$Na^+ + K^+$ (meq/L)	0.83	3.05

greater than total hardness, there is a large amount of sodium and potassium relative to calcium and magnesium.

If a water containing carbonate hardness is heated, carbon dioxide is driven off and calcium carbonate, magnesium carbonate or both precipitate as was shown for calcium carbonate precipitation in (9.2). Sulfate, chloride, sodium, and potassium are not precipitated by boiling, but the carbonate hardness can be removed from water by precipitation by boiling for a sufficient time. Because it can be removed by heating, carbonate hardness is called temporary hardness. Of course, boiling will not remove all of the hardness or alkalinity from a water with no noncarbonate hardness because calcium and magnesium carbonates have a degree of solubility.

Carbonate hardness is responsible for boiler scale. Hardness remaining in water after boiling is the permanent hardness. Because bicarbonate and carbonate precipitate with part of the calcium and magnesium during boiling, the permanent hardness also is known as noncarbonate hardness, and

$$\text{Noncarbonate hardness} = \text{total hardness} - \text{carbonate hardness}. \qquad (9.7)$$

Water Softening

Two methods have traditionally been used for softening water—the lime-soda ash process and the zeolite process (Sawyer and McCarty 1978). The lime-soda ash procedure involves treating water with calcium hydroxide to remove calcium carbonate and to convert magnesium bicarbonate to magnesium carbonate:

$$Ca(OH)_2 + CO_2 = CaCO_3\downarrow + H_2O \qquad (9.8)$$

$$Ca(OH)_2 + Ca(HCO_3)_2 = 2CaCO_3\downarrow + 2H_2O \qquad (9.9)$$

$$Ca(OH)_2 + Mg(HCO_3)_2 = MgCO_3 + CaCO_3\downarrow + 2H_2O. \qquad (9.10)$$

The magnesium carbonate is soluble, but it will react with lime to precipitate magnesium hydroxide:

$$Ca(OH)_2 + MgCO_3 = CaCO_3\downarrow + Mg(OH)_2\downarrow. \qquad (9.11)$$

Lime also removes any magnesium associated noncarbonate hardness:

$$Ca(OH)_2 + MgSO_4 = CaSO_4\downarrow + Mg(OH)_2\downarrow. \qquad (9.12)$$

Soda ash is then added to remove the calcium associated noncarbonate hardness:

$$Na_2CO_3 + CaSO_4 = Na_2SO_4 + CaCO_3\downarrow. \qquad (9.13)$$

The water contains excess lime after treatment and has a high pH. Carbon dioxide is added to remove the lime and reduce the pH to an acceptable level. The total hardness concentration following lime-soda ash treatment is 50–80 mg/L.

Zeolite is an aluminosilicate mineral of high cation exchange capacity that is used as an ion exchanger. Natural zeolite deposits occur in many locations and may be mined as a source of zeolites for water softening or other purposes, but synthetic zeolites also are widely used for the same purposes. The cation exchange sites on zeolites for softening water was saturated with sodium by exposure to a sodium chloride solution. Water is passed through a zeolite bed and calcium and magnesium ions in the water are exchanged with sodium on the zeolite resulting in softening of the water as illustrated below:

		Zeolite column				
Inflowing water		Zeolite		Zeolite		Outflowing water
$2Ca^{2+}$, Mg^{2+}, $5HCO_3^-$, Cl^-	$\rightarrow$	$6Na^+$	$\rightarrow$	$2Ca^{2+}$, Mg^{2+}	$\rightarrow$	$5HCO_3^-$, Cl^-, $6Na^+$

(9.14)

When the zeolite bed has exchanged so much sodium for calcium and magnesium that it becomes ineffective, its water softening capacity is regenerated by backwashing with a concentrated sodium chloride solution.

Calcium Carbonate Saturation

The degree of calcium carbonate saturation of water affects the likelihood of calcium carbonate precipitation or scaling by a water. Several indices of scaling potential have been formulated, and one of the most popular is the Langelier Saturation Index (Langelier 1936). The full explanation of this index is beyond the scope of this book but the equation is

$$\mathrm{LSI} = \mathrm{pH} - \mathrm{pH_{sat}} \tag{9.15}$$

where LSI = Langelier Saturation Index; pH = pH of water; pH_{sat} = pH at calcium carbonate saturation of the water. The equation for pH_{sat} is

$$\mathrm{pH_{sat}} = (9.3 + A + B) - (C + D) \tag{9.16}$$

where $A = (\log \mathrm{TDS} - 1)/10$, $B = [-13.12 \times \log\ (°C + 273)] + 34.55$, $C = \log\ (Ca^{2+} \times 2.5) - 0.4$, and $D = \log$ (alkalinity as $CaCO_3$).

For waters with TDS concentration <500 mg/L, a simplified equation (Gebbie 2000) may be used for pH_{sat}:

$$\mathrm{pH_{sat}} = 11.5 - \log\left(Ca^{2+}\right) - \log\,(\mathrm{Alkalinity}). \tag{9.17}$$

At LSI = 0, the water is saturated with calcium carbonate. Waters with a negative LSI are undersaturated with calcium carbonate and potentially corrosive (see Chap. 7), because the precipitation of calcium carbonate over a surface protects from corrosion. When the LSI is positive, the water is not corrosive, but it can cause scaling that can be problematic. Calculation of the LSI is illustrated in Ex. 9.3.

Ex. 9.3: *Calculation of the Langelier Saturation Index for a water at 20 °C with pH 7.7, 20 mg/L Ca^{2+}, 75 mg/L total alkalinity, and 375 mg/L total dissolved solids.*

Solution:
The pH_{sat} value is calculated with (9.16)

$A = (\log 375 - 1)/10 = 0.16$
$B = [-13.12 \times \log (20 + 273)] + 34.55 = 2.18$
$C = \log (20 \times 2.5) - 0.4 = 1.30$
$D = \log (75) = 1.88$

$$pH_{sat} = (9.3 + 0.16 + 2.18) - (1.30 + 1.88) = 8.46.$$

The LSI is computed using (9.15)

$$LSI = 7.7 - 8.46 = -0.76.$$

The simplified equation (9.17) could be used instead of (9.16) because TDS is <500 mg/L. The result is

$$pH_{sat} = 11.5 - \log 20 - \log 75 = 8.32$$

and

$$LSI = 7.7 - 8.32 = -0.62.$$

The calculation by either procedure suggests that the water would not cause scaling by either procedure.

Suppose that the water referred to in Ex. 9.3 is in a lake that develops a dense plankton bloom in summer and pH rises to 8.6 and the water temperature increases to 28 °C. The LSI would be +0.29 by the Langelier method for estimating pH_{sat} (9.16) and +0.28 by the Gebbie (2000) shortcut method for estimating pH_{sat} (9.17). At the higher pH and water temperature, the water would have potential to cause scaling.

Significance

The biological productivity of natural freshwaters will increase with greater hardness up to a concentration of 150–200 mg/L (Moyle 1946, 1956). However, in reality, hardness *per se* has less biological significance than does alkalinity. Productivity depends upon the availability of carbon dioxide, nitrogen, phosphorus, and other nutrients, a suitable pH range, and many other factors. As a rule, alkalinity tends to increase along with concentration of other dissolved ions where watershed soils are not highly acidic and fertile. Hardness and alkalinity often increase in proportion to each other, and in many waters of humid regions, they may be roughly equal. Nevertheless, for analytical reasons, it was easier in the past to measure total hardness than to determine total alkalinity. Thus, hardness became a common index of productivity. Today, there is no reason to follow this tradition, and indices of productivity in aquatic ecosystems should be based on total alkalinity and other variables.

Hardness is an important factor regarding the use of water for many purposes. Hardness of water is manifest most commonly by the amount of soap needed to produce suds. Hardness might be called the soap-wasting property of water, because no suds will be produced in a hard water until the minerals causing the hardness have been removed from the water by combining with the soap. The material that is removed by the soap is evident as an insoluble scum—familiar ring on the bathtub—that forms during bathing in some waters.

Waters with a hardness of less than 50 mg/L are considered soft. A hardness of 50–150 mg/L is not objectionable for most purposes, but the amount of soap needed increases with hardness. Laundries or other industries using large quantities of soap generally find it profitable to lower hardness concentrations to about 50 mg/L. Water having 100–150 mg/L hardness will deposit considerable scale in steam boilers. Hardness of more than 150 mg/L is decidedly noticeable. At levels of 200–300 mg/L or higher, it is common practice to soften water for household uses. Where municipal water supplies are softened, the hardness is reduced to about 85 mg/L. Further softening of a whole public water supply is not considered economical.

Where scale forms when water is heated, calcium carbonate scale deposits first, because it is more insoluble than magnesium carbonate. In the absence of carbon dioxide, water will carry only about 14 mg/L calcium carbonate in solution. Under the same conditions, the solubility of magnesium carbonate is more than five times as great, or about 80 mg/L.

References

Eaton AD, Clesceri LS, Rice EW, Greenburg AE (eds) (2005) Standard methods for the examination of water and wastewater. American Public Health Association, Washington

Gebbie P (2000) Water stability—what does it mean and how do you measure it? In: Proceedings of 63rd annual water industry engineers and operators conference, Warrnambool, Australia, pp 50–58. http://wioa.org.au/conference_papers/2000/pdf/paper7.pdf

Langelier WF (1936) The analytical control of anti-corrosion water treatment. J Am Water Works Assoc 28:1500–1521

Moyle JB (1946) Some indices of lake productivity. Trans Am Fish Soc 76:322–334

Moyle JB (1956) Relationships between the chemistry of Minnesota surface waters and wildlife management. J Wildl Manag 20:303–320

Sawyer CN, McCarty PL (1978) Chemistry for environmental engineering. McGraw-Hill, New York

Microorganisms and Water Quality 10

Abstract

Phytoplankton and bacteria have a greater effect on water quality than do other aquatic microorganisms. Phytoplankton are the main primary producers while bacteria are responsible for the majority of organic matter decomposition and nutrient recycling. An overview of microbial growth, photosynthesis, and respiration is provided, and methods for measuring primary production and respiration in water bodies are discussed. The combined physiological activities of producer and decomposer organisms in water bodies cause pH and dissolved oxygen concentration to increase and carbon dioxide concentration to decrease in daytime, while the opposite occurs during nighttime. In unstratified water bodies, aerobic conditions usually exist in the water column and at the sediment-water interface. Nevertheless, sediment typically is anaerobic at depths greater than a few centimeters in oligotrophic water bodies or greater than a few millimeters in eutrophic water bodies. In anaerobic sediment (or water), the metabolic activity of chemotrophic bacteria is important in decomposing organic compounds resulting from fermentation. Although chemotrophic bacteria are beneficial in assuring more complete decomposition of organic matter, toxic metabolic wastes—particularly nitrite and hydrogen sulfide—produced by these microorganisms can enter the water column. Blue-green algae—often called cyanobacteria—tend to dominate phytoplankton communities in eutrophic waters. Blue-green algae can cause surface scums and shallow thermal stratification, be toxic to other algae and aquatic animals, or produce taste and odor problems in public water supplies.

Keywords

Types of aquatic microorganisms • Aerobic and anaerobic respiration • Photosynthesis by aquatic plants • Carbon and oxygen cycles • Chlorination

C.E. Boyd, *Water Quality*, DOI 10.1007/978-3-319-17446-4_10

Introduction

Aquatic ecosystems contain a wide variety of microorganisms to include algae, bacteria, fungi, protozoa, rotifers, bryozoa, and arthropods. Algae are the primary producers, and bacteria and fungi are the decomposers. Microscopic animals feed on other small plants and animals and dead organic matter; they serve as a link between primary producers and larger animals in the food web of aquatic ecosystems. Although all of these organisms are important ecologically, phytoplankton and bacteria have a greater impact on water quality than do other microorganisms. Phytoplankton produce large amounts of organic matter through photosynthesis and release prodigious quantities of oxygen into water during the process. Bacteria decompose organic matter thereby transforming and recycling nutrients. Respiration by bacteria and respiration and photosynthesis by phytoplankton have a pronounced effect on pH and concentrations of carbon dioxide and dissolved oxygen in water. Certain bacteria and other microscopic organisms are pathogenic, and some species of algae can impart bad tastes and odors to water as well as to the flesh of fish and other aquatic food animals.

Although bacteria and phytoplankton are the focus of this chapter, much of the discussion applies also to other primary producer and decomposer organisms.

Bacteria

Bacteria are a diverse group of microscopic organisms. The nuclei of bacterial cells are not enclosed in a membrane—they are known as prokaryotic organisms as opposed to eukaryotic organisms in which the nucleus is bound by a membrane. Bacteria may be either unicellular or filamentous, and their cells usually are spherical or cylindrical (rods). They may live free in water, attached to surfaces, or in sediment. A few types are capable of locomotion. Some bacteria are pathogenic to plants, animals, and humans.

The food source for most bacteria is dead organic matter, while a few kinds of bacteria are capable of synthesizing organic matter. These two groups of bacteria are known as heterotrophic and autotrophic bacteria, respectively. Obligate aerobic bacteria cannot live without oxygen, obligate anaerobic bacteria cannot exist in environments with oxygen, and facultative anaerobic bacteria can do well with or without oxygen. Some species that can function without molecular oxygen obtain oxygen from nitrate, sulfate, carbon dioxide, or other inorganic compounds.

The major ecological role of bacteria is to decompose organic matter and recycle its essential inorganic components, e.g., carbon dioxide, water, ammonia, phosphate, sulfate, and other minerals.

Nutrients

Nutrients have three functions in growth and metabolism. They are raw materials for elaboration of biochemical compounds of which organisms are made. Carbon and nitrogen contained in organic nutrients are used in making protein, carbohydrate, fat, and other components of microbial cells. Nutrients supply energy for growth and chemical reactions. Organic nutrients are oxidized in respiration, and the energy released is used to drive chemical reactions to synthesize biochemical compounds necessary for growth and maintenance. Nutrients also serve as electron and hydrogen acceptors in respiration. The terminal electron and hydrogen acceptor for aerobic respiration is molecular oxygen. An organic metabolite or inorganic substance replaces oxygen as electron or hydrogen acceptor in anaerobic respiration.

Growth

Water and sediment contain species of microorganisms capable of decomposing almost any organic substance. Some substances decompose faster than others, but few organic compounds completely resist decay by microorganisms. Microbial activity is slow where organic matter is scarce, but actively growing microorganisms, resting spores, and other propagules are present almost everywhere. An increase in organic matter provides substrate for microbial growth, and the number of microorganisms increases.

Bacteria reproduce by binary fission. One cell divides into two cells, and the new cells continue to divide. The time between cell divisions is called the generation time or doubling time. The number of bacterial cells present after a given time (N_t) can be computed from the initial number of cells (N_o) and the number of generations (n)

$$N_t = N_o \times 2^n. \quad (10.1)$$

Microorganisms with a short generation time increase their numbers quickly (Ex. 10.1).

Ex. 10.1: *A water containing 10^3 bacterial cells per milliliter is treated with organic matter. The bacteria can double every 4 h. The number of bacterial cells after 36 h will be estimated.*

***Solution*:**
The number of generations is

$$n = \frac{36\,\text{h}}{4\,\text{h}} = 9.$$

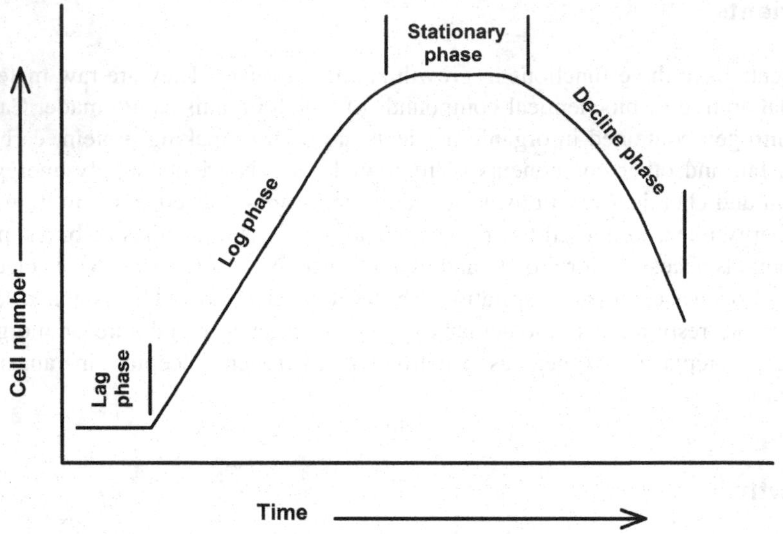

Fig. 10.1 Characteristic phases of growth in microbial cultures

The number of cells after 36 h can be calculated with (10.1).

$$N_t = 10^3 \times 2^9 = 512 \times 10^3 = 5.12 \times 10^5 cells/\text{mL} \ \ (a\ 512\,fold\ increase\ in\ 36\text{h}).$$

When bacteria are inoculated into fresh organic substrate, it takes a short time for them to adjust to the new nutrient supply. The period during which there is little or no increase in cell number is called the lag phase (Fig. 10.1). Rapid growth follows as microorganisms utilize the new substrate; this period is known as the logarithmic or log phase. After a period of rapid growth, a stationary phase is attained during which cell number remains relatively constant. As the substrate is used up and metabolic by-products accumulate, growth slows and a decline phase occurs in which the number of cells decreases.

The generation time for a bacterial population can be computed from data on cell numbers during the logarithmic phase. Equation 10.1 can be rewritten as

$$\log N_t = \log N_o + n \log 2 \tag{10.2}$$

which becomes

$$0.301\ n = \log N_t - \log N_o \tag{10.3}$$

and

$$n = \frac{\log N_t - \log N_o}{0.301}. \tag{10.4}$$

The generation time (g) is computed from the time interval (t) over which a given number of generations (n) occurred

$$g = \frac{t}{n} \text{ or } n = \frac{t}{g}. \tag{10.5}$$

Substituting $\frac{t}{g}$ for n in Eq. 10.4 gives

$$g = \frac{0.301\ t}{\log N_t - \log N_o}. \tag{10.6}$$

The utility of (10.6) is illustrated in Ex. 10.2.

***Ex. 10.2:** A bacterial culture increases from 10^4 cells/mL to 10^8 cells/mL in 12 h. The generation time will be calculated using* (10.6).

***Solution**:*

$$g = \frac{0.301(12)}{8-4} = 0.903\,\text{h} = 54\,\text{min}.$$

The generation time of bacteria differs among species, and for a given species, generation time varies with temperature, substrate availability, and other environmental factors. When bacteria are growing rapidly, they also are quickly decomposing substrate. For example, in a water where microorganisms are growing logarithmically, the evolution of carbon dioxide increases logarithmically.

Aerobic Respiration

An overview of respiration will be provided. The reader should note that aerobic respiration is much the same for all organisms, and aerobic respiration involves oxidation of organic compounds to carbon dioxide and water. The purpose of respiration is to release energy from organic compounds and store it in high-energy phosphate bonds of adenosine triphosphate (ATP)

$$\text{Adenosine diphosphate(ADP)} + PO_4 + \text{energy} \rightleftharpoons \text{ATP}. \tag{10.7}$$

Energy stored in ATP can be used to drive chemical reactions in cells. When energy is released from ATP, ADP is regenerated and used again.

In aerobic carbohydrate metabolism, the organic carbon is a glucose molecule that passes through glycolysis and the citric acid cycle (also called the Krebs cycle or the tricarboxylic acid cycle) before being completely oxidized to carbon dioxide. Glycolysis is a series of ordered reactions catalyzed by numerous enzymes—not indicated in the equations below—that transform one glucose molecule to two pyruvate molecules

$$C_6H_{12}O_6 \rightarrow 2C_3H_4O_3 + 4H^+. \quad (10.8)$$

Two moles of ATP are formed per mole of glucose in glycolysis, but no carbon dioxide is released. Glycolysis does not require oxygen.

In aerobic respiration each pyruvate molecule from glycolysis can react with coenzyme A (CoA) and nicotinamide adenine dinucleotide (NAD^+) to form one acetyl CoA molecule

$$\text{Pyruvate} + \text{CoA} + 2\text{NAD}^+ \rightleftharpoons \text{Acetyl CoA} + 2\text{NADH} + 2H^+ + CO_2. \quad (10.9)$$

The formation of acetyl CoA from pyruvate is an oxidative decarboxylation, and two H^+ ions, two electrons, and one CO_2 molecule are removed from each molecule of pyruvate—a total of four hydrogen ions and two carbon dioxide molecules per molecule of the original glucose entering glycolysis. Acetyl CoA formation links glycolysis to the citric acid cycle. Although glycolysis does not use oxygen, acetyl CoA production and the citric acid cycle require oxygen.

The first step in the citric acid cycle is the reaction of acetyl CoA and oxaloacetic acid to form citric acid. In succeeding reactions—also catalyzed by enzymes—a series of transformations of organic acids occur with release of eight H^+ ions and two CO_2 molecules for each molecule of citric acid entering the cycle. This accounts for the oxidation of the six organic carbon atoms present in one glucose molecule. One molecule of oxaloacetic acid also is regenerated in the cycle (recycled) and it can react again with acetyl CoA.

Hydrogen ions produced through oxidation of glucose in glycolysis and the citric acid cycle reduce the enzymes NAD, NADP (nicotinamide adenine dinucleotide phosphate), and FAD (flavin adenine dinucleotide). A representative reaction where H^+ is released in the citric acid cycle is

$$\underset{\text{(Isocitric acid)}}{C_6H_8O_7} \rightarrow \underset{\text{(Oxalosuccinic acid)}}{C_6H_6O_7} + 2H^+$$

and

$$\text{NADP}^+ + 2H^+ \rightleftharpoons \text{NADPH} + H^+. \quad (10.10)$$

Enzymes of the citric acid cycle are brought in contact with the electron transport system where they are reoxidized. Energy released in these oxidations is used for ATP synthesis, and the regenerated NAD, NADP, and FAD are used again as hydrogen acceptors. Reoxidation of enzymes is accomplished by a series of cytochrome enzymes that pass electrons from one cytochrome compound to another. In the electron transport system, ADP combines with inorganic phosphate to form ATP and H^+ ions combine with molecular oxygen to form water.

A total of 38 ATP molecules can be formed from 1 mol of glucose in aerobic respiration (glycolysis and citric acid cycle). The ATP molecules contain approximately one-third of the theoretical energy released during glucose oxidation. Energy not used to form ATP is lost as heat. In aerobic respiration, 1 mol of glucose consumes 6 mol of oxygen and releases 6 mol of CO_2 and water. This stoichiometry is illustrated in the summary equation for respiration.

$$C_6H_{12}O_6 + 6O_2 \rightarrow 6CO_2 + 6H_2O + \text{energy}. \quad (10.11)$$

Oxygen Stoichiometry in Aerobic Respiration

Organic matter contains a wide array of organic compounds, most of which are more complex than glucose. Bacteria excrete extracellular enzymes that break complex organic matter down into particles small enough to be absorbed. These fragments are broken down further inside the bacterial cell by enzymatic action until they are small enough to be used in glycolysis. These fragments pass through the respiratory reactions in the same way as glucose fragments.

The ratio of moles carbon dioxide produced to moles oxygen consumed is the respiratory quotient (RQ)

$$RQ = \frac{CO_2}{O_2}. \quad (10.12)$$

Reference to (10.11) reveals that in the oxidation of glucose, six oxygen molecules are consumed and six carbon dioxide molecules are released, i.e., $RQ = 1.0$. Other classes of organic matter do not contain the same ratio of carbon to oxygen to hydrogen as carbohydrate (Table 10.1), and RQ may be lesser or greater than 1.0. For example, consider the oxidation of a fat

$$C_{57}H_{104}O_6 + 80O_2 \rightarrow 57CO_2 + 52H_2O. \quad (10.13)$$

The respiratory quotient for the above reaction is $57CO_2/80O_2$ or 0.71. Organic compounds more highly oxidized (more oxygen relative to carbon) than carbohydrates will have a respiratory quotient greater than 1.0, and more reduced compounds such as fat and protein will have a respiratory quotient less than 1.0. It follows that the more reduced an organic substance, the more oxygen that is required to oxidize a unit quantity (Ex. 10.3).

Table 10.1 Average carbon, hydrogen, and oxygen content of three major classes of organic compounds

	% C	% H	% O
Carbohydrate	40	6.7	53.3
Protein	53	7	22
Fatty acids	77.2	11.4	11.4

Ex. 10.3: *Calculate the amount of oxygen required to completely oxidize the organic carbon in 1 g each of carbohydrate, protein, and fat.*

***Solution*:**
Assume that carbon concentrations are carbohydrate, 40 %; protein, 53 %; fat, 77.2 % (Table 10.1). One gram of mass represents 0.4 g C, 0.53 g C, and 0.772 g C for carbohydrate, protein, and fat, respectively. The stoichiometry for converting organic carbon to CO_2 is

$$C + O_2 \rightarrow CO_2.$$

The appropriate ratio for computing oxygen consumption is

$$\frac{\text{g C}}{12} = \frac{O_2\ used}{32}$$

$$O_2\ used = \frac{32\ (\text{g C})}{12}$$

$$O_2\ carbohydrate = \frac{(32)(0.4)}{12} = 1.07\,\text{g}$$

$$O_2\ protein = \frac{(32)(0.53)}{12} = 1.41\,\text{g}$$

$$O_2\ fat = \frac{(32)(0.772)}{12} = 2.06\ g.$$

Based on the relationship between the degree of reduction of organic compounds and their oxygen requirement for decomposition, one can see that there is not a direct proportionality between the weight of organic matter and of oxygen required for its decomposition. There is, however, a direct proportionality between the percentage carbon in organic matter and the amount of oxygen necessary to completely decompose a given weight of it. Of course, some complex organic matter is more resistant to decay than simple carbohydrates and the rate of decomposition is not always related to carbon concentration.

Anaerobic Respiration

Under anaerobic conditions, certain kinds of bacteria, yeast, and fungi, continue to respire, but in the absence of molecular oxygen, terminal electron acceptors for anaerobic respiration are organic or inorganic compounds. Carbon dioxide may be produced in anaerobic respiration, but other end products include alcohols, formate, lactate, propionate, acetate, methane, other organic compounds, gaseous nitrogen, ferrous iron, manganous manganese, and sulfide.

Fermentation is a common type of anaerobic respiration. Organisms capable of fermentation hydrolyze complex organic compounds to simpler ones that can be used in a process identical to glycolysis to produce pyruvate. In fermentation, pyruvate cannot be oxidized to carbon dioxide and water with molecular oxygen serving as the terminal electron and hydrogen acceptor. Hydrogen ions removed from organic matter during pyruvate formation are transferred via nicotinamide adenine dinucleotide to an intermediary product of metabolism.

Ethanol production from glucose has traditionally been used in general biology textbooks as an example of fermentation. In this process, glucose is converted to pyruvate, and four ATP molecules result from each molecule of glucose. Hydrogen ions removed from pyruvate are transferred to acetaldehyde with the release of carbon dioxide. The summary equations are as follows:

$$\underset{\text{(Glucose)}}{C_6H_{12}O_6} \rightarrow \underset{\text{(Pyruvate)}}{2C_3H_4O_3} + 4H^+ \tag{10.14}$$

$$2C_3H_4O_3 \rightarrow \underset{\text{(Acetaldehyde)}}{2C_2H_4O} + 2CO_2 \tag{10.15}$$

$$2C_2H_4O + 4H^+ \rightarrow \underset{\text{(Ethanol)}}{2CH_3CH_2OH}. \tag{10.16}$$

In addition to producing carbon dioxide, fermentation also may yield hydrogen gas (H_2). The hydrogen gas apparently is formed when NADH is oxidized at low hydrogen pressure with liberation of hydrogen gas.

$$NADH + H^+ \rightarrow NAD^+ + H_2. \tag{10.17}$$

Ethanol is only one of many organic compounds that can be produced by fermentation. For example, some microorganisms convert glucose to lactic acid

$$\underset{\text{Glucose}}{C_6H_{12}O_6} \rightarrow 2C_3H_4O_3 + 4H^+ \tag{10.18}$$

$$2C_3H_4O_3 + 4H^+ \rightarrow \underset{\text{(Lactic acid)}}{2C_3H_6O_3}. \tag{10.19}$$

In lactic acid production, carbon dioxide is not released as it is in ethanol production (10.15) and (10.16).

Fermentation does not oxidize organic matter completely. In ethanol production only one-third of the organic carbon in glucose is converted to carbon dioxide, and no carbon dioxide was produced in the fermentation of glucose to lactic acid. As a result, carbon dioxide and organic products accumulate in zones where fermentation is occurring. The end products of fermentation can be oxidized by microorganisms capable of using inorganic substances instead of molecular oxygen as electron acceptors.

Bacteria that use nitrate as an electron acceptor can hydrolyze complex compounds and oxidize the hydrolytic products to carbon dioxide. Nitrate is reduced to nitrite, ammonia, nitrogen gas, or nitrous oxide. In the zone where nitrate-reducing bacteria occur, a part of the organic carbon is oxidized completely to carbon dioxide and a portion is converted to organic fermentation products.

The iron- and manganese-reducing bacteria utilize oxidized iron and manganese compounds as oxidants in the same manner as nitrate-reducing bacteria use nitrate. They attack organic fermentation products and oxidize them to carbon dioxide. Ferrous iron (Fe^{2+}) and manganous manganese (Mn^{2+}) are released as by-products of respiration.

Sulfate-reducing bacteria and methane-producing bacteria cannot hydrolyze complex organic substances or decompose simple carbohydrates and amino acids originating from hydrolytic activity by other bacteria. They only utilize short-chain fatty acids and simple alcohols produced by fermentation as organic carbon sources. Sulfate-reducing bacteria use sulfate as an oxygen source to oxidize fermentation products to carbon dioxide. Sulfide is released as a by-product.

Fermentation products also can be utilized by bacteria that produce methane. In the most common method of methane formation, a simple organic molecule is fermented and carbon dioxide is utilized as the electron (hydrogen) acceptor as shown below:

$$CH_3COOH + 2H_2O \rightarrow 2CO_2 + 8H^+ \tag{10.20}$$

$$8H^+ + CO_2 \rightarrow CH_4 + 2H_2O. \tag{10.21}$$

Subtraction of the two reactions gives

$$CH_3COOH \rightarrow CH_4 + CO_2. \tag{10.22}$$

Some bacteria also can use carbon dioxide as an oxidant to convert hydrogen gas produced in fermentation to methane

$$4H_2 + CO_2 \rightarrow CH_4 + 2H_2O. \tag{10.23}$$

Methane production is an important process because the hydrogen that accumulates from fermentation must be disposed of or it will inhibit the fermentation process.

High concentrations of alternative electron acceptors such as nitrate, ferric iron, sulfate, and other oxidized inorganic compounds in water or sediment favor anaerobic decomposition. Complete decomposition of organic matter in aquatic environments requires both aerobic and anaerobic organisms. The anaerobic organisms are especially important in sediment, because organic matter tends to settle to the bottom of water bodies, and anaerobic conditions typically occur a few centimeters or millimeters below the sediment-water interface. In eutrophic waters, the hypolimnion will become anaerobic during thermal stratification.

Environmental Effects on Bacterial Growth

Major factors affecting growth and respiration of bacteria are temperature, oxygen supply, moisture availability, pH, mineral nutrients, and composition and availability of organic substrates. Temperature effect on growth is illustrated in Fig. 10.2.

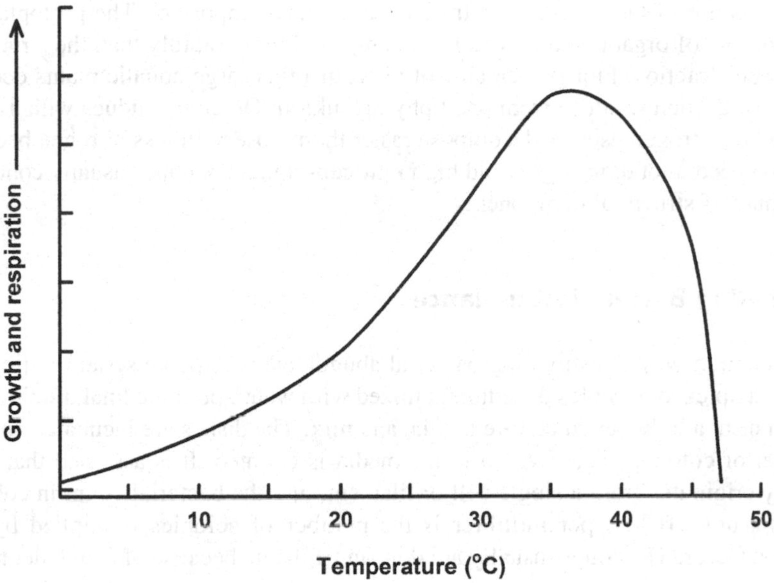

Fig. 10.2 Typical response of microbial growth and respiration to temperature

Growth is limited by low temperature, there is a narrow optimal temperature range, temperature can be too high for good growth, and temperature can reach the thermal death point.

According to van Hoff's law, the rate of many chemical reactions doubles with a 10 °C increase in temperature. Most physiological processes are chemical reactions and comply closely to van Hoff's law. The factor of increase in reaction rate that typically is about 2 for a 10 °C increase in temperature usually is called the Q_{10}. Thus, at suboptimal temperature, an increase of 10 °C normally doubles respiration and growth. Decomposition rate of organic matter also will double if the temperature is increased by 10 °C. Most bacteria in natural waters and sediment probably grow best at 30–35 °C.

A continuous supply of oxygen is needed to support aerobic bacterial activity, but under anaerobic conditions, facultative and anaerobic bacteria can decompose organic matter. Leachable and easily hydrolysable compounds from fresh organic matter decompose equally fast under aerobic and anaerobic conditions, but anaerobic decomposition of structural remains (complex macromolecules) of living things is less efficient than aerobic decomposition because of slower bacterial hydrolysis of these compounds (Kristensen et al. 1995).

Moisture is abundant in aquatic habitats, so aquatic decomposition is not hampered by lack of moisture. Bacteria thrive at pH 7–8, and decomposition is more rapid in neutral or slightly alkaline environments than in acidic ones. Bacteria must have inorganic nutrients such as sulfate, phosphate, calcium, potassium, etc., but these nutrients normally can be obtained from the substrate even if they are deficient in the water.

The nature of the organic substrate is particularly important. The protoplasmic components of organic matter can be decomposed more rapidly than the structural (cell wall) fraction. Fibrous remains of reeds or other large aquatic plants decompose slower than residues from dead phytoplankton. Organic residues with a large amount of nitrogen usually decompose faster than those with less nitrogen because bacteria need a lot of nitrogen, and high nitrogen-content residues usually contain a minimum of structural components.

Estimating Bacterial Abundance

The common way of estimating bacterial abundance is to make serial dilutions of water samples, or samples of sediment mixed with water, pour the final dilution into a petri dish, add liquefied culture media, and mix. The dishes are incubated and the number of colonies that develop in the media is counted. It is assumed that each colony originates from a single cell or filament, and the bacterial count in colony-forming units (CFU) per milliliter is the number of colonies multiplied by the dilution factor. The counts usually are done in triplicate because of a high degree of variation.

Metabolic activity of certain types of bacteria can be used as an index of their presence or abundance. For example, the formation of gas in a culture tube with lactose medium indicates the presence of lactose fermentating bacteria. Conversion of ammonia to nitrate reveals the presence of nitrifying bacteria. The rate of consumption of oxygen or the release of carbon dioxide can indicate the rate of decomposition of organic matter, and the abundance of the decomposing bacteria is related to the rate of carbon dioxide release.

Sediment Respiration

Decomposition occurs in both water and sediment, and the main effect on water quality of aerobic respiration is to remove dissolved oxygen and release carbon dioxide, ammonia, and other mineral substances. The flocculent layer of sediment just above the sediment-water interface and the upper few centimeters or millimeters of sediment contain a large amount of fresh, readily decomposable organic matter. Microbial activity in aquatic ecosystems usually is greatest in the flocculent layer and upper sediment layer. Uptake of oxygen results from respiration in the flocculent layer, and from respiration and chemical oxidation of reduced metabolites in the sediment. Sediment respiration rates as great as 20–30 g O_2/m^2 per day have been measured, but rates usually are 5 g O_2/m^2 per day or less. The removal of dissolved oxygen for use in sediment respiration is illustrated in Ex. 10.4.

Ex. 10.4: *The loss of dissolved oxygen from the water will be estimated for sediment respiration of 2 g $O_2/m^2/day$ in a 3-m deep body of water.*

Solution:
The water column above a 1 m² area of sediment will contain 3 m³ of water. Remembering that 1 mg/L is the same as 1 g/m³,

$$\frac{2\,\text{g}\,O_2/\text{m}^2}{3\,\text{m}^3\,\mathit{water}/\text{m}^2} = 0.67\,\text{g/m}^3\ \mathit{or}\ 0.67\,\text{mg/L}.$$

Some examples of oxygen uptake by non-biologically-mediated chemical reactions in sediment follow:

$$2Fe^{2+} + 0.5O_2 + 2H_2O \rightarrow Fe_2O_3 + 4H^+ \tag{10.24}$$

$$Mn^{2+} + 0.5O_2 + H_2O \rightarrow MnO_2 + 2H^+ \tag{10.25}$$

$$H_2S + 2O_2 \rightarrow SO_4^{2-} + 2H^+. \tag{10.26}$$

Usually, only the surface few millimeters of sediment are aerobic, and reduced microbial metabolites are at high concentration in pore water of sediment. There is opportunity for chemical oxidation of these reduced substances when oxygen enters the pore water. It is difficult to separate oxygen consumption by respiration from oxygen consumption in chemical oxidation, and sediment oxygen uptake measurements usually indicate the combined uptake by both processes.

Organic matter decomposition controls the redox potential in water and sediment. As long as dissolved oxygen is plentiful, only aerobic decomposition occurs. However, when dissolved oxygen concentration falls to 1 or 2 mg/L, certain bacteria begin to use oxygen from nitrate. As nitrate is used up, the redox potential declines, and when nitrate is depleted, the bacteria begin to use oxidized forms of iron and manganese in respiration. Redox continues to fall until sulfate and finally carbon dioxide become oxygen sources. Because the utilization of oxidized inorganic compounds is sequential, the different bacterial reductions occur sequentially in time or occur in different layers of sediment. For example, when a body of water stratifies thermally, it no longer obtains dissolved oxygen from the illuminated epilimnion in which photosynthesis occurs. Organic matter settles into the hypolimnion, and dissolved oxygen and redox potential decline as aerobic bacteria decompose organic matter. Simply put, once dissolved oxygen is depleted, nitrate supports respiration; when nitrate is gone, iron and manganese compounds become oxidants in respiration; sulfate and finally carbon dioxide serve as oxidants as redox falls.

Sediment becomes depleted of oxygen quicker than water, because dissolved oxygen cannot move downward rapidly in the pore water. Oxygen availability declines with sediment depth and zonation of the different processes develops (Fig. 10.3). This zonation will develop in sediment in bodies of water that are not chemically stratified, but in most unstratified bodies of water, the thin surface layer of the sediment remains aerobic as shown in Fig. 10.3. If organic matter inputs are

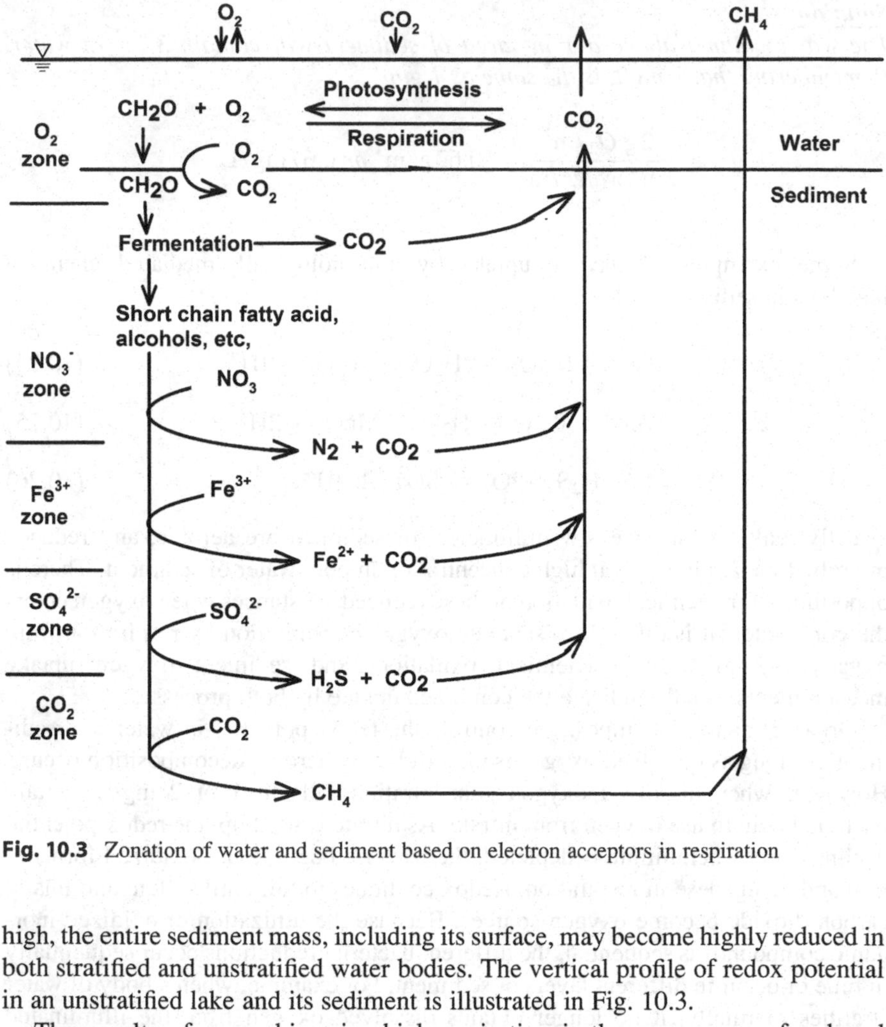

Fig. 10.3 Zonation of water and sediment based on electron acceptors in respiration

high, the entire sediment mass, including its surface, may become highly reduced in both stratified and unstratified water bodies. The vertical profile of redox potential in an unstratified lake and its sediment is illustrated in Fig. 10.3.

The results of anaerobic microbial respiration is the occurrence of reduced inorganic compounds to include nitrite, ferrous iron, manganous manganese, sulfide, and methane and many reduced organic compounds in waters of the hypolimnion of lakes, ponds, and other water bodies. When these reduced substances diffuse or are mixed into waters containing dissolved oxygen, they will be oxidized. If the input rate of reduced substances into aerobic water exceeds the oxidation rate, an equilibrium concentration of reduced substances may persist.

Some bacteria can oxidize ammonia to nitrate and thereby lower ammonia concentrations in water. This process will be discussed in Chap. 11.

Phytoplankton

The aquatic flora is very diverse and includes many thousand species. These include planktonic algae (phytoplankton) that are microscopic and suspended in the water, macrophytic algae that live on the bottom or form mats in the water, fungi which lack chlorophyll and are decomposers rather than producers like other plants, liverworts and mosses that lack the flowers and water conducting tissues of more advanced plants, and vascular plants that have water-conducting tissues throughout their bodies. Phytoplankton, macrophytic algae, and vascular plants usually are present in various proportions in most aquatic ecosystems, but phytoplankton are more biologically active and generally have a greater influence on water quality than do other plants.

Biology

The phytoplankton is distributed primarily among the phyla Pyrrophyta, Euglenophyta, Cyanophyta, Chlorophyta, and Heterokontophyta. The Pyrrophyta are mainly marine and include the dinoflagellates. The Euglenophyta—like the Pyrrophyta—contain flagellated motile organisms the best known of which are species of the genus *Euglena*. The Chlorophyta are known as the green algae, and they are mostly freshwater organisms. Members of this phylum include macroalgae such as the marine genus, *Ulva*, and the freshwater genus, *Spirogyra*, as well as many phytoplankton genera. Some common planktonic genera are *Scenedesmus*, *Chlorella*, and *Closterium*. Heterokontophyta is a large phylum of both marine and freshwater algae that includes the yellow-green algae (Xanthophyceae), golden algae (Chrysophyceae), brown algae (Phaeophyceae), and diatoms (Bacillariophyceae). The Cyanophyta are the blue-green algae. However, this group of algae is prokaryotic like bacteria, and many taxonomic authorities consider blue-green algae to be cyanobacteria. But, in this book these organisms will be referred to as blue-green algae. A more complete discussion of phytoplankton systematics is beyond the scope of this text, but there are many books and websites from which more information can be obtained.

Blue-green algae can have particularly large effects on water quality. They tend to be abundant in nutrient enriched waters, and many species of planktonic blue-green algae are considered undesirable. They form scums on the water surface, they are subject to sudden population crashes, a few species may be toxic to other aquatic organisms, and some species excrete odorous compounds into the water. These odorous compounds may cause a bad odor and taste in drinking water or impart bad flavor to fish and other aquatic food animals.

Phytoplankton requires sunlight for growth, so actively-growing phytoplankton are found only in illuminated water where there is more than about 1 % of incident light. The density of phytoplankton is slightly greater than that of water, so they tend to sink from the photic zone (illuminated surface stratum), but all species have adaptations such as small size, irregular morphology often with projecting surfaces, gas vacuoles, or means of motility to reduce their sinking rate and permit them to remain suspended even in waters of low turbulence. Nearly all waters contain phytoplankton, and waters with abundant nutrients will contain enough phytoplankton to cause turbidity and discoloration. Waters discolored by phytoplankton are said to have phytoplankton blooms and appear green, blue-green, red, yellow, brown, black, gray, or various other colors.

Reproduction in phytoplankton usually is by binary fission as described above for bacteria. The life span of phytoplankton is short, individual cells probably survive for no more than 1 or 2 weeks. As dead cells settle, they quickly rupture and the protoplasmic contents spill into the water. Natural waters usually contain much organic detritus originating from dead phytoplankton.

Most species of phytoplankton disperse readily. The atmosphere contains spores and vegetative bodies of many species. If a flask of sterile nutrient solution is opened to the atmosphere, algal communities will soon develop. Wading birds often spread phytoplankton from one body of water to another. This results because phytoplankton propagules stick to their external surfaces, and viable algal spores and vegetative bodies pass through their digestive tracts. Species of phytoplankton that colonize a habitat depend upon the suitability of the habitat into which their propagules happened to enter.

Suppose that a tank of nutrient-enriched water containing no algae is exposed outdoors. Species of phytoplankton in the air will fall into the tank, and the species entering would be primarily the result of chance. However, the species that persist in the tank will depend upon environmental conditions. Moreover, once the phytoplankters begin to grow they will go through a succession until a climax phytoplankton community is attained. This community eventually will make the environment unsuitable for its growth and crash (see Fig. 10.1), or the season may change making the conditions more suitable for other species. Thus, phytoplankton communities are continually changing in species composition.

Photosynthesis

Photosynthesis is the process by which green plants capture energy from photons of sunlight and use it to reduce carbon dioxide to organic carbon in form of carbohydrate. The basic features of the reaction of photosynthesis are depicted (Fig. 10.4).

In the light reaction of photosynthesis, sunlight strikes chlorophyll and other light sensitive pigments in plant cells, and photons of light are captured by the pigments. This energy is used in a reaction known as photolysis to split water molecules into molecular oxygen, hydrogen ions or protons, and electrons. Two water molecules yield one oxygen molecule, four hydrogen ions, and four electrons. The hydrogen ions and electrons then react to convert nicotinamide adenine dinucleotide phosphate ion ($NADP^+$) to its un-ionized form (NADPH). Energy captured by the pigments also converts ADP to ATP—a process called photophosphorylation.

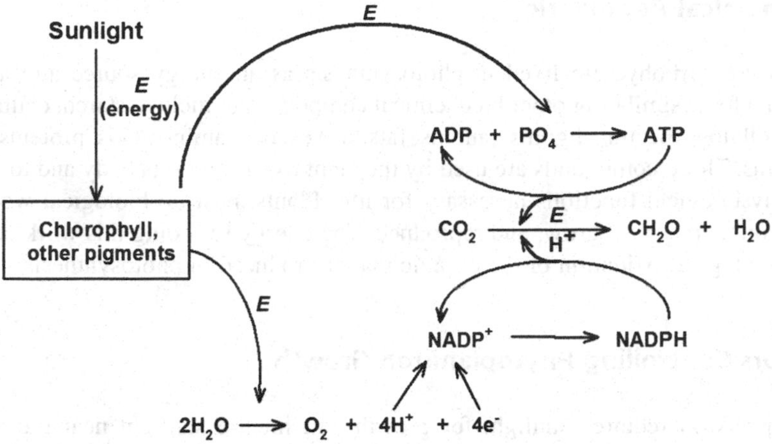

Fig. 10.4 Basic features of the photosynthesis process

The second phase of photosynthesis is not dependent on light, so it is termed the dark reaction despite occurring in the light. In a series of enzyme catalyzed reactions, energy from ATP and hydrogen ions and electrons from photolysis of water reduce carbon dioxide to carbohydrate (CH_2O) and water. Of course, $NAPD^+$ and ADP are regenerated and used again in the light reaction.

The summary equation for photosynthesis can be written as

$$2H_2O + CO_2 \xrightarrow{\text{light}} CH_2O + H_2O + O_2. \tag{10.27}$$

However, it is tradition to give the product of photosynthesis as glucose ($C_6H_{12}O_6$). This is done by multiplying (10.27) by six

$$12H_2O + 6CO_2 \xrightarrow{\text{light}} C_6H_{12}O_6 + 6H_2O + 6O_2. \tag{10.28}$$

Of course, (10.28) can be presented in more familiar form by subtracting the $6H_2O$ from both sides of the reaction equation to give

$$6H_2O + CO_2 \xrightarrow{\text{light}} C_6H_{12}O_6 + 6O_2. \tag{10.29}$$

Green plants can produce their own organic matter and are said to be autotrophic. The resources required for photosynthesis and plant production are carbon dioxide, water, sunlight, and inorganic nutrients. Molecular oxygen and hydrogen are available from the photolysis of water in photosynthesis. Other nutrients are primarily absorbed from the soil solution by terrestrial plants and from the water by aquatic plants. Essential mineral nutrients for the majority of plants are nitrogen, phosphorus, sulfur, potassium, calcium, magnesium, iron, manganese, zinc, copper, and molybdenum. Some plants require in addition one or more of the following: sodium, silicon, chloride, boron, and cobalt.

Biochemical Assimilation

Plants use carbohydrate fixed in photosynthesis as an energy source and a raw material for assimilating other biochemical compounds to include starch, cellulose, hemicellulose, pectins, lignins, tannins, fats, waxes, oils, amino acids, proteins, and vitamins. These compounds are used by the plant to construct its body and to carry out physiological functions necessary for life. Plants must do biological work to maintain themselves, grow, and reproduce. The energy for doing this work comes from biological oxidation of the organic matter produced in photosynthesis.

Factors Controlling Phytoplankton Growth

Phytoplankton requires sunlight for growth, but intense sunlight near the water surface may inhibit certain species. Some waters contain enough turbidity from suspended mineral or organic particles to greatly restrict light penetration and lessen the growth of phytoplankton. Phytoplankton growth responds favorably to warmth. Highest growth rates usually occur during spring and summer, but growth continues at a slower rate during colder months. Measurable rates of photosynthesis may even occur under clear ice.

The nutrient in shortest supply relative to the requirements of the plant will limit growth. For example, if phosphorus is the nutrient present in the shortest supply relative to plant needs, production will be limited to the amount possible with the ambient phosphorus concentration (Fig. 10.5). If phosphorus is added to the water, plant growth will increase. Growth will continue until the nutrient present in the second shortest supply becomes limiting. Further growth can be achieved by adding more of the second limiting nutrient. This is a case of multiple limiting factors (Fig. 10.6).

There is an important caveat, too much of a nutrient may limit the growth of some species. Thus, adding more phosphorus typically stimulates growth of some phytoplankton species while inhibiting the growth of other species. Thus, eutrophic water bodies may have high primary productivity, but relatively few species of phytoplankton.

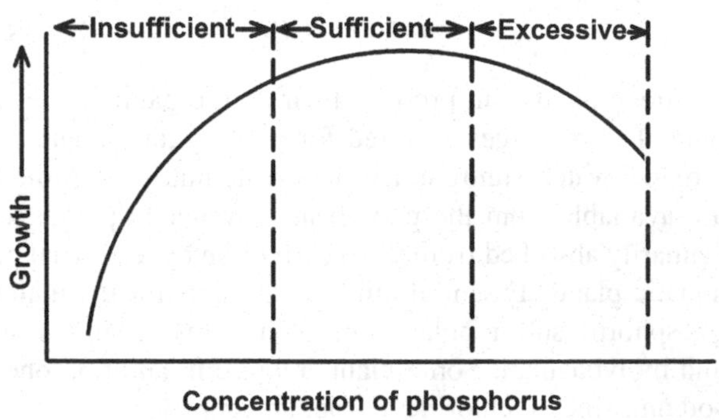

Fig. 10.5 Effect of a single limiting nutrient on plant growth

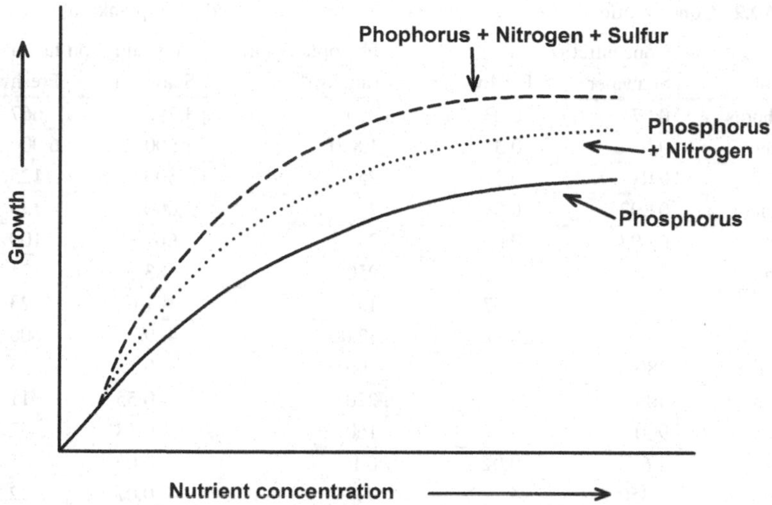

Fig. 10.6 Example of multiple limiting factors on plant growth

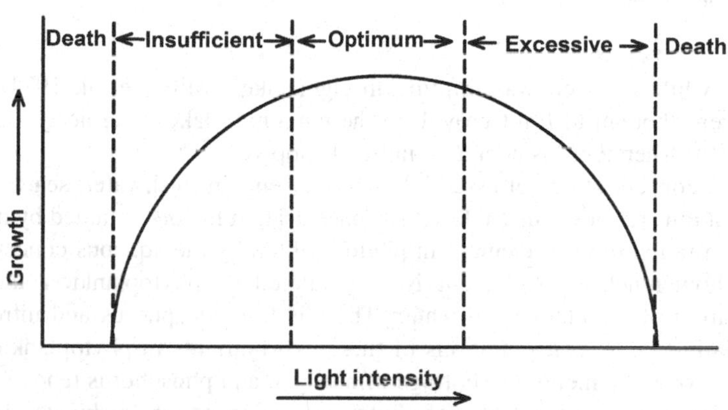

Fig. 10.7 Example of the response of a plant to light, illustrating the range of tolerance to a particular factor

The concept of limiting factors and ranges of tolerance also applies to other environmental factors, such as light intensity (Fig. 10.7). There is a range of tolerance within which plants can grow, and within this range light may be inadequate, optimum, or excessive for growth. Beyond the range of tolerance, plants will die. The ideas expressed in this paragraph are known as Liebig's "Law of the Minimum" and Shelford's "Law of Tolerance."

Shortage of any one or a combination of the essential nutrients may limit phytoplankton growth in natural waters. However, in most bodies of water, phosphorus, and to a lesser extent, nitrogen, are the most important limiting nutrients. A study of 49 American lakes showed phosphorus to limit phytoplankton growth in

Table 10.2 Concentrations of nutrients in seawater, freshwater, and phytoplankton

Element	Concentration (mg/L)		Phytoplankton	Concentration factors	
	Seawater	Freshwater	(mg/kg)[a]	Seawater	Freshwater
Phosphorus	0.07	0.03	230	3,286	7,667
Nitrogen	0.5	0.3	1,800	3,600	6,000
Iron	0.01	0.2	25	2,500	125
Manganese	0.002	0.03	4	2,000	133
Copper	0.003	0.03	2	667	100
Silicon	3	2	250[b]	83	125
Zinc	0.01	0.07	1.6	1.6	23
Carbon	28	20	12,000	429	600
Potassium	380	2	190	0.5	95
Calcium	400	20	220	0.55	11
Sulfur	900	5	160	0.18	32
Boron	4.6	0.02	0.1	0.02	5
Magnesium	1,350	4	90	0.07	22.5
Sodium	10,500	5	1,520	0.14	304

[a]Wet weight basis
[b]Concentration is much greater for diatoms

35 lakes, while nitrogen was limiting in eight lakes (Miller et al. 1974). Other factors were thought to limit growth in the remaining lakes. Turbidity may limit growth even when there is an ample nutrient supply.

Typical concentrations of essential plant nutrients in freshwater, seawater, and phytoplankton are shown in Table 10.2. Concentration factors obtained by dividing the concentration of each element in phytoplankton by the aqueous concentration indicate how much each element is accumulated by phytoplankton above the concentration of the element in water. There is less phosphorus and nitrogen in water relative to the concentrations of these two nutrients in phytoplankton than there is for other elements. A shortage of nitrogen and phosphorus tends to be the most common nutrient limitation of phytoplankton growth in freshwater, brackish water, and seawater. In spite of the common belief that phosphorus is not as likely to be a limiting factor in marine environments as in freshwater ones, a review of the literature on this topic (Elser et al. 2007) suggested that nitrogen and phosphorus limitations of freshwater, marine, and terrestrial ecosystems are similar. Of course, there are waters where nutrients other than phosphorus and nitrogen limit phytoplankton growth. Iron and manganese in particular may limit marine phytoplankton productivity.

Redfield (1934) observed that the average molecular ratio of carbon:nitrogen: phosphorus in marine phytoplankton was about 106:16:1 (weight ratio $\approx$ 41:7:1); this ratio became known as the Redfield ratio. However, this ratio does not imply that the addition of nitrogen and phosphorus into ecosystems at a ratio of 7:1 would be the most effective ratio in promoting phytoplankton growth. As will be seen in Chaps. 11 and 12, there is considerably more recycling of nitrogen than of phosphorus in aquatic ecosystems. Thus, a narrower nitrogen:phosphorus ratio than 7:1 would need to be added to realize a 7:1 ratio in the water.

There has been much debate over the concentrations of nitrogen and phosphorus necessary to cause phytoplankton blooms in water bodies. The response of phytoplankton to nitrogen and phosphorus differs among aquatic ecosystems, and development of noticeable phytoplankton blooms may require concentrations of 0.01–0.1 mg/L soluble inorganic phosphorus and 0.1–0.75 mg/L of inorganic nitrogen.

Many limnologists feel that carbon does not limit phytoplankton growth in natural waters, but algal culture studies indicate that the possibility of carbon as a limiting factor in aquatic ecosystems cannot be entirely dismissed (King 1970; King and Novak 1974; Boyd 1972). Moreover, several workers have observed that phytoplankton production and fish production increase as total alkalinity increased in natural waters up to 100–150 mg/L. This observation does not necessarily indicate that waters with higher alkalinities have higher concentrations of available carbon, and therefore, greater phytoplankton productivity. The studies were conducted in natural waters that were not fertilized intentionally or polluted through human activity. Water with higher alkalinity tends to have a greater complement of most plant nutrients than does water of low alkalinity. Correlations between alkalinity and phytoplankton productivity may have been related to differences in nitrogen and phosphorus availability rather than different concentrations of carbon dioxide or alkalinity *per se* (Boyd and Tucker 2014).

Phytoplankton productivity also may be regulated by the growth of macrophytes because these plants compete with phytoplankton for nutrients. Macrophytes floating on pond surfaces and those with leaves at the surface shade the water column and greatly restrict phytoplankton growth. Certain macrophytes are allelopathic; they excrete substances toxic to phytoplankton. Regardless of the reasons, additions of nutrients to water bodies containing extensive macrophyte communities often do not cause phytoplankton blooms; rather, they stimulate further macrophyte growth.

Estimating Phytoplankton Abundance

The number of individuals (single cells, filaments, or colonies) per milliliter or liter of water may be determined directly by microscopic examination. Different species have characteristic sizes, and the volume of phytoplankton cells per unit of water provides a better estimate of biomass than does enumeration of phytoplankton abundance. However, it is exceedingly difficult to measure the volume of phytoplankton in a sample. A formula must be concocted for estimating the value of each species present and multiplied by the number of that species per unit volume of water.

Indirect methods of assessing phytoplankton abundance are popular. Chlorophyll *a* determination on particulate matter removed from a water sample by filtration provides an index of phytoplankton abundance. Because particulate organic matter often consists mostly of phytoplankton, this variable is indicative of phytoplankton abundance in many waters. Suitable techniques for separating phytoplankton from other particulate matter are not available, so results must be assessed with caution. The Secchi disk visibility decreases with increasing plankton abundance, and in many bodies of water, the Secchi disk visibility can be a useful measure of phytoplankton abundance. Of course, one must consider how much of the turbidity appears to be from other sources when assessing plankton abundance with a Secchi disk.

Effects of Phytoplankton on Water Quality

The most profound effects of phytoplankton activity on water quality are changes in pH and concentrations of dissolved oxygen and carbon dioxide. Photosynthesis usually dominates over respiration during daylight in a water body, while the opposite occurs at night. In other words, there is a net increase in dissolved oxygen and a net decrease in carbon dioxide during the day and *vice versa* at night. Changes in carbon dioxide concentration affect pH. The pH increases when carbon dioxide decreases. Daily changes in carbon dioxide, dissolved oxygen, and pH are illustrated in Fig. 10.8. The magnitude of daily fluctuations in concentrations of the three variables will tend to increase as phytoplankton abundance increases.

Light availability decreases with depth, so the rate of phytoplankton photosynthesis will tend to decrease with depth. In the afternoon, dissolved oxygen and pH will tend to be higher in surface water than in deeper water while the opposite will be true for carbon dioxide.

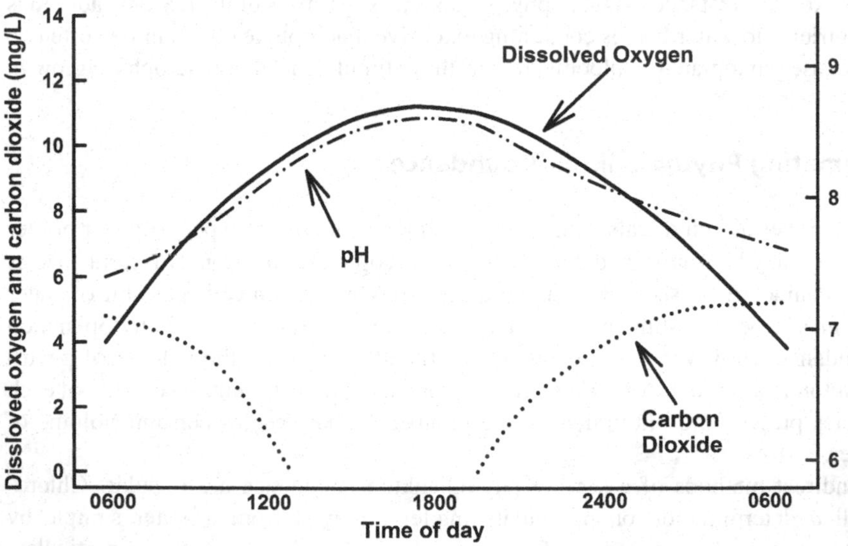

Fig. 10.8 Changes in pH and concentrations of dissolved oxygen and carbon dioxide over a 24-h period in a eutrophic water body

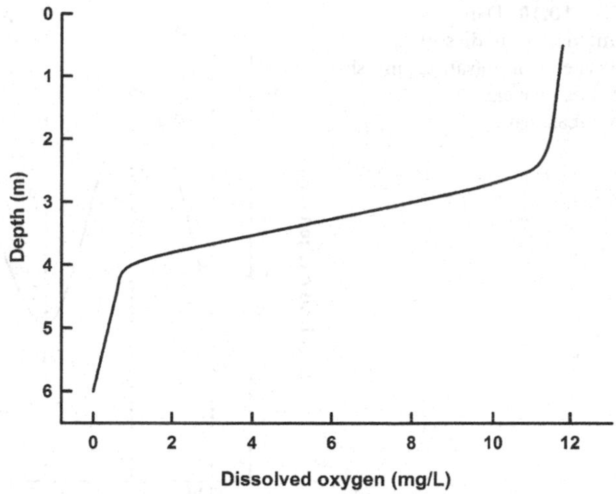

Fig. 10.9 Change in dissolved oxygen concentration with depth in a thermally-stratified, eutrophic lake

Stratification of dissolved oxygen concentration in a thermally stratified lake is illustrated in Fig. 10.9. The lake can be considered eutrophic because oxygen depletion in the hypolimnion is a criterion for separating eutrophic (nutrient rich) and oligotrophic (nutrient poor) lakes.

Phytoplankton blooms sometimes die suddenly. Sudden, massive mortality of phytoplankton—often called die-offs—can cause severe depression of dissolved oxygen concentrations. Die-offs are characterized by sudden death of all or a great portion of the phytoplankton followed by rapid decomposition of dead algae. Dissolved oxygen concentrations decline drastically, and they may fall low enough to cause fish kills. The reasons for phytoplankton die-offs have not been determined exactly, but they usually involve dense surface scums of blue-green algae. Die-offs often occur on calm, bright days when dissolved oxygen concentrations are high, carbon dioxide concentrations are low, and pH is high. This combination has been suggested to kill blue-green algae through a photo-oxidative process (Abeliovich and Shilo 1972; Abeliovich et al. 1974).

Events surrounding a complete die-off of a dense population of the blue-green algae *Anabaena variabilis* in a eutrophic pond at Auburn, Alabama were documented (Boyd et al. 1975). The pond contained a uniform density of *A. variabilis* throughout the water column on windy days in March and April. In late April, a succession of clear, calm days resulted in a surface scum of phytoplankton on April 29. On the afternoon of April 29, the phytoplankton died, and the pond water was brown and turbid with decaying algae on April 30. No living *A. variabilis* filaments and few individuals of other algal species were observed in water samples taken between April 30 and May 5. Between May 5 and 8, a new phytoplankton community developed that consisted primarily of desmids. Dissolved oxygen concentrations quickly dropped to 0 mg/L following death of the *A. variabilis* population, and dissolved oxygen remained at or near this concentration for nearly a week until the new phytoplankton community developed.

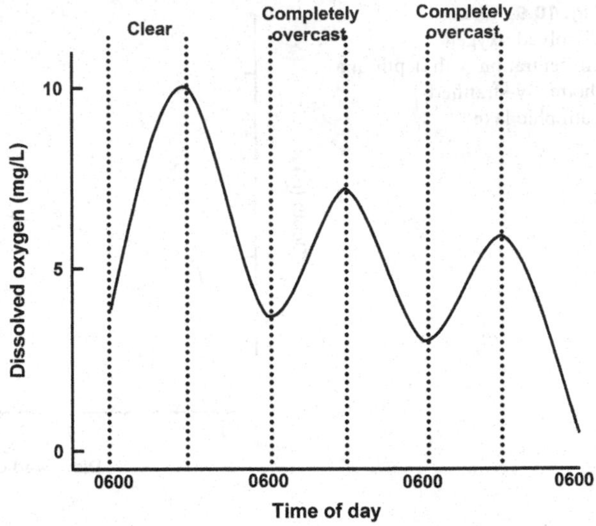

Fig. 10.10 Daily fluctuations in dissolved oxygen concentrations in fish ponds on clear and overcast days

All phytoplankton die-offs are not as spectacular as the one described above, but they are rather common events.

Weather profoundly influences dissolved oxygen concentrations. On clear days, water bodies normally have dissolved oxygen concentrations near saturation at dusk. On cloudy days, photosynthesis is limited by insufficient light, and dissolved oxygen concentrations often are lower than normal at dusk. The probability of dissolved oxygen depletion is greater during nights following cloudy days than during nights following clear days (Fig. 10.10). Thermal destratification or overturns of water bodies may occur during unusually cool spells, heavy winds, and heavy rains. Fish kills may occur following overturns because sudden mixing of large volumes of oxygen-deficient hypolimnetic water with epilimnic water can result in rapid oxygen depletion. Weather-related problems with low dissolved oxygen are most common in water bodies with abundant phytoplankton.

Phytoplankton also influences water quality by removing nutrients from the water and by serving as a source of organic matter. Ammonia nitrogen concentrations and soluble reactive phosphorus concentrations in water bodies usually decline when phytoplankton are growing rapidly.

Measuring Photosynthesis and Respiration

The oxygen light–dark bottle technique is a relatively simple procedure for measuring photosynthesis and respiration rates. In its simplest application, this procedure involves filling three bottles [usually 300-mL biochemical oxygen demand (BOD) bottles]—two transparent bottles and an opaque one—with water from the water body of interest. The dissolved oxygen concentration is measured immediately in one of the transparent bottles (initial bottle). The other transparent bottle, called the light bottle, and the opaque bottle, called the dark bottle, are incubated in the water body. After a period of incubation, the bottles are removed for measurement of dissolved oxygen concentration.

The initial bottle provides an estimate of the concentration of dissolved oxygen in water of the light and dark bottles at the beginning of incubation. During incubation, there is both photosynthesis and respiration in the light bottle, so part of the oxygen produced in photosynthesis is used in respiration. The gain in dissolved oxygen in the light bottle represents the amount of net photosynthesis. No light enters the dark bottle; photosynthesis does not occur, but respiration removes dissolved oxygen. The loss of dissolved oxygen in the dark bottle represents respiration of phytoplankton and other microorganisms. The total amount of oxygen produced in photosynthesis (gross photosynthesis) is the sum of dissolved oxygen produced in net photosynthesis and used in respiration.

If the objective is to determine the influence of depth on photosynthesis and respiration or to obtain average rates for the water column, bottles must be filled with water from several depths and then incubated at those same depths. This arrangement is necessary because light penetration decreases with depth and photosynthesis rates are dependent upon light intensity. Both temperature and the abundance of planktonic organisms also differ with depth and influence photosynthesis and respiration.

Light intensity varies with time of day and influences photosynthetic rate. A 24-h incubation would be ideal, but sedimentation of organisms within bottles, depletion of dissolved oxygen in the dark bottle, or supersaturation of dissolved oxygen in the light bottle can alter respiration and photosynthesis rates. These "bottle effects" make it necessary to restrict incubation to a few hours. Incubations from dawn (provided the initial dissolved oxygen is high enough to sustain respiration in the dark bottle) until noon are more feasible than incubations from noon until dusk, because at noon waters are already high in dissolved oxygen concentration. If the bottles are incubated at dawn and removed at noon, the incubation period will represent one-half of a photoperiod. Another alternative is to incubate the bottles for a specific time interval during the day and measure the solar radiation for the incubation interval and for the entire photoperiod. The fraction of the daily radiation occurring during the incubation period can be determined by dividing radiation during the incubation period by total radiation for the day. The total amount of photosynthesis is assumed to equal the amount during the incubation period times the ratio of daily radiation to radiation during incubation.

It is possible to calculate three variables, net photosynthesis, gross photosynthesis, and water column respiration from light–dark bottle data. Of course, net photosynthesis will be underestimated because organisms other than phytoplankton contribute to respiration and detract from net photosynthesis. Equations for these calculations follow:

$$NP = LB - IB \tag{10.30}$$

$$R = IB - DB \tag{10.31}$$

$$GP = NP + R \tag{10.32}$$

$$\text{or} \quad GP = LB - DB \tag{10.33}$$

where NP = net photosynthesis, GP = gross photosynthesis, R = respiration, IB = initial bottle, LB = light bottle, and DB = dark bottle. All variables are reported in milligrams per liter of dissolved oxygen. Where bottles are incubated at different depths, results can be averaged to obtain values for the water column. Results can be adjusted to a 24-h basis by considering the length of the incubation period and the amount of radiation during the incubation period. The computations are illustrated in Ex. 10.5.

Ex. 10.5: *The following data were collected on averages for light–dark bottle incubations in a 1.75-m deep column of water.*

IB = 4.02 mg/L
LB = 6.75 mg/L
DB = 2.98 mg/L
Incubation period = dawn to noon (6.5 h)

NP, GP, and R will be calculated for a 24-h day.

Solution:
For the incubation period,

$$NP = 6.75 - 4.02 = 2.73\,\text{mg/L}$$

$$R = 4.02 - 2.98 = 1.04\,\text{mg/L}$$

$$GP = 6.75 - 2.98 = 3.77\,\text{mg/L}.$$

Assuming a clear day, the values may be doubled to provide totals for the photoperiod. Respiration will continue throughout the 11-h period of darkness, so nighttime respiration will be only 11/13 of daytime values. Nighttime respiration must be subtracted from daytime NP in order to estimate 24-h NP.

Photosynthesis and respiration rates may be expressed per unit area (usually square meter). This is done by multiplying the average values for NP, GP, and R in the entire water column by water depth. Values for NP, GP, and R also can be expressed in terms of carbon. In the general expressions for respiration and photosynthesis, one mole of oxygen can be equated to one mole of carbon dioxide or carbon. The ratio of the atomic weight of carbon (C) to the molecular weight of oxygen (O_2), 12/32 or 0.375, can be used to convert oxygen concentrations to carbon concentrations.

The Compensation Depth

Photosynthesis rate decreases with increasing depth because of diminishing light. In most water bodies there will be a depth at which oxygen produced by photosynthesis will be equal to oxygen used in respiration. At this depth—called the compensation depth or point—net photosynthesis will be zero. Above the compensation depth, more dissolved oxygen will be produced by phytoplankton than used in respiration by microorganisms in the water column. Below the compensation depth, more dissolved oxygen will be used in respiration than produced by photosynthesis.

In a stratified body of water, the compensation depth usually corresponds to the thermocline, and no oxygen is produced in the hypolimnion. Because of stratification, an oxygen debt develops in the hypolimnion. When destratification occurs, water from the hypolimnion is mixed with epilimnetic water, and the accumulated oxygen debt must be satisfied. This can cause a decrease in dissolved oxygen concentration at the time of overturns that is proportional to the size of the hypolimnetic oxygen debt.

In bodies of water that do not stratify thermally, there is daily mixing of bottom and surface waters. If the oxygen deficit below the compensation depth is great in comparison to the oxygen surplus above the compensation depth, low dissolved oxygen concentration can result. In an unstratified water body, the difference between the oxygen surplus above the compensation depth and the oxygen deficit below this point (Fig. 10.11) represents the oxygen available for larger organisms in the water column whose respiration is not included in light–dark bottle measurements and sediment respiration.

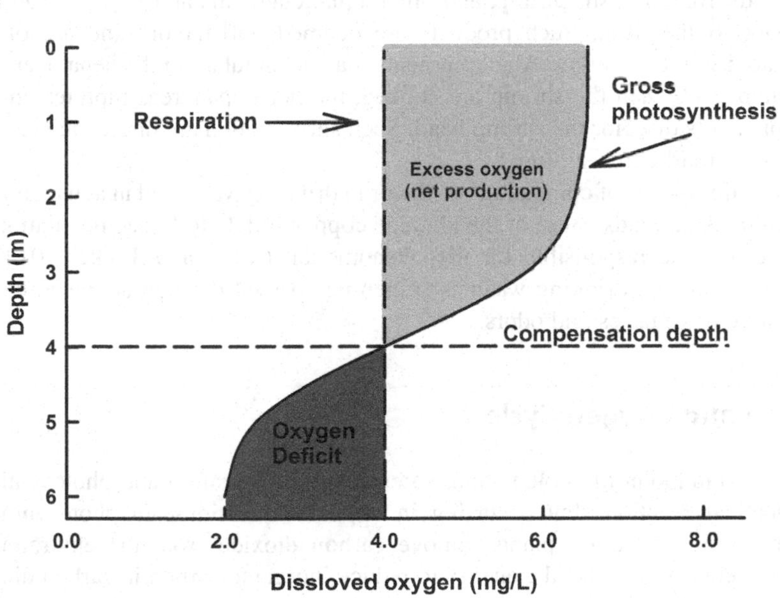

Fig. 10.11 Illustration of the compensation depth in a water body

Harmful Algae

Phytoplankton blooms can be harmful to aquatic animals if their standing crops become great enough to cause dissolved oxygen depletion. A few species of algae may be directly toxic to aquatic animals or even to humans and livestock. These algae include certain unicellular marine algae known as prymnesiophytes, and some blue-green algae, dinoflagellates, diatoms, and chloromonads. Toxic algae—especially those that are responsible for the red tide phenomenon in marine waters—can cause huge fish kills over wide areas.

Toxins passed along the food chain may represent a health threat to humans who consume certain aquatic products. The best example is shell fish poisoning in which bivalve molluscs such as oysters, mussels, clams, and scallops filter toxic algae—certain dinoflagellates, diatoms, and blue-green algae—from the water and accumulate the algal toxins in their tissues. There are four types of shellfish poisoning: amnesic (ASP), diarrheal (DSP), neurotoxic (NSP), and paralytic (PSP) that may result in humans from eating improperly or uncooked shellfish. The effects vary ranging from an unpleasant bought with diarrhea (DSP) or unusual sensations (NSP), to amnesia and possible permanent cognitive damage (ASP), or even death (PSP has a mortality rate of 10–12 %). Some blue-green algae are known to cause allergic reactions, mostly skin rashes, in humans. Livestock and wildlife have been killed by drinking water from pools infested with toxic algae.

Algae, and blue-green algae in particular, are known to produce compounds that can cause bad taste and odor in drinking water supplies. The two most common offensive compounds are geosmin and 2-methylisoborneol. These same compounds can be adsorbed by fish, shrimp, and other aquatic animals and impart a bad taste and odor to the flesh. Such products are deemed "off-flavor" and are of low acceptability in the market. Algal pigments may accumulate in the hepatopancreas of shrimp, and when the shrimp are cooked, the hepatopancreas ruptures and the algal pigments discolor the shrimp head. Such shrimp often are unacceptable in the "heads-on" market for shrimp.

The major means of combating off-flavor in drinking water and in aquaculture of food animals in ponds is use of the algicide copper sulfate to lessen populations of blue-green algae responsible for the phenomenon (Boyd and Tucker 2014). In extreme situations, drinking water is sometimes passed through activated carbon filters to remove tastes and odors.

Carbon and Oxygen Cycle

From the standpoint of ecology and water quality, respiration and photosynthesis are identical reactions but occurring in opposite directions. In photosynthesis, phytoplankton and other plants remove carbon dioxide from the environment, trap solar energy, and use the energy to reduce inorganic carbon in carbon dioxide to organic carbon in carbohydrate and release oxygen into the environment. In respiration, organisms oxidize organic carbon to inorganic carbon of carbon dioxide

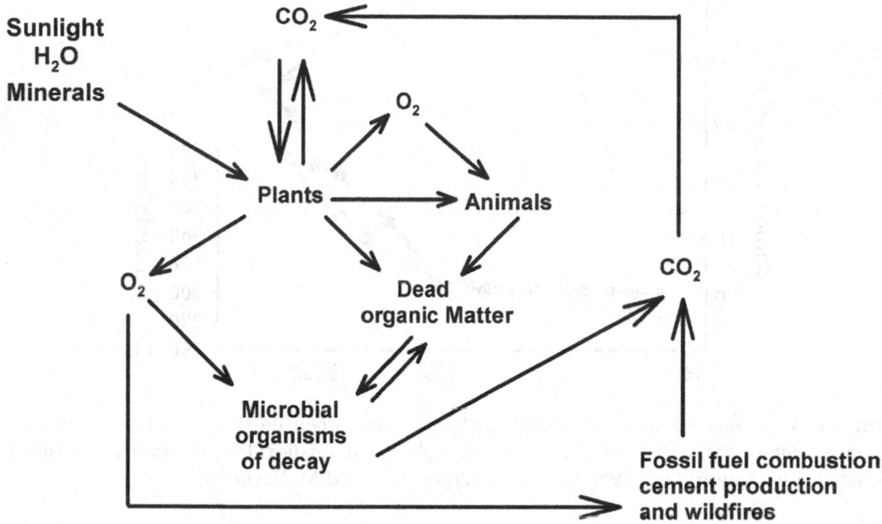

Fig. 10.12 Carbon and oxygen cycles

with the consumption of oxygen from the environment and the release of energy and carbon dioxide. Plants generally produce more organic matter and oxygen than they use—this is net photosynthesis available to animals, bacteria, and other heterotrophic organisms for food. The surplus oxygen is used by aerobic, heterotrophic organisms as an oxidant in respiration. Organic matter in ecosystems is the remains of organisms. It consists primarily of plant remains, because production of plant biomass greatly exceeds that of animal biomass in ecosystems.

The dynamics of carbon and oxygen are closely entwined in nature. When organic carbon is fixed, oxygen is released, and oxygen is consumed when organic carbon is mineralized. All life depends upon the input of energy from the sun and upon the cyclic transformations of carbon and oxygen. The global carbon and oxygen cycles are depicted in simple form in Fig. 10.12. The food web which consists of all of the transfers of organic matter among producer, consumer, and decomposer organisms may be prepared by expanding the carbon cycle to include all the pathways by which food (organic matter) and energy move through ecosystems.

The carbon cycle has been greatly altered since the beginning of the industrial age about 1750 by use of fossil fuels, deforestation, cement manufacturing and other anthropogenic sources of carbon dioxide. The carbon dioxide concentration in the atmosphere has increased from about 280 ppm to about 400 ppm today (Fig. 10.13). This increase is considered by most scientists as the major reason for global warming, climate change, sea-level rise, and acidification of the ocean.

In the past century, average global surface temperature increased by about 0.78 °C and mean sea level rose by 17 cm because of melting of polar ice and thermal expansion of ocean water as a result of warming. Climate change is

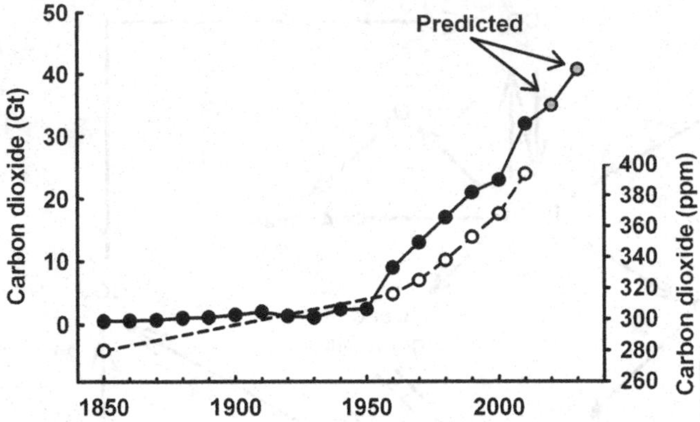

Fig. 10.13 Annual estimates of global, anthropogenic carbon dioxide emissions at 10-year intervals (*dots*) (1850–2030) and changes in atmospheric carbon dioxide concentrations measured at Mauna Loa, Hawaii from 1958 to 2012 (*circles*) (Boyd and McNevin 2014)

occurring resulting in more extreme weather. The extreme weather index maintained by the US National Oceanographic and Atmospheric Administration fluctuated but showed no clear pattern of increase from 1920–1970, but since, it has exhibited an upward trend. Average ocean pH declined from 8.12 in 1988 to 8.09 in 2008, and it is resulting in thinner calcium carbonate shells on many marine invertebrates. The changes are expected to accelerate during the remainder of the twenty-first century (Boyd and McNevin 2014).

Human Pathogens

Water supplies for human use may be contaminated with pathogens. Some of the most serious waterborne infections such as typhoid fever, dysentery, and cholera are not common in developed countries, but waterborne diseases are still of major significance in the developing world. Organisms responsible for waterborne diseases include viruses, bacteria, protozoans, mites, and worms.

Waters that are contaminated with human pathogens usually are contaminated by human fecal material. Coliform bacteria are used to identify waters that may have received human wastes, and water containing significant numbers of coliforms is a health hazard because of the possibility of waterborne diseases.

Coliform organisms are aerobic and facultative anaerobic, gram-negative, nonspore-forming, rod-shaped bacteria that ferment lactose with gas formation within 48 h at 35 °C. Some of these organisms can be found in soil and on vegetation, but fecal coliforms *Escherichia coli* usually originate from feces of warm-blooded animals. The ratio of fecal coliforms to fecal streptococci is sometimes used to differentiate waters contaminated with human fecal coliforms from those contaminated with fecal coliforms of other warm-blooded animals. The fecal coliform:fecal streptococci ratio for humans is above 4.0, but for other warm-blooded animals, the ratio is 1.0 or less (Tchobanoglous and Schroeder 1985). Fecal coliforms should not be present in drinking water, but in some cases, public health authorities may permit up to 10 total coliform organisms/100 mL provided there are no fecal coliforms. When water contains coliforms, it should be disinfected before it is consumed by humans.

Chlorination

Chlorination is the most common method of disinfecting public water supplies. The common, commercial chlorine disinfectants are chlorine gas (Cl_2), sodium hypochlorite (NaOCl), and calcium hypochlorite [$Ca(OCl)_2$]. But, chloramines—especially monochloramine (NH_2Cl)—are being used increasingly to disinfect drinking water.

Chlorine reacts with water to form hydrochloric acid (HCl) and hypochlorous acid (HOCl)

$$Cl_2 + H_2O \rightarrow HOCl + H^+ + Cl^-. \tag{10.34}$$

Hydrochloric acid dissociates completely, but hypochlorous acid partially dissociates according to the following equation

$$HOCl \rightleftharpoons H^+ + OCl^- \quad K = 10^{-7.53}. \tag{10.35}$$

Chlorination of water may yield four chlorine species: chloride that is not a disinfectant, and chlorine, hypochlorous acid, and hypochlorite ion that have disinfecting power and are called free chlorine residuals. The disinfecting powers of chlorine and hypochlorous acid are about 100 times greater than that of hypochlorite (Snoeyink and Jenkins 1980). The dominant free chlorine residual in water depends on pH rather than type of chlorine compound applied (Fig. 10.14). Chlorine occurs only at very low pH, HOCl is the dominant residual between pH 2 and 6; HOCl and OCl^- both occur in significant portions between pH 6 and 9, but HOCl declines relative to OCl^- as pH increases; OCl^- is the dominant residual above pH 7.53 (Fig. 10.14).

Disinfection at pH 7 normally requires about 1 mg/L of free chlorine residuals. The ratio HOCl:OCl^- decreases as pH increases, ranging from 32:1 at pH 6 to 0.03:1 at pH 9. Greater concentrations of free chlorine residuals are required as the pH rises, because disinfection power decreases as the proportion of HOCl in the free chlorine residual declines. Free chlorine residuals participate in many reactions that diminish their disinfecting power. Energy from sunlight drives the reaction in which hypochlorous acid is reduced to nontoxic chloride.

$$2HOCl \xrightarrow{\text{sunlight}} 2H^+ + 2Cl^- + O_2. \tag{10.36}$$

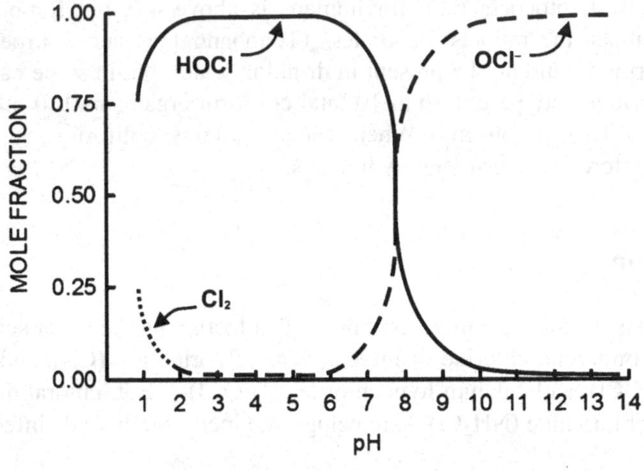

Fig. 10.14 Effect of pH on distribution of free chlorine residuals in water

Free chlorine residuals oxidize organic matter, nitrite, ferrous iron, and sulfide, and disinfecting power is lost because free chlorine residuals are reduced to chloride. Free chlorine residuals combine with organic nitrogen compounds, phenols, and humic acid to produce organochlorine compounds. At least one group of these compounds—the trihalomethanes—is suspected to be a carcinogen in humans (Jimenez et al. 1993).

A common extraneous reaction of chlorine is formation of chloramines

$$\text{Monochloramine} \quad NH_3 + HOCl \rightarrow NH_2Cl + H_2O \tag{10.37}$$

$$\text{Dichloroamine} \quad NH_2Cl + HOCl \rightarrow NHCl_2 + H_2O \tag{10.38}$$

$$\text{Trichloramine} \quad NHCl_2 + HOCl \rightarrow NCl_3 + H_2O. \tag{10.39}$$

Because these extraneous reactions diminish concentrations of free chloride residuals, they must be considered in establishing the chlorine dose for disinfection. Water samples typically are treated with the chlorinating agent until the dose necessary to provide the desired concentration of free chlorine residuals is established.

Chloramines are becoming popular for chlorination for two reasons. They are less likely to react with organic matter to form trihalomethanes and other byproducts, and they have a longer residual life in water distribution systems than traditional chlorination compounds.

Significance

The biological aspects of water quality are extremely important, because photosynthesis and respiration by microorganisms are major factors controlling dissolved oxygen dynamics and redox potential in waters and sediment of aquatic ecosystems. Dissolved oxygen concentration is probably the most important variable related to the overall well-being of aquatic ecosystems. In highly eutrophic ecosystems, excessive phytoplankton production results in high concentrations of organic matter, wide daily fluctuations in dissolved oxygen concentrations, and low redox potential at the sediment-water interface. Low dissolved oxygen concentration is stressful to aquatic species, and only those species most tolerant to low dissolved oxygen concentrations can prosper in eutrophic water bodies. Usually, species diversity and stability decline in eutrophic ecosystems in spite of the high level of primary productivity.

Techniques relying on the presence, absence, or relative abundance of indicator species have been developed for assessing the trophic status of ecosystems and to indicate relative degrees of pollution in aquatic ecosystems. However, these procedures are difficult and time consuming. A less tedious technique for assessing the trophic status of water bodies is measurement of primary productivity. Wetzel (1975) ranked lake trophic status according to the average net primary productivity as follows: ultra-oligotrophic, <50 mg C/m^2/day; oligotrophic, 50–300 mg C/m^2/day; mesotrophic, 250–1,000 mg C/m^2/day; eutrophic, >1,000 mg C/m^2/day.

Simpler measures of trophic status also can be used. For example, the depletion of dissolved oxygen in the hypolimnion of a water body during stratification indicates eutrophic conditions. The clarity of water as determined by Secchi disk visibility in waters where plankton is the main source of turbidity can indicate trophic status. Oligotrophic bodies of water may have Secchi disk visibilities of 4 or 5 m or more while Secchi disk visibility is less than 1.0 m in highly eutrophic waters. Of course, Secchi disk visibility is not a good index of productivity where light penetration is restricted by suspended soil particles or humic substances. Chlorophyll *a* concentrations are not difficult to measure, and they provide a good index of phytoplankton abundance and trophic status of ponds, lakes, reservoirs, and coastal waters. The relationship between chlorophyll *a* concentration and the trophic status of lakes and reservoirs is provided (Table 10.3).

Table 10.3 Relationship between chlorophyll *a* concentrations and conditions in reservoirs and lakes

Chlorophyll *a* (μg/L)		
Annual mean	Annual maximum	Conditions
<2	<5	Oligotrophic, aesthetically pleasing, very low phytoplankton levels
2–5	5–15	Mesotrophic, some algal turbidity, reduced aesthetic appeal, oxygen depletion not likely
5–15	15–40	Mesotrophic, obvious algal turbidity, reduced aesthetic appeal, oxygen depletion likely
>15	>40	Eutrophic, high levels of phytoplankton growth, significantly reduced aesthetic appeal, serious oxygen depletion in bottom waters, reduction in other uses

Microorganisms in water bodies also can have a great influence on human and animal health. Better sanitary conditions to prevent the contamination of public water supplies with human wastes and methods for disinfecting drinking water have been major milestones in the continuing effort to improve public health. Of course, many diseases of fish and wildlife also can be spread by water.

The presence of excessive algae discolors water and creates surface scums that detract from aesthetic value, and turbidity created by planktonic algae restricts underwater visibility and makes waters less desirable for swimming and other water sports. Bad tastes and odors are imparted to drinking waters by some phytoplankton species that occur in public water supplies. The same algal compounds that cause bad tastes and odors in drinking water can be absorbed by fish and other aquatic organisms. Absorption of algal compounds taints the flesh of aquatic food animals and makes them less desirable to the consumer or even harmful to the health of the consumer. Shellfish poisoning is caused by an algal toxins and can lead to various painful symptoms or even death in humans.

References

Abeliovich A, Shilo M (1972) Photo-oxidative death in blue-green algae. J Bacteriol 11:682–689

Abeliovich A, Kellenberg D, Shilo M (1974) Effects of photo-oxidative conditions on levels of superoxide dismutase in *Anacystis nidulans*. Photochem Photobiol 19:379–382

Boyd CE (1972) Sources of CO_2 for nuisance blooms of algae. Weed Sci 20:492–497

Boyd CE, McNevin AA (2015) Aquaculture, resource use, and the environment. Wiley-Blackwell, Hoboken

Boyd CE, Tucker CS (2014) Handbook for aquaculture water quality. Craftmaster, Auburn

Boyd CE, Prather EE, Parks RW (1975) Sudden mortality of a massive phytoplankton bloom. Weed Sci 23:61–67

Elser JJ, Bracken M, Cleland EE et al (2007) Global analysis of nitrogen and phosphorus limitation of primary producers in freshwater, marine and terrestrial ecosystems. Ecol Lett 10:1135–1142

Jimenez MCS, Dominguez AP, Silverio JMC (1993) Reaction kinetics of humic acid with sodium hypochlorite. Water Res 27:815–820

King DL (1970) The role of carbon in eutrophication. J Water Pollut Control Fed 42:2035–2051

King DL, Novak JT (1974) The kinetics of inorganic carbon-limited algal growth. J Water Pollut Control Fed 46:1812–1816

Kristensen E, Ahmed SI, Devol AH (1995) Aerobic and anaerobic decomposition of organic matter in marine sediment: Which is fastest? Limnol Oceanogr 40:1430–1437

Miller WE, Maloney TE, Greene JC (1974) Algal productivity in 49 lake waters as determined by algal assays. Water Res 8:667–679

Redfield AC (1934) On the proportions of organic deviations in sea water and their relation to the composition of plankton. In: Daniel RJ (ed) James Johnstone memorial volume. University Press of Liverpool, Liverpool, pp 177–192

Snoeyink VL, Jenkins D (1980) Water chemistry. Wiley, New York

Tchobanoglous G, Schroeder ED (eds) (1985) Water quality: characteristics, modeling, modification. Adison-Wesley, Reading

Wetzel RG (1975) Limnology. WB Saunders, Philadelphia

Nitrogen

11

Abstract

The atmosphere is a vast storehouse of nitrogen—it consists of 78 % by volume of this gas. Atmospheric nitrogen is converted by electrical activity to nitrate (NO_3^-) that reaches the earth's surface in rainfall. Atmospheric nitrogen also can be fixed as organic nitrogen by bacteria and blue-green algae, and it can be reduced to ammonia (NH_3) by industrial nitrogen fixation. Plants use ammonium (NH_4^+) or nitrate as nutrients for making protein that is passed through the food web. Elevated concentrations of ammonium and nitrate contribute to eutrophication of water bodies. Because it has several valence states, nitrogen undergoes oxidations and reductions most of which are biologically mediated. Nitrogen in organic matter is converted to ammonia (and ammonium) by decomposition. Organic matter with high nitrogen content typically decomposes quickly with release of appreciable ammonia nitrogen ($NH_3 + NH_4^+$). In aerobic zones, nitrifying bacteria oxidize ammonia nitrogen to nitrate, while in anaerobic zones, nitrate is reduced to nitrogen gas by denitrifying bacteria. Ammonia and ammonium exist in a temperature and pH dependent equilibrium—the proportion of NH_3 increases with greater temperature and pH. Elevated concentrations of un-ionized ammonia can be toxic to aquatic organisms. Nitrite sometimes reaches high concentrations even in aerobic water and is potentially toxic to aquatic animals.

Keywords

Nitrogen cycle • Mineralization of organic nitrogen • Nitrification and denitrification • Ammonia toxicity • Nitrogen compounds in water

C.E. Boyd, *Water Quality*, DOI 10.1007/978-3-319-17446-4_11

Introduction

Nitrogen gas (N_2) comprises 78.08 % by volume of the atmosphere. It also is an important constituent of protein in living and dead organic matter—protein contains an average of 16 % nitrogen. Nitrogen is an essential nutrient for all organisms, but it can be converted to amino acids, the basic units of protein, only by microorganisms capable of converting nitrogen gas to ammonia and by plants that use nitrate (NO_3^-) or ammonium (NH_4^+) as nitrogen sources.

Animals and saprophytic microorganisms obtain protein from their diets. They can transform amino nitrogen in their dietary protein into proteins characteristic of their species. Certain essential amino acids that cannot be synthesized by animals or saprophytic microbes must be in their diet—these essential amino acids are made by plants and nitrogen-fixing microorganisms.

Un-ionized ammonia (NH_3) and nitrite (NO_2^-) can be toxic to aquatic organisms. Plants require nitrogen for growth, but excessive concentrations of ammonium and nitrate in water bodies contribute to dense phytoplankton blooms causing eutrophication. Microbial oxidation of ammonia nitrogen ($NH_3 + NH_4^+$)—usually called nitrification—removes dissolved oxygen from the water and produces acidity. Excessive nitrogen gas in water can cause gas bubble trauma in fish and some other aquatic animals as discussed in Chap. 6.

Nitrogen exists in an unusually large number of valence states (Table 11.1). Organisms of decay convert organically-bound nitrogen to ammonia nitrogen during decomposition of organic matter. Ammonia can be oxidized to a variety of nitrogen species, which, in turn, also can be reduced. These oxidations and reductions are mostly microbial transformations that have a great influence on the forms and concentrations of nitrogen in water.

The purpose of this chapter is to discuss nitrogen dynamics in aquatic ecosystems and to consider the role of nitrogen in water quality.

Table 11.1 Valence states of forms of nitrogen

Compound or ion	Formula	Valence
Amino nitrogen	R-NH_2 or R-NH-R	−3
Ammonia and ammonium	NH_3 and NH_4^+	−3
Hydrazine	N_2H_4	−2
Hydroxylamine	H_2NOH	−1
Nitrogen	N_2	0
Nitrous oxide	N_2O	+1
Nitric oxide	NO	+2
Nitrite	NO_2^-	+3
Nitrogen dioxide	NO_2	+4
Nitrate	NO_3^-	+5

R = organic moiety

The Nitrogen Cycle

The nitrogen cycle (Fig. 11.1) usually is presented as a global cycle, but most of the components of this cycle function in much smaller systems. For example, many of the steps depicted in Fig. 11.1 occur in a small pond or even in an aquarium.

The depiction of the nitrogen cycle (Fig. 11.1) shows that the ultimate, natural source of plant available nitrogen is atmospheric and biological nitrogen fixation. But, mankind has discovered how to fix atmospheric nitrogen by an industrial process providing most of the nitrogen fertilizers used in agriculture. Protein made by plants—the primary producers—passes through the food web providing amino nitrogen required by animals. Fecal material of animals and dead plants and animals become a pool of organic matter that is decomposed by bacteria and other organisms of decay. Ammonia released by decomposition of organic matter and excreted by animals is oxidized to nitrate by nitrifying bacteria, and nitrate is reduced to nitrogen gas and returned to the atmosphere by denitrifying bacteria—this completes the cycle.

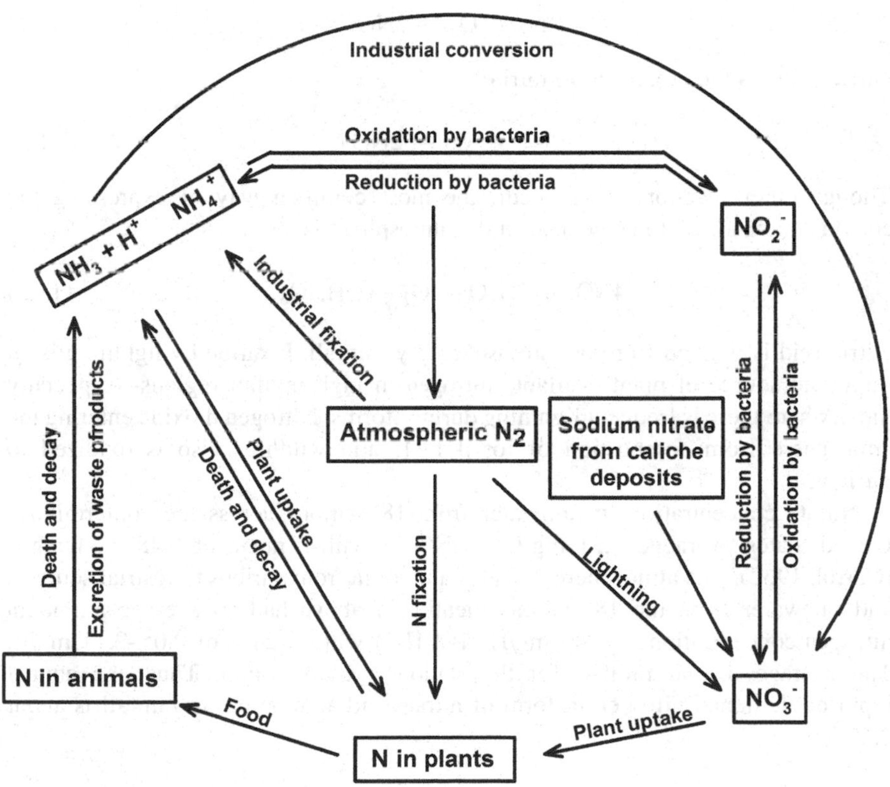

Fig. 11.1 The nitrogen cycle

Most of the earth's nitrogen is in the atmosphere, or contained in organic matter including fossil fuels and living biomass. A small quantity is present in deposits of a sodium nitrate bearing mineral—called caliche—found in the Atacama Desert of Chile and a few other arid places. Most nitrogen transformations are biologically mediated, but gaseous forms of nitrogen can freely exchange between air and water.

Transformations of Nitrogen

Atmospheric Fixation by Lightning

Nitrogen gas in the atmosphere must be transformed to ammonia nitrogen or nitrate to be useful to plants. Some nitrogen gas is oxidized to nitric acid by lightning. This process occurs when the heat from lightning breaks the triple bond between nitrogen atoms allowing them to react with molecular oxygen in the atmosphere forming nitric oxide

$$N_2 + O_2 \rightarrow 2NO. \tag{11.1}$$

Nitric oxide is then oxidized to nitrite

$$2NO + O_2 \rightarrow 2NO_2. \tag{11.2}$$

Though other reactions may occur, the most common way of expressing the conversion of nitrite to nitric acid in the atmosphere is

$$4NO_2 + 2H_2O + O_2 \rightarrow 4HNO_3. \tag{11.3}$$

Nitric acid is washed from the atmosphere by rainfall. Fixation by lightning is an important source of plant available nitrogen in high rainfall regions—especially those where there is frequent lightning during storms. Nitrogen dioxide entering the atmosphere from combustion of fossil fuels and wildfires also is oxidized to nitric acid.

Nitrate concentration in rainwater from 18 stations across the conterminous United States averaged 2.31 mg/L (as NO_3^-) with a range of 0.48–5.34 mg/L (Carroll 1962). The atmosphere contains ammonia from various terrestrial sources, and rainwater from the 18 stations mentioned above had an average ammonia nitrogen concentration of 0.43 mg/L (as NH_4^+) with a range of 0.05–2.11 mg/L. The average annual rainfall for the stations was 875 mm. Thus, the annual input of elemental nitrogen in form of nitrate and ammonium in rainfall is about 7.49 kg/ha.

Biological Fixation

Nitrogen gas is soluble in water; its concentration at equilibrium for different temperatures and salinities are found in Table 6.2. Nitrogen is not highly reactive, but certain species of blue-green algae and bacteria are able to absorb molecular nitrogen from water, transform it to ammonia, and combine ammonia with intermediate carbohydrate compounds to make amino acids. The summary for the reaction producing ammonia nitrogen is

$$N_2 + 8H^+ + 8e^- + 16ATP \rightleftharpoons 2NH_4^+ + 16\,ADP + 16\,PO_4^{2-}. \qquad (11.4)$$

The reaction requires two enzymes both of which require iron and one of which contains molybdenum as cofactors. The ammonia nitrogen produced in (11.4) is used to synthesize amino acids that are combined into protein. The process requires metabolic energy, but the process *per se* does not require molecular oxygen, and can occur in either aerobic or anaerobic environments.

Nitrogen fixation rates in aquatic ecosystems usually are in the range of 1–10 kg/ha per year, but larger rates have been observed. The greatest rates of nitrogen fixation in aquatic environments are for wetlands where certain trees and shrubs have symbiotic nitrogen-fixing bacteria associated with their roots. Large amounts of nitrogen also may be fixed by blue-green algae in rice paddies.

Blue-green algae capable of fixing nitrogen have heterocysts; these large, thick-walled, spherical cells found in filaments of *Nostoc*, *Anabaena*, *Gloeocapsa*, and a few other genera are the site of nitrogen fixation. The ability of some blue-green algae to fix nitrogen gives them an advantage over other algae when nitrate and ammonium concentrations are low; however, nitrogen fixation by phytoplanktic blue-green algae tends to decline as nitrate and ammonia nitrogen concentrations increase in the water. This results because it is more energy efficient to use nutrients already present than to reduce N_2. There also tends to be a decrease in nitrogen fixation in water bodies as the ratio of total N:total P in water increases. At a total N: total P ratio of 13 or more, nitrogen fixation stops (Findlay et al. 1994). This suggests that adding phosphorus to waters will increase the rate of nitrogen fixation.

Industrial Fixation

Industrial nitrogen fixation is accomplished by the Haber-Bosch process. The process involves bringing hydrogen gas made from natural gas or petroleum and atmospheric nitrogen together at high pressure (15–25 M Pa) and temperature (300–550 °C) in the presence of a catalyst (K_2O, CaO, SiO_2, or Al_2O_3). One molecule of N_2 reacts with three molecules of H_2 to form two molecules of NH_3. Ammonia can be used directly or oxidized to nitrate by an industrial process. Most fertilizer nitrogen is from the Haber-Bosch process.

Plant Uptake of Nitrate and Ammonium

Ammonia, ammonium, nitrite, and nitrate nitrogen are said to be combined forms of nitrogen. Plants can absorb both ammonium and nitrate from the water. Most species apparently prefer to use ammonium, because it is energetically less demanding to do so. Nitrate must be reduced to ammonia by the nitrate reductase pathway before it can be used for amino acid synthesis. The process is illustrated below:

$$NO_3 \rightarrow NO_2 \rightarrow NH_3 \quad (11.5)$$

$$\text{Carbohydrates} + NH_3 \rightarrow \text{amino acids.} \quad (11.6)$$

The nitrate reductase enzyme contains molybdenum and the reduction requires metabolic energy from ATP.

Rapidly growing phytoplankton absorb large amounts of combined nitrogen from water because they need nitrogen to make their protein (Ex. 11.1). Most species contain from 5 to 10 % of their dry weight as nitrogen.

Ex. 11.1: *Phytoplankton net productivity in a body of water is 2 g carbon/m²/day in the 2-m deep photic zone. Assuming the phytoplankton is 50 % carbon and 8 % nitrogen (dry weight), daily nitrogen removal from the water will be estimated.*

Solution:

$$2\,\text{g}\,C/\text{m}^2/\text{day} \div 0.50\,\text{g}\,C/\text{g}\,dry\ wt = 4\,\text{g}\,dry\ wt/\text{m}^2/\text{day}$$

$$4\,\text{g}\,dry\ wt/\text{m}^2/\text{day} \times 0.08\,\text{g}\,N/\text{g}\,dry\ wt = 0.32\,\text{g}\,N/\text{m}^2/\text{day}$$

This amount is equivalent to 3.2 kg N/ha/day.

In terms of concentration, phytoplankton removal equaled 0.16 mg N/L/day

$$0.32\,\text{g}\,N/\text{m}^2/\text{day} \div 2\,\text{m}^3/\,\text{m}^2 = 0.16\,\text{g}\,N/\text{m}^3\ or\ 0.16\,\text{mg/L}.$$

Phytoplankton uptake is a major factor affecting combined nitrogen concentrations in water. Aquatic macrophyte stands also may contain large amounts of nitrogen (Ex. 11.2). Stands of macrophytes can remove nitrogen from the water and hold it in their biomass throughout the growing season, and thereby reduce the availability of nitrogen to phytoplankton.

Ex. 11.2: *A body of water 10-ha in area and 1-m average depth contains a 2-ha stand of rooted aquatic plants along the shallow edges. Assuming a standing crop of macrophytes of 800 g/m² (dry wt) with 3 % nitrogen, estimate the amount of nitrogen in macrophyte biomass.*

Solution:

$$800\,\text{g/m}^2 \times 10{,}000\,\text{m}^2/\text{ha} \times 2\,\text{ha} \times 0.03\,\text{g}\,N/\text{g}\ \textit{dry wt} = 480{,}000\,\text{g}\,N.$$

In terms of concentration, nitrogen in macrophyte biomass is equal to a concentration of 4.8 mg/L if returned to the water as combined nitrogen

$$\frac{480{,}000\,\text{g}\,N}{10\,\text{ha} \times 1\,\text{m} \times 10{,}000\,\text{m}^2/\text{ha}} = 4.8\,\text{g}\,N/\text{m}^3 \ \textit{or}\ 4.8\,\text{mg/L}.$$

Fate of Nitrogen in Plants

Phytoplankton and other plants may be consumed by animals or they may die and become organic matter to be decomposed by microorganisms. When animals eat plants, at least 10 units of plant dry weight usually are necessary to produce 1 unit of dry animal biomass and the ratio is often greater and especially for fibrous plants of low nitrogen content. The nitrogen in plant material consumed by animals and not converted to nitrogen in biomass is excreted in feces or other metabolic wastes (Ex. 11.3).

Ex. 11.3: *Suppose the conversion of phytoplankton to animal tissue (dry weight basis) is 15:1, and the plants contain 6 % nitrogen while the animals are 11 % nitrogen. The nitrogen waste will be estimated for 1,000 g of animal biomass.*

Solution:

To obtain 1,000 g of animal biomass require 15,000 g of plants. The nitrogen budget is
Plant nitrogen $15{,}000\,\text{g} \times 0.06\,\text{g}\,N/\text{g} = 900\,\text{g}\,N$
Animal nitrogen $1{,}000\,\text{g} \times 0.11\,\text{g}\,N/\text{g} = 110\,\text{g}\,N$
Waste nitrogen $(900 - 110)\,\text{g}\,N = 790\,\text{g}\,N.$

Of course, when animals die, their bodies become organic matter to be decomposed by bacteria resulting in nitrogen recycling.

Mineralization of Organic Nitrogen

When bacteria and fungi decompose organic matter, part of the nitrogen in organic matter is converted to organic nitrogen in microbial biomass and the remainder is released to the environment mainly in the form of ammonia—a process called mineralization of nitrogen. The ability of saprophytic microorganisms to convert substrate organic carbon to microbial carbon is known as the carbon assimilation efficiency (CAE). Aerobic bacteria typically convert 5–10 % of substrate carbon to biomass with the rest being released as carbon dioxide or waste carbon compounds. The CAE for anaerobic bacteria of 2–5 % is less than that of aerobic bacteria. Fungi have a CAE of 40–55 % (Hoorman and Islam 2010). The dry matter carbon and nitrogen contents of bacterial biomass are about 50 % and 10 %, respectively. Fungi contain as much carbon as bacteria, but only about 5 % or half as much nitrogen. Based on these facts, a nitrogen budget for decomposition of organic matter will be made to illustrate nitrogen mineralization. In this budget (Ex. 11.4), it will be assumed that bacteria completely and quickly use all of the substrate leaving only living microbial biomass and wastes.

Ex. 11.4: *The amount of nitrogen mineralized by bacteria with a CAE of 7.5 % in completely decomposing 1,000 g (dry wt) of substrate containing 42 % organic carbon and 4 % nitrogen will be determined.*

Solution:

Substrate C: $1{,}000\,\text{g} \times 0.42 = 420\,\text{g}\,C$
Substrate N: $1{,}000\,\text{g} \times 0.04 = 40\,\text{g}\,N$
Bacterial biomass produced:

$$420\,\text{g}\,C \times 0.075\,\text{g}\,\textit{bacterial}\,C/\text{g}\,\textit{substrate}\,C = 31.5\,\text{g}\,\textit{bacterial}\,C$$

$$31.5\,\text{g}\,\textit{bacterial}\,C \div 0.5\,\text{g}\,C/\text{g}\,\textit{dry bacteria} = 63\,\text{g}\,\textit{bacteria}$$

$$\textit{Bacterial nitrogen}: 63\,\text{g}\,\textit{bacteria} \times 0.1\,\text{g}\,N/\text{g}\,\textit{bacteria} = 6.3\,\text{g}\,\textit{bacterial}\,N$$

Nitrogen mineralized:

$$40\,\text{g}\,\textit{substrate}\,N - 6.3\,\text{g}\,\textit{bacterial}\,N = 33.7\,\text{g}\,N\,\textit{released to environment.}$$

Using the same methods illustrated in Ex. 11.4, only 21.1 g of nitrogen would be mineralized by fungi with a CAE of 45 %.

The ratio of carbon to nitrogen (C:N ratio) in organic matter is a key factor controlling decomposition and nitrogen mineralization rates. Organic matter normally contains 40–50 % carbon on a dry matter basis. Nitrogen content is more variable ranging from less than 0.4–10 % or more. As the nitrogen percentage in organic matter increases, there tends to be an increase in the rate of decomposition and in the proportion of organic nitrogen mineralized. This results because a high nitrogen content is associated with organic matter that contains less decay-resistant structural compounds and more easily degradable protoplasmic material. Also, a high nitrogen content assures that there is more than enough nitrogen to sustain the microbial biomass that develops during decay.

In a substrate with less nitrogen than needed to effect its complete and rapid decomposition by microorganisms, one or both of two phenomena occur during decomposition: (1) decomposition is slow and microorganisms must die so that their nitrogen can be mineralized and used again to decompose substrate;

(2) ammonia nitrogen and nitrate in the environment can be removed and used by microorganisms to decompose the nitrogen-deficient residue. The later process is called nitrogen immobilization, and it results in a decrease in concentrations of ammonia nitrogen and nitrate in the zone where decomposition is occurring. Where nitrogen is available for immobilization, decomposition of low-nitrogen content residues often is greatly accelerated.

The influence of the C:N ratio on nitrogen balance during decomposition can be illustrated by solving the nitrogen budget (see Ex. 11.4 for procedure) for two 1,000 g residues, one with 42 % carbon and 1 % N and a second with 42 % carbon and 0.2 % N, decomposed by bacteria with a CAE of 7.5 %. Decomposition of the first residue would mineralize 3.7 g nitrogen—the second residue with 0.2 % nitrogen contains only 2 g nitrogen. This is 4.3 g less nitrogen than needed by the bacteria decomposing the residue.

In reality, high nitrogen content will not assure that an organic residue will quickly decompose. Most organic matter consists of a variety of different compounds, and some of these substances will be much slower to decay than others. Nevertheless, organic matter with a narrow C:N ratio usually will decompose faster and release more nitrogen to the environment than organic substances with a wide C:N ratio.

Nitrification

In nitrification, chemoautotrophic bacteria oxidize ammonia nitrogen to nitrate. The reaction occurs in two steps:

$$NH_4^+ + 1\tfrac{1}{2}\,O_2 \rightarrow NO_2^- + 2H^+ + H_2O \quad (11.7)$$

$$NO_2^- + 0.5\,O_2 \rightarrow NO_3^-. \quad (11.8)$$

Bacteria of the genus *Nitrosomonas* conduct the first oxidation, and the second oxidation is caused by bacteria of the genus *Nitrobacter*. These two genera of bacteria usually occur together in the same environment. Nitrite produced in the first reaction is quickly oxidized to nitrate in the second reaction, and nitrite seldom accumulates. The two reactions can be added to give the overall nitrification equation

$$NH_4^+ + 2O_2 \rightarrow NO_3^- + 2H^+ + H_2O. \quad (11.9)$$

Energy is released when ammonium is oxidized to nitrate, and *Nitrosomonas* and *Nitrobacter* have mechanisms for capturing in ATP a part of this energy and using it to reduce carbon dioxide to organic carbon. This result is similar to photosynthesis in that organic matter is synthesized, but light is not required and the biochemical pathway differs from the one of photosynthesis. The ratio of organic carbon synthesized to energy released through ammonia oxidation is quite low, and nitrification is not a significant, primary source of organic carbon in ecosystems.

Nitrification is most rapid at temperatures of 25–35 °C and at pH values between 7 and 8. Nitrification can occur down to pH 3 or 4 and at low temperatures, but rates will be slow.

Some bacteria can carry out anaerobic oxidation of ammonia. According to van der Graaf et al. (1995), the process probably occurs by the reaction

$$5NH_4^+ + 3NO_3^- \rightarrow 4N_2 + 9H_2O + 2H^+. \quad (11.10)$$

Nitrate is the oxygen source for oxidation of ammonium to nitrogen gas. This reaction could just as aptly be called denitrification because nitrate also is reduced to nitrogen gas.

Nitrification by both aerobic and anaerobic processes produces hydrogen ion, so the nitrification process contributes to acidity (Ex. 11.5). Aerobic nitrification requires an abundant supply of molecular oxygen (Ex. 11.5), while anaerobic oxidation of ammonium (11.10) requires oxygen from nitrate.

Ex. 11.5: *The amount of oxygen consumed and the amount of acidity produced (in terms of alkalinity loss) in the oxidation of 1 mg/L ammonia nitrogen in aerobic nitrification will be estimated.*

Solution:

From (11.9), *one nitrogen consumes two oxygen molecules and produces two hydrogen ions. Thus,*

$$\text{(i)} \quad \begin{matrix} 1\,\text{mg/L} \\ N \\ 14 \end{matrix} = \begin{matrix} x \\ 2O_2 \\ 64 \end{matrix} \quad x = 4.57\,\text{mg/L } \textit{of dissolved oxygen consumed.}$$

(ii) *Two hydrogen ions react with one calcium carbonate (alkalinity is expressed as* $CaCO_3$*)*

$$CaCO_3 + 2H^+ = Ca^{2+} + CO_2 + H_2O$$

and

$$\begin{matrix} 1\,\text{mg/L} \\ N \\ 14 \end{matrix} = 2H^+ = \begin{matrix} x \\ CaCO_3 \\ 100 \end{matrix}$$

$$x = 7.14\,\text{mg/L } \textit{of total alkalinity neutralized.}$$

Rates of nitrification differ greatly among waters and over time in the same body of water. They tend to be much greater in summer than in winter, and they increase when ammonia nitrogen concentrations increase. An abundant supply of dissolved oxygen is necessary for rapid nitrification. Nitrification can occur both in the water column and in aerobic sediment. Gross (1999) found that nitrification produced an average of 240 mg NO_3-N/m^2/day and 260 mg NO_3-N/m^2/day in the water column and sediment, respectively, of shallow, eutrophic ponds. However, Hargreaves (1995) reported that nitrification in the aerobic layers of sediment could be several times greater than in the water column.

Denitrification

Under anaerobic conditions, some bacteria use oxygen from nitrate as a substitute for molecular oxygen to oxidize organic matter (respire). One equation for denitrification is

$$6NO_3^- + 5CH_3OH \rightarrow 5CO_2 + 3N_2 + 7H_2O + 6OH^-. \tag{11.11}$$

Methanol is the carbon source for denitrification in (11.11)—this is the carbon source frequently used in the denitrification process in sewage treatment plants. But, in nature many types of organic matter can be used. The process is called denitrification because nitrate nitrogen is converted to nitrogen gas and lost from the ecosystem by diffusion into the atmosphere. Denitrification yields one mole of hydroxyl ion for each mole of nitrate reduced; thus, it adds alkalinity to the water.

$$OH^- + CO_2 \rightarrow CO_3^-. \tag{11.12}$$

Nitrification and denitrification can be thought of as coupled, because nitrate produced in nitrification can be denitrified. The loss of alkalinity caused by nitrification is not completely restored even if nitrification and denitrification are perfectly coupled. Each mole of nitrogen oxidized in nitrification releases two hydrogen ions, while each mole of nitrogen denitrified releases only one hydroxyl ion. The potential acidity of nitrification is 7.14 mg/L equivalent $CaCO_3$ while the potential alkalinity of denitrification is 3.57 mg/L equivalent $CaCO_3$.

Denitrification does not always lead to formation of N_2. It can end with nitrite

$$NO_3^- + 2H^+ \rightarrow NO_2^- + H_2O. \tag{11.13}$$

Ammonia can be an end product when nitrate is reduced to nitrite followed by conversion of nitrite to hyponitrite ($H_2N_2O_2$) and hyponitrite to hydroxylamine (H_2NOH)

$$2NO_2^- + 5H^+ \rightarrow H_2N_2O_2 + H_2O + OH^- \tag{11.14}$$

$$H_2N_2O_2 + 4H^+ \rightarrow 2H_2NOH \tag{11.15}$$

$$H_2NOH + 2H^+ \rightarrow NH_3 + H_2O. \tag{11.16}$$

Another pathway results in formation of nitrous oxide as follows:

$$H_2N_2O_2 \rightarrow N_2O + H_2O. \quad (11.17)$$

However, N_2 formation is the most common pathway.

Denitrification rates can be very high in aquatic ecosystems where nitrate is abundant and there is a readily available source of organic matter. The denitrification rate during the period June to October in small, eutrophic ponds at Auburn, Alabama, in which fish were fed a high-protein content ration ranged from 0 to 60 mg N/m^2/day (average = 38 mg N/m^2/day). On an annual basis, the average rate is equivalent to 139 kg N/ha (Gross 1999). Nitrogen inputs to most aquatic ecosystems are not great enough to support such large amounts of denitrification.

Nitrite from denitrification or nitrification may sometimes accumulate in water to concentrations of 1 mg/L or more. When this phenomenon occurs, there may be toxicity to fish and other aquatic animals.

Ammonia-Ammonium Equilibrium

Ammonia nitrogen exists in water as ammonia (NH_3) and ammonium ion (NH_4^+). The un-ionized form is a gas, and it also is potentially toxic to fish and other aquatic life at relatively low concentration. Ammonium is not appreciably toxic.

Because ammonia hydrolyses to ammonium in aqueous solution, the two forms exist in a pH and temperature dependent equilibrium

$$NH_3 + H_2O = NH_4^+ + OH^- \quad K_b = 10^{-4.75}. \quad (11.18)$$

That also may be expressed as

$$NH_4^+ = NH_3 + H^+. \quad (11.19)$$

Ammonia acts as a base in (11.18), and ammonium is an acid in (11.19). Because $K_aK_b = K_w = 10^{-14}$ (at 25 °C), $K_a = 10^{-14}/K_b$, and K_a for (11.19) is $10^{-9.25}$. The K_a and K_b values mentioned above are for 25 °C, but values of K_a and K_b for other temperatures obtained from Bates and Pinching (1949) are presented (Table 11.2) along with K_w values for different temperatures.

Table 11.2 Equilibrium constants for ammonium dissociation (K_a), the hydrolysis of ammonia (K_b), and dissociation of water (K_w) (Modified from Bates and Pinching 1949)

Temperature (°C)	K_a	K_b	K_w
0	$10^{-10.08}$	$19^{-4.86}$	$10^{-14.94}$
5	$10^{-9.90}$	$10^{-4.83}$	$10^{-14.73}$
10	$10^{-9.73}$	$10^{-4.80}$	$10^{-14.53}$
15	$10^{-9.56}$	$10^{-4.78}$	$10^{-14.35}$
20	$10^{-9.40}$	$10^{-4.77}$	$10^{-14.17}$
25	$10^{-9.25}$	$10^{-4.75}$	$10^{-14.00}$
30	$10^{-9.09}$	$10^{-4.74}$	$10^{-13.83}$
35	$10^{-8.95}$	$10^{-4.73}$	$10^{-13.68}$
40	$10^{-8.80}$	$10^{-4.73}$	$10^{-13.53}$

The ratio of NH_3-N to NH_4-N can be calculated most conveniently from (11.19) as follows:

$$\frac{(NH_3\text{-N})(H^+)}{(NH_4^+\text{-N})} = K_a$$

$$\frac{(NH_3\text{-N})}{(NH_4^+\text{-N})} = \frac{K_a}{(H^+)}.$$

By assigning a value of 1.0 for (NH_4^+-N) concentration, the proportion of the total ammonia nitrogen present in NH_3-N form may be estimated as illustrated in Ex. 11.6 by aid of the equation.

$$\frac{(NH_3\text{-N})}{(NH_4^+\text{-N})} = \left(\frac{10^{-9.25}}{(H^+)}\right) \div \left(1 + \frac{(10^{-9.25})}{(H^+)}\right). \tag{11.20}$$

Ex. 11.6: *Calculation of the proportion of the TAN concentration in NH_3-N form will be calculated for pH 8.0 and 9.0 at temperatures of 20 °C and 30 °C.*

Solution:

At 20 °C and pH 8.0 (K_a from Table 11.2)

$$\frac{(NH_3\text{-}N)}{(NH_4^+\text{-}N)} = \left(\frac{10^{-9.40}}{10^{-8.0}}\right) \div \left(1 + \frac{10^{-9.40}}{10^{-8.0}}\right) = 0.0398 \div 1.0398 = 0.0383.$$

At 20 °C and pH 9.0

$$\frac{(NH_3\text{-}N)}{(NH_4^+\text{-}N)} = \left(\frac{10^{-9.40}}{10^{-9.0}}\right) \div \left(1 + \frac{10^{-9.40}}{10^{-9.0}}\right) = 0.398 \div 1.398 = 0.2847.$$

At 30 °C and pH 8.0 (K_a from Table 11.2)

$$\frac{(NH_3\text{-}N)}{(NH_4^+\text{-}N)} = \left(\frac{10^{-9.09}}{10^{-8.0}}\right) \div \left(1 + \frac{10^{-9.09}}{10^{-8.0}}\right) = 0.0813 \div 1.0813 = 0.075.$$

At 30 °C and pH 9.0

$$\frac{(NH_3\text{-}N)}{(NH_4^+\text{-}N)} = \left(\frac{10^{-9.09}}{10^{-9.0}}\right) \div \left(1 + \frac{10^{-9.09}}{10^{-9.0}}\right) = 0.813 \div 1.813 = 0.448.$$

The proportion of NH_3 increases with both increasing temperature and pH, but the increase is greater with pH (Ex. 11.6). At 20 °C the proportion of NH_3-N was 0.0382 at pH 8.0, but at the same pH and 30 °C, the proportion increased to 0.075—roughly a twofold increase. But, when the pH increased from 8.0 to 9.0 at 20 °C, the proportion of NH_3 rose from 0.0382 to 0.2847—an increase of more than sevenfold.

The K_a values for selected temperatures have been used to estimate the proportions of the total ammonia nitrogen concentration consisting of NH_3-N at different pHs (Table 11.3). The usual analytical methods for ammonia nitrogen concentration in water measure both NH_3-N and NH_4^+-N. Nevertheless, the concentration of potentially toxic NH_3-N can be estimated easily for a particular pH and water temperature by multiplying total ammonia nitrogen concentration by the appropriate factor from Table 11.3 as illustrated in Ex. 11.7. If the critical NH_3-N concentration—the concentration that is potentially toxic to fish is known, a table

Table 11.3 Decimal fractions (proportions) of total ammonia existing as un-ionized ammonia in freshwater at various pH values and temperatures (Boyd 1990)

	Temperature (°C)								
pH	16	18	20	22	24	26	28	30	32
7.0	0.003	0.003	0.004	0.004	0.005	0.006	0.007	0.008	0.009
7.2	0.004	0.005	0.006	0.007	0.008	0.009	0.011	0.012	0.015
7.4	0.007	0.008	0.009	0.011	0.013	0.015	0.017	0.020	0.023
7.6	0.011	0.013	0.015	0.017	0.020	0.023	0.027	0.031	0.036
7.8	0.018	0.021	0.024	0.028	0.032	0.036	0.042	0.048	0.057
8.0	0.028	0.033	0.038	0.043	0.049	0.057	0.065	0.075	0.087
8.2	0.044	0.051	0.059	0.067	0.076	0.087	0.100	0.114	0.132
8.4	0.069	0.079	0.090	0.103	0.117	0.132	0.149	0.169	0.194
8.6	0.105	0.120	0.136	0.154	0.172	0.194	0.218	0.244	0.276
8.8	0.157	0.178	0.200	0.223	0.248	0.276	0.306	0.339	0.377
9.0	0.228	0.255	0.284	0.313	0.344	0.377	0.412	0.448	0.490
9.2	0.319	0.352	0.386	0.420	0.454	0.489	0.526	0.563	0.603
9.4	0.426	0.463	0.500	0.534	0.568	0.603	0.637	0.671	0.707
9.6	0.541	0.577	0.613	0.645	0.676	0.706	0.736	0.763	0.792
9.8	0.651	0.684	0.715	0.742	0.768	0.792	0.815	0.836	0.858
10.0	0.747	0.774	0.799	0.820	0.840	0.858	0.875	0.890	0.905
10.2	0.824	0.844	0.863	0.878	0.892	0.905	0.917	0.928	0.938

similar to Table 11.3 may be constructed in which the table entries represent the maximum acceptable total ammonia nitrogen concentrations. For instance, if the critical concentration is 0.10 mg/L NH_3-N, at pH 8 and 25 °C, the total ammonia nitrogen concentration should not exceed 1.92 mg/L (0.1 mg/L ÷ 0.052).

Ex. 11.7: *A water sample contains 1 mg/L total ammonia nitrogen. The concentration of un-ionized ammonia nitrogen will be estimated for pH 7, 8, 9, and 10 and 30°C.*

Solution:

From Table 11.3, *the proportions of un-ionized ammonia are 0.008, 0.075, 0.448, and 0.890 in order of increasing pH. Thus,*

$pH\ 7\colon 1\,\mathrm{mg/L} \times 0.008 = 0.008\,\mathrm{mg}\ NH_3\text{-}N/\mathrm{L}$
$pH\ 8\colon 1\,\mathrm{mg/L} \times 0.075 = 0.075\,\mathrm{mg}\ NH_3\text{-}N/\mathrm{L}$
$pH\ 9\colon 1\,\mathrm{mg/L} \times 0.448 = 0.448\,\mathrm{mg}\ NH_3\text{-}N/\mathrm{L}$
$pH\ 10\colon 1\,\mathrm{mg/L} \times 0.890 = 0.890\,\mathrm{mg}\ NH_3\text{-}N/\mathrm{L}.$

Ammonia Diffusion

The potential for diffusion of ammonia from the surface of a water body into the air obviously will be greatest when the pH is high. Weiler (1979) reported ammonia diffusion losses up to 10 kg N/ha per day when wind velocity, pH, and total ammonia nitrogen concentrations were high. In most waters, the ammonia nitrogen concentration will not be great enough to support such large ammonia diffusion rates. Gross et al. (1999) reported diffusion losses of 9–71 mg N/m^2/day from small ponds with 0.05–5.0 mg/L of total ammonia nitrogen, afternoon pH of 8.3–9.0, and water temperature of 21–29 °C. The effect of ammonia diffusion to the atmosphere on the total ammonia concentration in water is illustrated (Ex. 11.8).

Ex. 11.8: *The lowering of ammonia nitrogen concentration by diffusion of ammonia at 10 mg NH_3-N/m^2/day will be calculated for a 1-ha water body that has an average depth of 2 m.*

Solution:

$$\frac{10\,\mathrm{mg}\ NH_3\text{-}N/\mathrm{m}^2/\mathrm{day}}{2\mathrm{m}^3/\mathrm{m}^2} = 5\,\mathrm{mg}\ NH_3\text{-}N/\mathrm{m}^3/\mathrm{day} = 0.005\,\mathrm{mg}\ NH_3\text{-}N\ /\mathrm{L}/\mathrm{day}.$$

The decrease in ammonia nitrogen concentration is small for 1 day, but over a year, the total loss of nitrogen could be substantial. If the loss rate used above is expanded for 1 year, the nitrogen loss would be

$$10\,\mathrm{mg}\ NH_3\text{-}N/\mathrm{m}^2/\mathrm{day}\ \times 365\ \mathrm{day/year} = 3.65\ \mathrm{g/m}^2.$$

This is 36.5 kg N/ha per year.

Nitrogen Inputs and Outputs

Aquatic ecosystems have various inputs and outputs of nitrogen. The inputs are rainfall, inflowing water containing nitrogen from natural sources and pollution, nitrogen added intentionally (as in aquaculture), and nitrogen fixation. The outputs are outflowing water, harvest of aquatic products, intentional withdrawal of water, seepage, diffusion of ammonia into the atmosphere, and denitrification. The relative importance of each gain and loss varies greatly from one body of water to another.

Nitrogen is stored in aquatic ecosystems primarily as organic matter in bottom sediments, but smaller amounts are found in the water to include nitrogen gas, nitrate, nitrite, ammonia, and nitrogen in dissolved and particulate organic matter. When nitrogen in organic matter is mineralized, it is subject to the various processes and transformations described above, and it can be lost from the ecosystem. However, there is a tendency for recycling of nitrogen in aquatic ecosystems, and in most, an equilibrium among inputs, outputs, and stored nitrogen is reached. Pollution can quickly disrupt this equilibrium and cause higher nitrogen concentrations in the water and sediment.

Concentrations of Nitrogen in Water and Sediment

Nitrogen gas is not as biologically active as dissolved oxygen, so its concentrations usually are near saturation (Table 6.2). Ammonia nitrogen and nitrate nitrogen concentrations normally are less than 0.25 mg/L in unpolluted waters. They may be above 1 mg/L in polluted waters, and in highly polluted waters concentrations of 5–10 mg/L are not uncommon. Nitrite nitrogen is seldom present at concentrations above 0.05 mg/L in oxygenated water, but it may reach concentrations of several milligrams per liter in polluted waters with low dissolved oxygen concentration. There is less information on dissolved organic nitrogen, for the usual procedure is to determine total organic nitrogen. Measurements of total organic nitrogen include dissolved and particulate organic nitrogen, and they usually are below 1 or 2 mg/L in relatively unpolluted natural waters. Effluents and polluted waters may have much higher concentrations of total organic nitrogen.

Concentrations of nitrogen also are highly variable in sediment. As a general rule, organic matter is about 5 % nitrogen, and sediments in aquatic ecosystems usually contain 1–10 % organic matter—0.05–0.5 % nitrogen.

Significance

As a Nutrient

Plant growth often increases in response to increased nitrogen concentrations, because a shortage of available nitrogen is a limiting factor in many aquatic ecosystems. In aquaculture, fertilizers that contain nitrogen and phosphorus often are applied to ponds to increase phytoplankton productivity and enhance the base of the food web to support greater fish or shrimp production. In most instances, however, addition of nitrogen to water bodies is considered to be nutrient pollution, because it can contribute to dense phytoplankton blooms, i.e., cause eutrophication. The concentration of nitrogen necessary to cause excessive phytoplankton blooms in natural waters is difficult to ascertain, because many factors in addition to nitrogen concentration influence phytoplankton productivity. Concentrations of 0.1–0.75 mg/L of ammonia nitrogen plus nitrate nitrogen in freshwaters—even less in brackish and marine waters—have been adequate to cause phytoplankton blooms. Limitation of ammonia nitrogen and nitrate nitrogen concentrations and loads in effluents is an important tool in water pollution and eutrophication control.

As a Toxin

Ammonia is the major nitrogenous excretory product of aquatic animals, and like other excretory products, is toxic if it cannot be excreted. A high concentration of ammonia in the water makes it more difficult for organisms to excrete ammonia. As a result the concentration of ammonia in the blood of fish and other aquatic animals increases with greater environmental ammonia concentration.

Although the mechanism of ammonia toxicity has not been elucidated entirely, a number of physiological and histological effects of high ammonia concentration in the blood and tissues of fish have been identified as follows: elevation of blood pH; disruption of enzyme systems and membrane stability; increased water uptake; increased oxygen consumption; gill damage; histological lesions in various internal organs.

The concentration of ammonia in most water bodies fluctuates daily because the proportion of ammonia to total ammonia nitrogen varies with pH and temperature (Table 11.3). Most data on ammonia toxicity were based on exposure to constant ammonia concentration, and less is known about the effects of fluctuating ammonia concentrations than about exposure to a constant concentration on aquatic animals. Moreover, the potential for toxicity at a given ammonia nitrogen concentration depends on several factors in addition to pH and temperature. Toxicity increases with decreasing dissolved oxygen concentration, but this effect often is negated because high carbon dioxide concentration decreases ammonia toxicity. There is evidence that ammonia toxicity decreases with increasing concentrations of salinity and calcium. Fish also tend to increase their tolerance for ammonia when acclimated to gradually increasing ammonia concentrations over several weeks or months.

Most ammonia toxicity data come from LC50 tests on fish—the LC50 is the concentration required to kill 50 % of the animals exposed. The 96-h LC50 concentrations for un-ionized ammonia nitrogen to various species of fish range from about 0.3 to 3.0 mg/L (Ruffier et al. 1981; Hargreaves and Kucuk 2001). Coldwater species usually are more susceptible to ammonia than warmwater species. Un-ionized ammonia should not cause lethal or sublethal effects at concentrations below 0.005–0.01 mg/L for coldwater species or below 0.01–0.05 mg/L for warmwater species. Corresponding concentrations of total ammonia nitrogen depend upon pH and temperature. For example, the concentrations of ammonia nitrogen necessary to give 0.05 mg/L un-ionized ammonia nitrogen at selected pH values and 26 °C are as follows: pH 7, 8.33 mg/L; pH 8, 0.88 mg/L; pH 9, 0.13 mg/L. Ammonia toxicity obviously is of greater concern in waters where the pH is well above neutral than in neutral or acidic waters. The main effect of ammonia on aquatic life probably is stress rather than mortality. Several studies have shown that ammonia concentrations well below lethal concentrations lead to poor appetite, slow growth, and greater susceptibility to disease.

Nitrite is absorbed from the water by fish and other organisms. In the blood of fish, nitrite reacts with hemoglobin to form methemoglobin. Methemoglobin does not combine with oxygen like hemoglobin, so high nitrite concentrations in water can result in a functional anemia known as methemoglobinemia. Blood that contains significant methemoglobin is brown, so the common name for nitrite toxicity in fish is "brown blood disease." Nitrite also can bind hemocyanin, the oxygen binding pigment in crustacean blood, to lower the ability of their blood to transport oxygen.

Nitrite is transported across the gill by lamellar chloride cells. These same cells apparently transport chloride and they cannot distinguish between nitrite and chloride. Thus, the rate of nitrite absorption declines, at least in freshwater fish, as the concentration of chloride increases relative to nitrite. In freshwater, a chloride to nitrite ratio of 6–10:1 will prevent methemoglobinemia at nitrite concentrations up to at least 5 or 10 mg/L (Boyd and Tucker 1998). In brackishwater, high concentrations of calcium and chloride tend to minimize problems of nitrite toxicity in fish (Crawford and Allen 1977).

Because the toxic effects of nitrite are much worse in waters of low dissolved oxygen concentration and also depend upon chloride, calcium, and salinity concentrations, it is virtually impossible to make recommendations on lethal, sublethal, and safe concentrations. The 96-h LC50 values (as nitrite) are in the range of 0.66–200 mg/L for freshwater fish and crustaceans and 40–4,000 mg/L for brackishwater and marine species (Boyd 2014). Safe concentrations would likely be about 0.05 times the 96-h LC50 concentration for a species.

As Cause of Gas Bubble Trauma

Because nitrogen is the most abundant atmospheric gas, it is the major gas occurring in water supersaturated with air. Nitrogen usually is the primary gas contributing to gas bubble trauma in aquatic animals exposed to gas supersaturated water. Common reasons for air supersaturation are sudden warming of water and water entraining air when it falls over high dams into streams below. Air leaks on the suction sides of pumps also can cause air supersaturation of discharge water (Chap. 6).

As Atmospheric Pollutant

Nitrous oxide that can be produced by denitrification in anaerobic environments is not generally considered a water quality issue, because it diffuses from water bodies into the atmosphere. However, nitrous oxide is one of the greenhouse gases that can increase the heat retaining capacity of the atmosphere to cause global warming. The global warming potential of nitrous oxide is about 289 times greater over a 20-year span than that of carbon dioxide—the standard for greenhouse global warming potential (Boyd and McNevin 2014).

Nitrogen dioxide (NO_2) and related nitrogen oxides—often referred to collectively as NO_x—are produced by combustion of fossil fuels and by wildfires. In the atmosphere, nitrogen oxides are oxidized to nitric acid and contribute to the acid rain phenomenon.

References

Bates RG, Pinching GD (1949) Acid dissociation constant of ammonium ion at 0° to 50°C, and the base strength of ammonia. J Res Natl Bur Stan 42:419–420

Boyd CE (1990) Water quality in ponds for aquaculture. Alabama Agricultural Experiment Station, Auburn University, Alabama

Boyd CE (2014) Nitrite toxicity affected by species susceptibility, environmental conditions. Glob Aquac Advocate 17:34–37

Boyd CE, McNevin AA (2014) Aquaculture, resource use, and the environment. Wiley-Blackwell, Hoboken, in press

Boyd CE, Tucker CS (1998) Pond water quality management. Kluwer Academic, Boston

Carroll D (1962) Rainwater as a chemical agent of geologic processes—a review. United States Geological Survey water-supply paper 1535 G, United States Government Printing Office, Washington, DC

Crawford RE, Allen GH (1977) Seawater inhibition of nitrite toxicity to Chinook salmon. Trans Am Fish Soc 106:105–109

Findlay DL, Hecky RE, Hendzel LL, Stainton MP, Regehr GW (1994) Relationship between N_2-fixation and heterocyst abundance and its relevance to the nitrogen budget of lake 227. Can J Fish Aquat Sci 51:2254–2266

Gross A (1999) Nitrogen cycling in aquaculture ponds. Dissertation, Auburn University, Alabama

Gross A, Boyd CE, Wood CW (1999) Ammonia volatilization from freshwater fish ponds. J Environ Qual 28:793–797

Hargreaves JA (1995) Nitrogen biochemistry of aquaculture pond sediments. Dissertation, Louisiana State University, Baton Rouge

Hargreaves JA, Kucuk S (2001) Effects of diel un-ionized ammonia fluctuation on juvenile striped bass, channel catfish, and blue tilapia. Aquaculture 195:163–181

Hoorman JJ, Islam R (2010) Understanding soil microbes and nutrient recycling. Fact sheet SAG-16-10, The Ohio State University Extension, Columbus

Ruffier PJ, Boyle WC, Kleinschmidt J (1981) Short-term acute bioassays to evaluate ammonia toxicity and effluent standards. J Water Pollut Control Fed 53:367–377

van der Graaf AA, Mulder A, de Bruijin P, Jetten MSM, Robertson LA, Kuenen JG (1995) Anaerobic oxidation of ammonium is a biologically mediated process. Appl Environ Microbiol 61:1246–1251

Weiler RR (1979) Rate of loss of ammonia from water to the atmosphere. J Fish Res Bd Can 36:685–689

Phosphorus

12

Abstract

Phosphorus usually is the most important nutrient limiting phytoplankton productivity in both aquatic and terrestrial ecosystems. Phosphorus occurs naturally in most geological formations and soils in varying amounts and forms; the main source of agricultural and industrial phosphate is deposits of the mineral apatite—known as rock phosphate. Municipal and agricultural pollution is a major source of phosphorus to many water bodies. Most dissolved inorganic phosphorus in aquatic ecosystems is an ionization product of orthophosphoric acid (H_3PO_4). At the pH of most water bodies, HPO_4^{2-} and H_2PO_4 are the forms of dissolved phosphate. Despite its biological significance, the dynamics of phosphorus in ecosystems are dominated by chemical processes. Phosphate is removed from water by reactions with aluminum, and to a lesser extent, with iron in sediment. In alkaline environments, phosphate is precipitated as calcium phosphate. Aluminum, iron and calcium phosphates are only slightly soluble, and sediments act as sinks for phosphorus. Concentrations of inorganic phosphorus in water bodies seldom exceed 0.1 mg/L, and total phosphorus concentration rarely is greater than 0.5 mg/L. In anaerobic zones, the solubility of iron phosphates increases; sediment pore water and hypolimnetic water of eutrophic lakes may have phosphate concentrations above 1 mg/L. Phosphorus is not toxic at elevated concentration, but along with nitrogen, it can lead to eutrophication.

Keywords

Biological role of phosphorus • Phosphorus chemistry • Sediment phosphorus • Phosphorus dynamics in water bodies • Phosphorus and eutrophication

C.E. Boyd, *Water Quality*, DOI 10.1007/978-3-319-17446-4_12

Introduction

Phosphorus is an extremely important element for all living things. It is contained in deoxyribonucleic acid (DNA) that contains the genetic code (or genome) that instructs organisms how to grow, maintain themselves, and reproduce. Phosphorus also is a component of ribonucleic acid (RNA) that provides the information needed for protein synthesis in organisms, i.e., DNA is responsible for RNA production which in turn controls protein synthesis. Phosphorus is a component of adenine diphosphate (ADP) and adenine triphosphate (ATP) that are responsible for energy storage and use at the cellular level. Phosphorus is a component of many other biochemical compounds such as phospholipids important in all cellular membranes. Moreover, calcium phosphorus comprises bone and teeth of vertebrates.

Phosphorus—like carbon and nitrogen—is incorporated into plants and supplied to animals and the microorganisms of decay via food webs. The storage form of phosphorus in plant tissues is phytic acid, a saturated cyclic acid with the formula $C_6H_{18}O_{24}P_6$—a six carbon ring with a phosphate radical ($H_2PO_4^-$) attached to each carbon. This compound is particularly abundant in bran and cereal grains. Phytic acid is not readily digestible by non-ruminant animals, but the enzyme phytase produced in the rumen by microorganisms allows ruminant animals to digest phytic acid. Thus, most animals get their phosphorus from other phosphorus compounds in plants rather than from phytic acid. But, microorganisms of decay can degrade phytic acid with the release of phosphate.

Phosphorus concentrations typically are low in natural waters, but it is required in relatively large amounts by plants. Phosphorus is generally the most important nutrient controlling plant growth in aquatic ecosystems, and phosphorus pollution of natural waters is considered a primary cause of eutrophication. Phosphorus is absorbed strongly by bottom sediment where it is bound in iron, aluminum, and calcium phosphate compounds and adsorbed onto iron and aluminum oxides and hydroxides. The solubilities of the mineral forms are regulated by pH, and there are few situations in nature in which phosphorus minerals are highly soluble. The phosphorus in organic matter is mineralized, but when it is, it usually will be adsorbed by sediment unless it is quickly absorbed by plants or bacteria. Because sediment tends to be a sink for phosphorus, high rates of plant growth in aquatic ecosystems require a continuous input of phosphorus.

Phosphorus has several valence states ranging from −3 to +5, but most phosphorus found in nature has a valence state of +5. In contrast with nitrogen and sulfur, oxidation and reduction reactions mediated by chemotropic bacteria are not an important feature of its cycle. Despite its tremendous biological importance, the phosphorus cycle is largely a chemical cycle instead of a biologically-driven one. Of course, microbial decomposition is an important factor in releasing phosphorus from organic matter.

This chapter will discuss the sources and reactions of phosphorus in aquatic ecosystems and consider the importance of phosphorus in water quality.

Phosphorus in Aquatic Ecosystems

Unlike carbon, oxygen, and nitrogen, there is not a well-defined global phosphorus cycle. This results primarily because the major sources of phosphorus are phosphorus-bearing minerals, e.g. iron, aluminum, and calcium phosphates that occur widely in soils at relatively low concentrations and in massive deposits of calcium phosphates of high phosphorus content at a few locations. These massive deposits of calcium phosphate consist of the mineral apatite that is commonly known as rock phosphate. Rock phosphate is mined and processed to make highly soluble calcium phosphate compounds for use as agricultural, industrial, and household phosphates. In relatively unpolluted natural waters, the primary source of phosphorus is runoff from watershed soils and dissolution of sediment phosphorus, because the atmospheric is not a significant source of phosphorus. Phosphorus concentrations in such waters reflect concentrations and solubilities of phosphorus minerals in soils and sediment.

Phosphorus has many uses: in agriculture, processing of food, manufacturing of beverages, other industries, and the home. The greatest agricultural use is for phosphate fertilizers and in some pesticides; fertilizers also are used on lawns, gardens, and golf courses. Phosphoric acid is included in soft drinks to give them a sharper flavor and to inhibit the growth of microorganisms on sugar present in the beverages. Phosphorus is used for acidification, for buffering, as an emulsifier, and for flavor intensification in food. It is important in an industry as a component of surfactants, lubricants, metal treating processes, component of matches, and as a fire retardant. Trisodium phosphate may be used as a cleaning agent and water softener in both industry and in the home. Thus, it is not surprising that agricultural runoff, effluent from industrial operations, and municipal sewage contains elevated concentrations of phosphorus and can lead to higher phosphorus concentrations in water bodies into which they are discharged.

Increased phosphorus concentrations in natural waters resulting from human activities normally stimulate aquatic plant growth and especially phytoplankton growth. If phosphorus additions to natural water are too great, eutrophication occurs with excessive phytoplankton blooms or nuisance growths of aquatic macrophytes. Phosphorus additions to natural waters are considered a form of water pollution. However, phosphorus often is applied to aquaculture ponds in fertilizers to increase natural productivity that is the base of the food web for fish production.

The dynamics of phosphorus in aquatic ecosystems are illustrated in Fig. 12.1. Dissolved and particulate phosphorus enters water bodies from the watershed. Dissolved inorganic phosphorus is absorbed by plants and incorporated into their biomass. Plant phosphorus is passed on to animals via the food web, and when plants and animals die, microbial activity mineralizes the phosphorus from their remains. If not absorbed by plants, dissolved inorganic phosphorus is strongly adsorbed by sediment. There is an equilibrium between inorganic phosphorus bound in sediment and phosphorus dissolved in water, but the equilibrium is shifted greatly towards sediment phosphorus. Rooted aquatic macrophytes can utilize sediment phosphorus that would otherwise not enter the water column, because

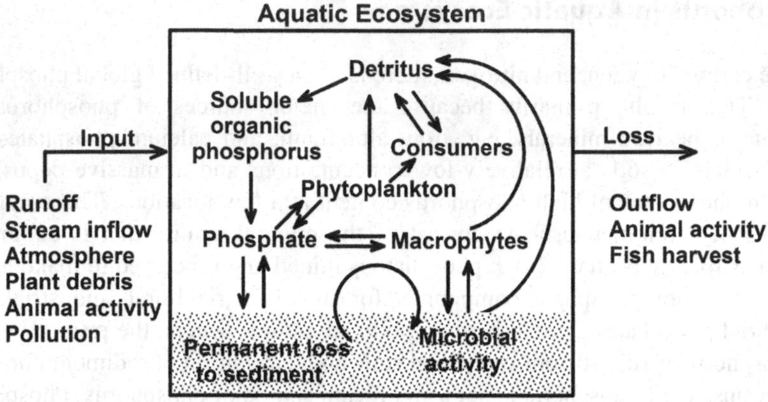

Fig. 12.1 A qualitative model of the phosphorus cycle in an aquatic ecosystem

their roots extract phosphorus dissolved in sediment pore water. Phosphorus is lost from aquatic ecosystems in outflowing water, intentional withdrawal of water for human use, and harvest of aquatic products. The sediment of water bodies tends to be a phosphorus sink, and it may increase in phosphorus content over time. As with nitrogen, in unpolluted natural waters, there tends to be an equilibrium among phosphorus inputs, outputs, and storage. Additions of phosphorus through water pollution can disrupt this equilibrium and cause undesirable changes in aquatic ecosystems.

Phosphorus Chemistry

To lead the reader to an understanding of why phosphorus concentrations in water tend to be low and controlled by interactions between water and sediment, a discussion of phosphorus chemistry is provided.

Dissociation of Orthophosphoric Acid

Inorganic phosphorus in soils, sediment, and water normally can be considered an ionization product of orthophosphoric acid (H_3PO_4) that dissociates as follows:

$$H_3PO_4 = H^+ + H_2PO_4^- \qquad K_1 = 10^{-2.13} \tag{12.1}$$

$$H_2PO_4^- = H^+ + HPO_4^{2-} \qquad K_2 = 10^{-7.21} \tag{12.2}$$

$$HPO_4^{2-} = H^+ + PO_4^{3-} \qquad K_3 = 10^{-12.36}. \tag{12.3}$$

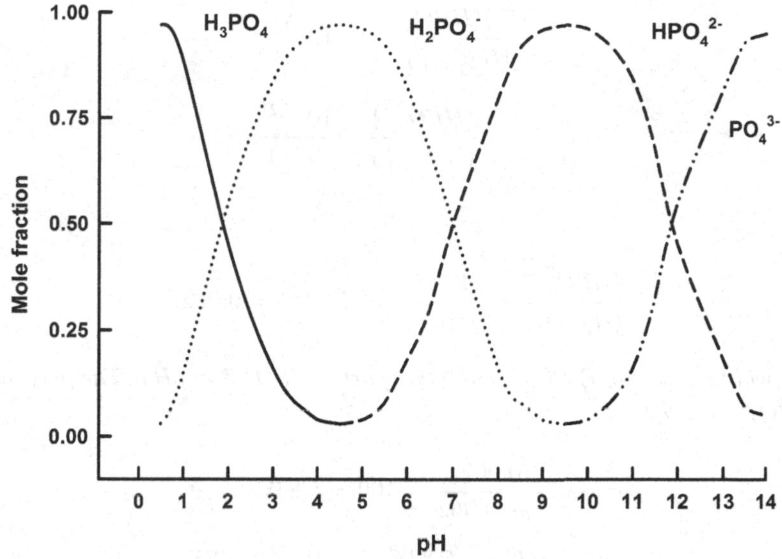

Fig. 12.2 Effects of pH on relative proportions (mole fractions) of H_3PO_4, $H_2PO_4^-$, HPO_4^{2-}, and PO_4^{3-} in an orthophosphate solution

When the hydrogen ion concentration equals the equilibrium constant for one of the steps in the dissociation, then the two phosphate ions involved in that step will have exactly equal concentrations. For example, in (12.1), when pH = 2.13 [$(H^+) = 10^{-2.13}$], the ratio $K_1/(H^+)$ is $10^{-2.13}/10^{-2.13}$ or 1.0. This means that the ratio of $(HPO_4^{2-})/(H_3PO_4)$ also equals 1.0, and the concentrations of the two forms of phosphorus are equal. The calculation of the proportions of the two forms of phosphorus in any one of the three dissociations of phosphoric acid may be done is illustrated in Ex. 12.1. This procedure can be applied across the entire pH range to provide the data needed for graphically depicting the effect of pH on proportions of H_3PO_4, $H_2PO_4^-$, HPO_4^{2-}, and PO_4^{3-} (Fig. 12.2). There is no un-ionized H_3PO_4 except in highly acidic solutions, and PO_4^{3-} predominates only in highly basic solutions. Within the pH range of most natural waters, dissolved phosphate will exist as $H_2PO_4^-$ and HPO_4^{2-}. At pH values below 7.21, there will be more $H_2PO_4^-$, and HPO_4^{2-} will dominate at pH 7.22 and above.

***Ex. 12.1:** The percentages of $H_2PO_4^-$ and HPO_4^{2-} at pH 6 will be estimated.*

***Solution**:*

The mass action expression for (12.2) *allows an expression for the ratio of HPO_4^{2-} to $H_2PO_4^-$,*

$$\frac{(H^+)(HPO_4^{2-})}{(H_2PO_4^-)} = 10^{-7.21}$$

$$\frac{(HPO_4^{2-})}{(H_2PO_4^-)} = \frac{10^{-7.21}}{(H^+)}$$

and

$$\frac{(HPO_4^{2-})}{(H_2PO_4^-)} = \frac{10^{-7.21}}{10^{-6}} = 10^{-1.21} = 0.062.$$

There will be 1 part of $H_2PO_4^-$ *and 0.062 part of* HPO_4^{2-} *at pH 6. The percentage of* HPO_4^{2-} *will be*

$$\frac{0.062}{1 + 0.062} \times 100 = 5.83\%$$

$$and \quad 100\% - 5.83\% = 94.17\% \; H_2PO_4^-.$$

Polyphosphates also enter waters naturally or in pollution. The parent form of polyphosphate can be considered a compound such as polyphosphoric acid with the formula $H_6P_4O_{13}$. Polyphosphate has a greater proportion of phosphorus than does orthophosphate, e.g., orthophosphate (PO_4) is 32.6 % phosphorus while the polyphosphate P_4O_{13} is 37.3 % phosphorus. However, when introduced into water, polyphosphate soon hydrolyzes to orthophosphate.

Phosphorus-Sediment Reactions

Inorganic phosphorus reacts with iron and aluminum in acidic sediments or waters to form slightly soluble compounds. Representative aluminum and iron phosphate compounds are variscite, $AlPO_4{\cdot}2H_2O$, and strengite, $FePO_4{\cdot}2H_2O$. The dissolution of these compounds is pH-dependent

$$AlPO_4 \cdot 2H_2O + 2H^+ = Al^{3+} + H_2PO_4^- + 2H_2O \qquad K = 10^{-2.5} \qquad (12.4)$$

$$FePO_4 \cdot 2H_2O + 2H^+ = Fe^{3+} + H_2PO_4^- + 2H_2O \qquad K = 10^{-6.85}. \qquad (12.5)$$

Decreasing pH favors the solubility of iron and aluminum phosphates as illustrated in Ex. 12.2.

Ex. 12.2: *The solubility of phosphorus from variscite will be estimated for pH 5 and pH 6.*

Solution:

(i) *From* (12.4),

$$\frac{(Al^{3+})(H_2PO_4^-)}{(H^+)^2} = 10^{-2.5}$$

and

$$(H_2PO_4^-) = \frac{(H^+)^2\,(10^{-2.5})}{(Al^{3+})}$$

Letting $x = (Al^{3+}) = (H_2PO_4^-)$, *for pH 5*

$$x = \frac{(10^{-5})^2\,(10^{-2.5})}{x}$$

$$x^2 = 10^{-12.5}$$

$$x = 10^{-6.25}.$$

At pH 5,

$$(H_2PO_4^-) = 10^{-6.25}\ \text{M}\ \textit{or}\ 5.62 \times 10^{-7}\ \text{M}.$$

There are 31 g of phosphorus per mole of $H_2PO_4^-$, *so*

$$5.62 \times 10^{-7}\,\text{M} \times 31\,\text{g}\,P/\text{mol} = 1.74 \times 10^{-5}\,\text{g}\,P/\text{L}$$

or 0.017 mg/L of phosphorus.

(ii) *Repeating the above for pH = 6 or* $(H^+) = 10^{-6}$, *we get 0.0017 mg P/L.*

Calculation of the solubility of strengite for the same pH values used in Ex. 12.2, gives phosphorus concentrations of 0.00012 mg/L at pH 5 and 0.000012 mg/L at pH 6. Strengite is less soluble than variscite at the same pH by a factor of nearly 150.

Acidic and neutral sediments usually contain appreciable iron and aluminum minerals, and aluminum tends to control phosphate solubility in aerobic water and sediments. To illustrate, two representative iron and aluminum compounds found in sediment are gibbsite, $Al(OH)_3$ and iron (III) hydroxide, $Fe(OH)_3$. The dissolution of these two representative compounds, like other iron and aluminum oxides and hydroxides, is pH dependent

$$Al(OH)_3 + 3H^+ = Al^{3+} + 3H_2O \qquad K = 10^9 \tag{12.6}$$

$$Fe(OH)_3 + 3H^+ = Fe^{3+} + 3H_2O \qquad K = 10^{3.54}. \tag{12.7}$$

The solubilities of the two minerals increase with decreasing pH as illustrated in Ex. 12.3.

Ex. 12.3: *The solubilities of gibbsite and iron (III) hydroxide will be estimated for pH 5 and 6.*

Solution:
From (12.6) *and* (12.7)

$$pH = 5 \quad \frac{(Al^{3+})}{(H^+)^3} = 10^9$$

$$and \quad (Al^{3+}) = (10^{-5})^3\,(10^9) = 10^{-6}\,\mathrm{M}$$

$$\frac{(Fe^{3+})}{(H^+)^3} = 10^{3.54}$$

$$and \quad (Fe^{3+}) = (10^{-5})^3\,(10^{3.54}) = 10^{-11.46}\,\mathrm{M}.$$

$$pH = 6 \quad (Al^{3+}) = (10^{-6})^3\,(10^9) = 10^{-9}\,\mathrm{M}$$

$$and \quad (Fe^{3+}) = (10^{-6})^3\,(10^{3.54}) = 10^{-14.46}\,\mathrm{M}.$$

The Al^{3+} concentration is 0.027 mg/L and .027 μg/L at pH 5 and pH 6, respectively. The iron concentration is more than four orders of magnitude lower than that of aluminum at each pH—iron is essentially imperceptible below pH 6.

Ex. 12.3 clearly shows that the amounts of iron and aluminum in solution in sediment pore water and available to precipitate phosphorus as iron and aluminum phosphates will increase as pH decreases. But, at the same pH, aluminum compounds are more soluble than iron compounds.

Despite iron and aluminum phosphate compounds being more soluble at lower pH (Ex. 12.2), iron and aluminum oxides and hydroxides tend to be much more abundant in sediment than are aluminum and iron phosphates. Thus, when phosphorus is added to acidic sediment there is adequate Al^{3+} and Fe^{3+}—especially Al^{3+} because of the greater solubility of its hydroxides and oxides compared to those of iron (Ex. 12.3)—present to precipitate it. Thus, in reality, the availability of phosphorus from sediment tends to decrease with decreasing pH.

When a highly soluble source of phosphorus is added to sediment at pH of 7 or below, the phosphorus will react with aluminum and iron and precipitate. This phenomenon is illustrated in Ex. 12.4.

***Ex. 12.4:** The solubility of phosphorus from a highly soluble monocalcium phosphate [$Ca(H_2PO_4)_2$] in a system at equilibrium with gibbsite will be calculated.*

***Solution**:*

Gibbsite provides Al^{3+} to solution, and Al^{3+} can react with phosphorus to precipitate $AlPO_4{\cdot}2H_2O$ (variscite). The solubility of gibbsite in water of pH 5 is 10^{-6} M (see Ex. 12.3). Dissolution of variscite may be written as

$$AlPO_4 \cdot 2H_2O + 2H^+ = Al^{3+} + H_2PO_4^- + 2H_2O$$

$$for\ which\ K = 10^{-2.5}.\ Thus,$$

$$\frac{(Al^{3+})(H_2PO_4^-)}{(H^+)^2} = 10^{-2.5}$$

$$(H_2PO_4^-) = \frac{(H^+)^2\,10^{-2.5}}{(Al^{3+})}$$

$$= \frac{(10^{-5})^2\,10^{-2.5}}{10^{-6}} = 10^{-6.5}\ \mathrm{M}.$$

Expressed in terms of phosphorus, $10^{-6.5}$ M is 0.00000032 M × 30.98 g P/mol or 0.01 mg P/L. In this example, phosphate solubility is controlled by variscite, because variscite is less soluble than monocalcium phosphate.

Phosphate can be absorbed by iron and aluminum oxides as follows:

$$\mathrm{H_2PO_4^- + Al(OH)_3 = Al(OH)_2H_2PO_4 + OH^-} \qquad (12.8)$$

$$\mathrm{H_2PO_4^- + FeOOH = FeOH_2PO_4 + OH^-}\ . \qquad (12.9)$$

In soil and sediment—especially in tropical areas—some of the clay fraction is in the form of iron and aluminum hydroxides. Clays are colloidal and have a large surface area; they can bind large amounts of phosphorus.

Silicate clays also can fix phosphorus. Phosphorus is substituted for silicate in the clay structure. Clays also have some ability to adsorb anions, because they have a small number of positive charges on their surfaces. However, absorption by silicate minerals and anion exchange is less important than phosphorus removal by aluminum and iron in acidic soils and sediments.

Primary phosphate compounds in neutral and basic sediment are calcium phosphates. The most soluble calcium phosphate compound is monocalcium phosphate, $Ca(H_2PO_4)_2$. This is the form of phosphorus normally applied in fertilizer. In neutral or basic soils, $Ca(H_2PO_4)_2$ is transformed through dicalcium, octacalcium, and tricalcium phosphates to apatite. Apatite is not very soluble under neutral or alkaline conditions. A representative apatite, hydroxyapatite, dissolves as follows:

$$Ca_5(PO_4)_3OH + 7H^+ = 5Ca^{2+} + 3H_2PO_4^- + H_2O \qquad K = 10^{14.46}. \quad (12.10)$$

A high concentration of Ca^{2+} and a high pH favors formation of hydroxyapatite from dissolved phosphate in water or sediment pore water. Apatite is not appreciably soluble at pH 7 and above even at low calcium concentration as shown in Ex. 12.5.

Ex. 12.5: *The solubility of phosphorus from hydroxyapatite in water of pH 7 and pH 8 and 15 mg/L calcium* $(10^{-3.43}\ M)$ *will be estimated.*

Solution:
Assuming that the reaction controlling phosphorus concentration is (12.10),

$$\frac{(Ca^{2+})^5 (H_2PO_4^-)^3}{(H^+)^7} = 10^{14.46}$$

at pH 7

$$(H_2PO_4^-)^3 = -\frac{(10^{-7})^7 (10^{14.46})}{(10^{-3.43})^5} = (10^{-17.39})\,\mathrm{M}$$

$$(H_2PO_4^-) = 10^{-5.80}\,\mathrm{M}\ or\ 0.05\,\mathrm{mg}\ P/\mathrm{L}.$$

Repeating the calculation for pH 8 gives

$$(H_2PO_4^-) = 10^{-8.13}\,\mathrm{M}\ or\ 0.0002\,\mathrm{mg}\ P/\mathrm{L}.$$

At a higher calcium concentration, the phosphorus concentration would be less, e.g. at pH 7 and 20 mg Ca^{2+}/L ($10^{-3.30}$ M), the phosphorus concentration would be only 0.03 mg/L—40 % less than at 15 mg/L Ca^{2+}.

The maximum availability of phosphorus in aerobic soil or sediment typically occurs between pH 6 and 7 (Fig. 12.3). In this pH range there is less Al^{3+} and Fe^{3+} to react with phosphorus and a smaller tendency of aluminum and iron oxides to adsorb phosphorus than at lower pH. Also, at a pH of 6–7 the activity of calcium is normally lower than at higher pH. Nevertheless, in the pH range of 6–7, most of the phosphorus added to aquatic ecosystems still is rendered insoluble through adsorption by colloids or precipitation as insoluble compounds.

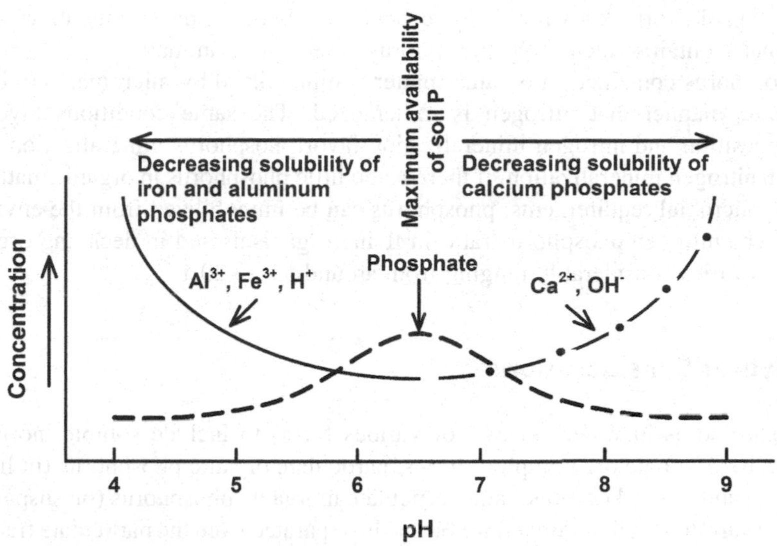

Fig. 12.3 Schematic showing effects of pH on the relative concentrations of dissolved phosphate in aerobic soil or sediment

Iron phosphates contained in sediment become more soluble when the redox potential falls low enough for ferric iron to be reduced to ferrous iron. Phosphorus concentrations in pore water of anaerobic sediment may be quite high (Masuda and Boyd 1994a). This pool of phosphorus is largely unavailable to the water column because iron phosphate reprecipitates when ferrous iron and phosphate diffuse into the aerobic layer normally existing at the sediment-water interface. The aerobic layer at the interface is lost during thermal stratification of eutrophic lakes and ponds. Diffusion of iron and phosphorus from anaerobic sediment can lead to high iron and phosphate concentrations in the hypolimnion. Concentrations of 10–20 mg/L iron and 1–2 mg/L soluble orthophosphate are not unusual. When thermal destratification occurs, hypolimnetic waters mix with surface waters, and concentrations of phosphorus increase briefly in surface waters. However, in oxygenated water, phosphorus concentrations quickly decline. Phosphorus either precipitates directly as iron phosphate (12.5) or it is adsorbed onto the surface of the floc of iron (III) hydroxide precipitating from the oxygenated water.

Organic Phosphorus

The dry matter of plants commonly contains 0.05–0.5 % phosphorus, while that of vertebrate animals such as fish may contain 2–3 % phosphorus or more. Crustaceans typically contain about 1 % phosphorus in their dry matter. As mentioned earlier in this chapter, phosphorus is present in plants, animals, and bacteria as phospholipids, nucleic acids, and other biochemicals. However, in plants some phosphorus is stored in an organic form known as phytic acid, and bone made mainly of calcium phosphate contains much of the phosphorus in vertebrate animals.

Phosphorus contained in organic matter is mineralized by microbial activity in the same manner that nitrogen is mineralized. The same conditions favoring decomposition and nitrogen mineralization favor phosphorus mineralization. Just as with nitrogen mineralization, if there is too little phosphorus in organic matter to satisfy microbial requirements, phosphorus can be immobilized from the environment. The nitrogen:phosphorus ratio in living organisms and in decaying organic residues varies considerably ranging from around 5:1 to 20:1.

Analytical Considerations

The phosphorus in water consists of various forms to include soluble inorganic phosphorus, soluble organic phosphorus, particulate organic phosphorus (in living plankton and in dead detritus), and particulate inorganic phosphorus (on suspended mineral particles). The soluble fraction can be separated from the particulate fraction by filtration through a membrane or glass fiber filter. However, common analytical methods do not distinguish perfectly between soluble inorganic and soluble organic phosphorus, and a portion of the soluble organic phosphorus will be included in measurements of soluble inorganic phosphorus. Therefore, when the phosphorus concentration is measured directly in filtrates of water, the resulting phosphorus fraction is called soluble reactive phosphorus. Digestion of a raw water sample in acidic persulfate releases all of the bound phosphorus, and analysis of the digestate gives total phosphorus. Most of the information on phosphorus concentrations in natural waters is for soluble reactive phosphorus and total phosphorus.

Phosphorus in sediment may be extracted with various solutions to give different fractions. A common way of fractioning sediment phosphorus is a sequential extraction with 1 M ammonium chloride to remove loosely-bound phosphorus, 0.1 N sodium hydroxide to remove iron and aluminum-bound phosphorus, and 0.5 N hydrochloric acid to remove calcium-bound phosphorus (Hieltjes and Liklema 1982). Other extractants also are used to remove phosphorus from sediment samples. A widely used method of soil phosphorus analysis in soil testing laboratories is to measure phosphorus extracted by dilute (0.05–0.1 N) hydrochloric acid, sulfuric acid, or a mixture of these two acids. Soil can be digested in perchloric acid to release bound phosphorus for total phosphorus analysis.

Phosphorus Dynamics

Concentrations in Water

Phosphorus concentrations in surface waters generally are quite low. Total phosphorus seldom exceeds 0.5 mg/L except in highly eutrophic waters or in wastewaters. There generally is much more particulate phosphorus than soluble

Table 12.1 Distribution of forms of soil and water phosphorus for a fish pond at Auburn, Alabama

Phosphorus pool	Phosphorus fraction	Amount (g/m^2)	(%)
Pond water[a]	Total phosphorus	0.252	0.19
	Soluble reactive phosphorus	0.019	0.01
	Soluble nonreactive phosphorus	0.026	0.02
	Particulate phosphorus	0.207	0.16
Soil[b,c]	Total phosphorus	132.35	99.81
	Loosely-bound phosphorus	1.28	0.96
	Calcium-bound phosphorus	0.26	0.20
	Iron- and aluminum-bound phosphorus	17.30	13.05
	Residual phosphorus[d]	113.51	85.60
Pond	All	132.60	100.00

[a]Average pond depth = 1.0 m
[b]Soil depth = 0.2 m
[c]Soil bulk density = 0.797 g/cm^3
[d]Phosphorus removed by perchloric acid digestion

reactive phosphorus. For example, Masuda and Boyd (1994b) found that water in eutrophic aquaculture ponds contained 37 % dissolved phosphorus and 63 % particulate phosphorus. However, most of the dissolved phosphorus was non-reactive organic phosphorus, and only 7.7 % of the total phosphorus was soluble reactive phosphorus. Typically, 10 % or less of the total phosphorus will be soluble reactive phosphorus and readily available to plants. Most surface waters contain less than 0.05 mg/L soluble reactive phosphorus, and most unpolluted water bodies only contain 0.001–0.005 mg/L of this fraction.

Sediment contains much more phosphorus than water. Total phosphorus concentrations found in the literature ranged from less than 10 mg/kg to more than 3,000 mg/kg. However, most of the phosphorus is tightly bound and not readily soluble in water. Phosphorus concentrations in the sediment of a eutrophic fishpond (Masuda and Boyd 1994b) are illustrated in Table 12.1. Notice that 85.6 % of the phosphorus was not removable by normal extracting agents and had to be released by perchloric acid digestion.

Plant Uptake

Phytoplankton can absorb phosphorus from water very quickly. In a water with a dense bloom of phytoplankton, phosphorus additions of 0.2–0.3 mg/L were completely removed within a few hours (Boyd and Musig 1981). Macrophytes also can remove phosphorus from water very quickly, and rooted macrophytes can absorb phosphorus from anaerobic zones in the sediment (Bristow and Whitcombe 1971). Plant uptake is a major factor controlling concentrations of soluble reactive phosphorus in water, and much of the total phosphorus in water is contained in phytoplankton cells. Macrophyte communities can store large amounts of phosphorus in their biomass.

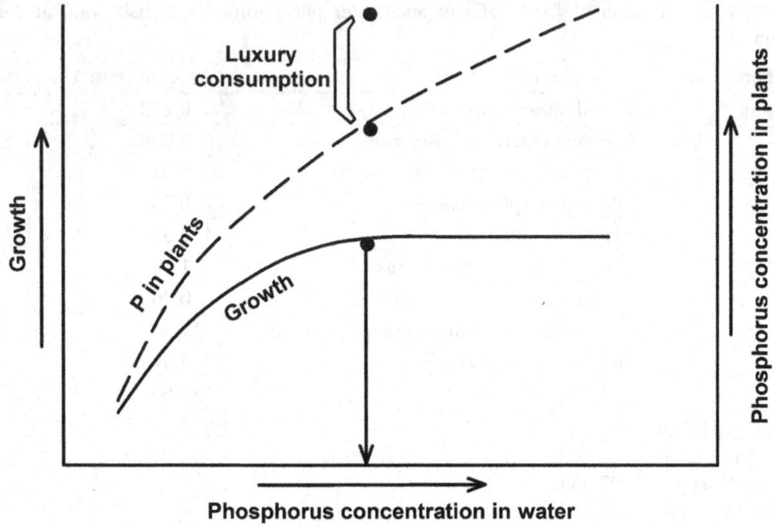

Fig. 12.4 Luxury consumption of phosphorus by phytoplankton

Some plants can absorb more phosphorus than they need immediately, and they store it for use later. The absorption of phosphorus and other nutrients in excess of the amount required for growth has been demonstrated in many plant species including species of phytoplankton. This phenomenon is termed luxury consumption and is illustrated in Fig. 12.4.

The ability to absorb and store more nutrients than needed at the moment is of competitive advantage for plants. The phosphorus can be removed from the environment thereby depriving competing plants of it. In phytoplankton, phosphorus in cells can be passed on to succeeding generations when cells divide and multiply. In larger plants, phosphorus can be translocated from storage sites in older tissue to rapidly growing meristematic tissues.

Exchange Between Water and Sediment

If some sediment is placed in a flask of distilled water and agitated until equilibrium phosphorus concentration is attained, very little phosphorus usually will be present in the water. For example, in a series of soil samples containing from 100 to 3,400 mg/kg of total phosphorus, water extractable phosphorus concentrations ranged from undetectable to 0.16 mg/L (Boyd and Munsiri 1996). The correlation between total phosphorus and water extractable phosphorus was weak ($r = 0.581$), but the correlation between dilute acid (0.075 N) extractable phosphorus and water soluble phosphorus was much stronger ($r = 0.920$).

It can be shown by successive extractions of a sediment with water that there is a continued release of phosphorus to the water for many extractions (Fig. 12.5).

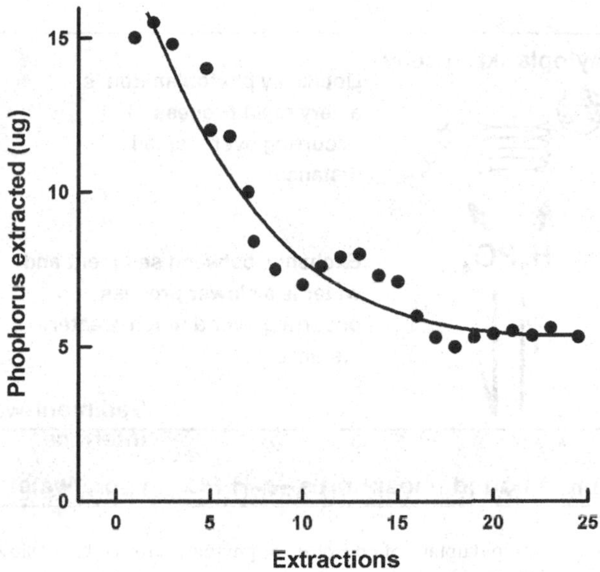

Fig. 12.5 Quantities of phosphorus removed from a mud by consecutive extractions with phosphorus free water

The amount released, however, declines with the number of extractions. Because of the relationship shown in Fig. 12.5, the sediment is a reserve of phosphorus available when plant removal causes phosphorus concentrations in the water to fall below the equilibrium concentration. Nevertheless, the concentrations of phosphorus at equilibrium normally are quite low, and phosphate additions are necessary to stimulate abundant phytoplankton growth.

Macrophytes—especially rooted, submerged macrophytes—grow quite well in waters that are low in phosphorus because they can absorb phosphorus and other nutrients from sediment. Sediment is not a readily available source of phosphorus or other nutrients for phytoplankton because of the difficult logistics of nutrient movement from sediment pore water to the illuminated zone where phytoplankton grow (Fig. 12.6).

When phosphorus enters water through intentional additions as in fishponds or through pollution, stimulation of phytoplankton growth creates turbidity and shades the deeper waters. Restriction in light caused by nutrient enrichment may eliminate many species of macrophytes from aquatic communities in eutrophic water bodies.

The relative concentrations of phosphorus in sediment, sediment pore water, at the sediment-water interface, and in surface water are illustrated in Fig. 12.7 for a small, eutrophic fishpond. There was roughly an order of magnitude difference in sediment phosphorus and pore water phosphorus concentrations, and another order of magnitude difference between phosphorus concentrations in pore water and at the water at the sediment-water interface. Pore water is anaerobic and phosphorus in pore water tends to precipitate at the aerobic interface and little enters the pond water.

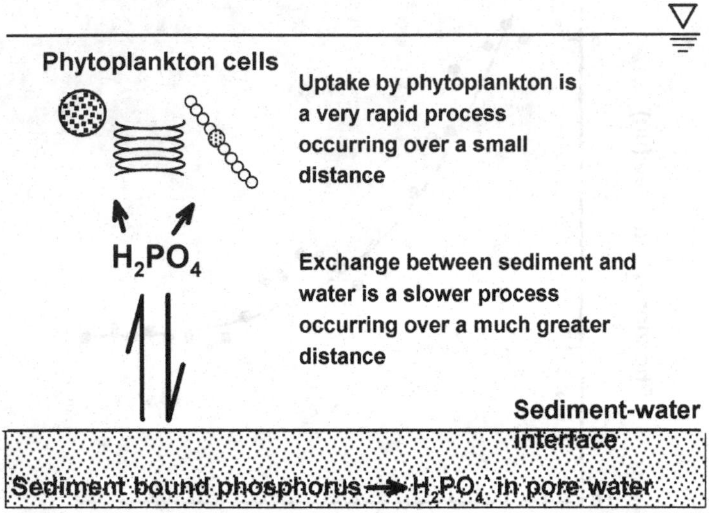

Fig. 12.6 Illustration of rapid uptake of phosphate by phytoplankton cells and slower exchange of phosphate between sediment and water

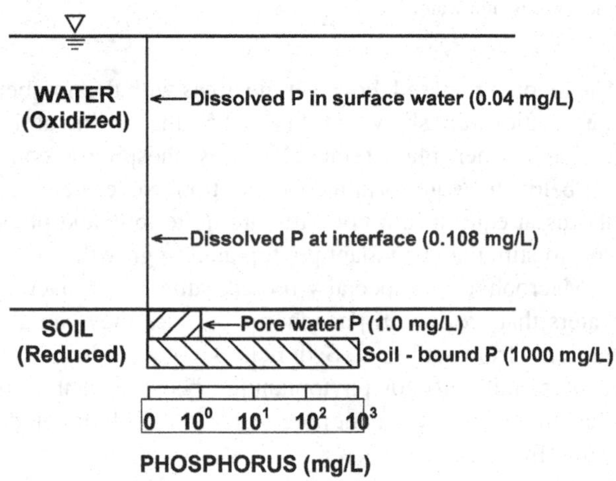

Fig. 12.7 Concentrations of phosphorus bound in soil, dissolved in pore water, and dissolved in overlaying pond water

Even when anaerobic conditions exist at the sediment-water interface, phosphorus must diffuse from the pore water into the open water, and diffusion is a relatively slow process. Once phosphorus enters the open water, it can be mixed throughout the water body rather quickly by turbulence.

In pond aquaculture, phosphorus often is added to increase dissolved inorganic phosphorus concentrations. A portion of the added phosphorus is quickly absorbed by phytoplankton. The part that is not removed by plants will tend to accumulate in the sediment, and most of the phosphorus removed by plants also will eventually reach the sediment. Turbulence will allow soluble phosphorus to reach the sediment more quickly when compared to the movement of sediment-bound phosphorus into

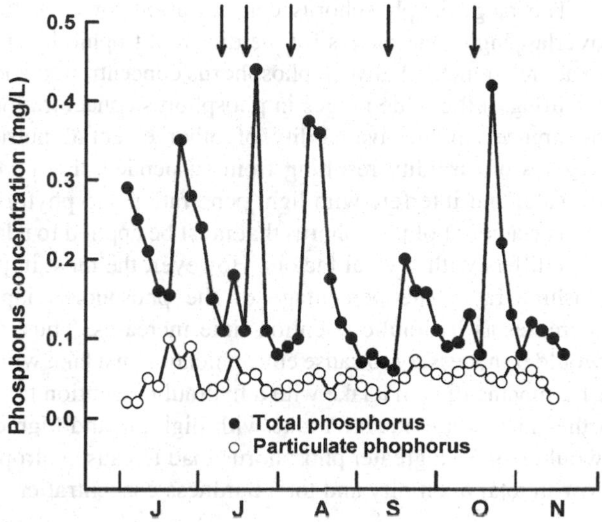

Fig. 12.8 Average concentrations of total and particulate phosphorus in two fertilized fish ponds. *Vertical arrows* indicate fertilizer application dates

the water. The removal of fertilizer phosphorus from fish ponds (Fig. 12.8) illustrates the rapidity of phosphorus removal from water.

The ability of sediment to hold phosphorus is usually quite large. Fish ponds on the Auburn University E. W. Shell Fisheries Center at Auburn, Alabama that had received an average phosphorus input of 4.1 g P/m^2/year (41 kg/ha/year) for 22 years were only about half-saturated with phosphorus and still rapidly adsorbed phosphorus from the water (Masuda and Boyd 1994b). Nevertheless, sediment can become saturated with phosphorus or have a very low capacity to adsorb phosphorus, e.g., sandy sediment. In such bodies of water, additions of phosphorus are particularly effective in stimulating phytoplankton growth.

Sediment is not always necessary for phosphorus removal from the water. In waters with significant concentrations of calcium and pH of 7–9, phosphate will precipitate directly from the water as calcium phosphate.

Eutrophication

Average, annual total phosphorus concentrations and their ranges in lakes classified to trophic status by different investigators were summarized by Wetzel (2001) as follows:

Status	n	Mean (mg/L)	Range (mg/L)
Oligotrophic	21	0.008	0.003–0.018
Mesotrophic	19	0.267	0.011–0.096
Eutrophic	71	0.084	0.016–0.386
Hyper-eutrophic	2	0.975	0.750–1.200

The range in phosphorus concentration for each trophic status level and the overlapping of the ranges for the different trophic levels reveal that there is not an exact relationship between phosphorus concentration and trophic status. One factor resulting in the wide ranges in phosphorus concentration for each trophic status is differences in the availability of other essential nutrients. Another is different degrees of turbidity resulting from suspended clay particles or humic substances in water that interfere with light penetration and phytoplankton photosynthesis.

The amount of phosphorus that must be applied to a lake to cause eutrophication also differs with several factors. However, the most important one is the hydraulic flushing rate. The percentage of the phosphorus input (load) that is retained increases as hydraulic retention time increases. Thus, a greater phosphorus input would be necessary to cause eutrophication in a lake with a hydraulic retention time of 2 months than in a lake with a hydraulic retention time of 8 months assuming all other factors are equal. A lake with high pH and high calcium concentration also would require a greater phosphorus load to cause eutrophication than would a lake low in total alkalinity and total hardness concentration.

Interaction with Nitrogen

Nitrogen and phosphorus are key nutrients regulating aquatic plant productivity. But, the amounts and ratios of these two nutrients vary among species. Redfield (1934) reported that marine phytoplankton contained on a weight basis about seven times more nitrogen than phosphorus. This value is often used as the average N:P ratio in plants, but the ratios for individual species vary from 5:1 to 20:1. However, in most ecosystems, an increase in phosphorus concentration will cause a greater response in plant growth than will an increase in nitrogen concentration. This results because phosphorus is quickly removed from the water and bound in the sediment. There is limited recycling of sediment bound phosphorus, and to maintain sufficient phosphorus in the water to promote high rates of plant growth, there must be a continuous external source of phosphorus.

Nitrogen also is removed from water bodies by various processes, but as much as 10–20 % of added nitrogen is present in organic matter deposited in sediment. Sediment organic matter is decomposed, and nitrogen is continually mineralized. The internal recycling of nitrogen in aquatic ecosystems is much greater than it is for phosphorus. As a result, in order to achieve a nitrogen to phosphorus ratio of 7:1 (the Redfield ratio) in the water, it likely would require an addition of these two elements in a lower N:P ratio.

Significance

The popular literature on lake eutrophication often refers to phosphorus as a toxic element. Phosphorus is not actually a toxic element. But, high phosphorus concentrations in aquatic ecosystems cause excessive aquatic plant productivity and eutrophication that can result in low dissolved oxygen concentration and fish kills. Phosphorus and nitrogen are considered to be the two key nutrients associated with eutrophication, and like nitrogen, it is difficult to establish the concentration of phosphorus that will cause excessive plant growth. Also, it is difficult to establish phosphorus loads necessary to cause eutrophication in natural waters. Nevertheless, there is ample evidence that concentrations of total phosphorus of 0.005–0.05 mg/L can cause phytoplankton blooms in many lakes. Thus, limitations on phosphorus concentrations in effluents have been one of the main tools in combating eutrophication.

References

Boyd CE, Munsiri P (1996) Phosphorus adsorption capacity and availability of added phosphorus in soils from aquaculture areas in Thailand. J World Aquacult Soc 27:160–167

Boyd CE, Musig Y (1981) Orthophosphate uptake by phytoplankton and sediment. Aquaculture 22:165–173

Bristow JM, Whitcombe M (1971) The role of roots in the nutrition of aquatic vascular plants. Am J Bot 58:8–13

Hieltjes AHM, Liklema L (1982) Fractionation of inorganic phosphate in calcareous sediments. J Environ Qual 9:405–407

Masuda K, Boyd CE (1994a) Chemistry of sediment pore water in aquaculture ponds built on clayey, Ultisols at Auburn, Alabama. J World Aquacult Soc 25:396–404

Masuda K, Boyd CE (1994b) Phosphorus fractions in soil and water of aquaculture ponds built on clayey, Ultisols at Auburn, Alabama. J World Aquacult Soc 25:379–395

Redfield AC (1934) On the proportions of organic deviations in sea water and their relation to the composition of plankton. In: Daniel RJ (ed) James Johnstone memorial volume. University Press of Liverpool, Liverpool

Wetzel RG (2001) Limnology, 3rd edn. Academic, New York

Sulfur

13

Abstract

Although sulfur is important as a nutrient, hydrogen sulfide is an odorous, toxic compound, and sulfur dioxide is an air pollutant responsible for acid rain. Plants primarily use sulfate as a sulfur source, and sulfur containing amino acids in plants are important to animal nutrition. Sulfur compounds undergo oxidations and reductions in the environment. The most famous sulfur oxidizing bacteria are of the genus *Thiobacillus*; they oxidize elemental sulfur, sulfides, and other reduced sulfur compounds releasing sulfuric acid into the environment—acid sulfate soils and acidic mine drainage result from sulfur oxidation. Bacteria of the genus *Desulfovibrio* use certain sulfur compounds as electron and hydrogen acceptors in respiration allowing them to decompose organic matter in anaerobic environments. Ferrous iron and other metals may react with sulfide in anaerobic sediments to form metallic sulfides, e.g., iron sulfide (iron pyrite). If such sediment later is exposed to oxygen, sulfides will be oxidized resulting in sulfuric acid production. Sulfide in anaerobic zones sometimes diffuses or is mixed into overlaying aerobic water at a rate exceeding the oxidation rate of sulfide—toxicity to aquatic animals can result. Sulfide toxicity is favored by low pH, because un-ionized hydrogen sulfide (H_2S) is the toxic form.

Keywords

Sulfur cycle • Sulfur transformations • Acid-sulfate soils • Hydrogen sulfide toxicity • Sulfur concentrations

Introduction

Sulfur is present in the environment in various forms to include the following: deposits of metal sulfides, sulfates and elemental sulfur; dissolved sulfate in water bodies; sulfur dioxide and other gaseous forms of sulfur in the atmosphere; sulfur in living organisms and dead organic matter. Sulfur is found in fossil fuels; when these

C.E. Boyd, *Water Quality*, DOI 10.1007/978-3-319-17446-4_13

fuels are combusted, sulfur gases are released into the air. Sulfate-bearing minerals such as epsomite ($MgSO_4 \cdot 7H_2O$), varite ($BaSO_4$), and gypsum ($CaSO_4 \cdot 2H_2O$) tend to be soluble, and sulfate leaches readily from the land. Because sulfate is common in the earth's crust, ocean water has a high sulfate concentration.

Sulfur is an essential element for plants, animals, and bacteria; one essential function is as a component of two amino acids (cysteine and methionine). Sulfur also is contained in the vitamins biotin and thiamine, and in iron-sulfur clusters such as the ferredoxins that influence transfer of electrons in cells. Moreover, sulfur plays a role in catalyzing many biochemical reactions.

Sulfur, like nitrogen, commonly occurs in the environment in several oxidation states (valences). As a result, bacteria can oxidize and reduce sulfur compounds in the same manner as they do nitrogen compounds, and oxidations and reductions of sulfur may occur by chemical reactions not involving biological activity. Sulfate is the form of sulfur used most commonly by plants, although some plants can fix atmospheric sulfur by a method analogous to nitrogen fixation. Sulfur oxidations often produce sulfuric acid; reductions yield sulfides. Anthropogenic emissions of sulfur dioxide are the main cause of the acid rain phenomenon, sulfide oxidation is responsible for acid mine drainage, and oxidation of sulfur or sulfide in soil can cause extreme acidity.

There is a sulfur cycle in nature, and most aspects of this cycle operate in aquatic ecosystems. Sulfate influences biological activity by acting as a nutrient, but sulfide can be toxic to aquatic organisms. The purpose of this chapter is to discuss sulfur chemistry and its influence on water quality.

The Sulfur Cycle

The main features of the sulfur cycle are illustrated in Fig. 13.1. Natural sources of sulfur in the atmosphere include volcanic activity which releases hydrogen sulfide (H_2S) and sulfur dioxide (SO_2), sulfate originating from evaporation of sea spray or suspension of dust containing gypsum ($CaSO_4 \cdot 2H_2O$) from arid lands, hydrogen sulfide released by microbial decomposition of organic matter, and sulfur dioxide released by combustion of fuels and by wildfires. Human activities release around 100 million metric tons of sulfur into the air annually (Smith et al. 2011). The amount of sulfur emissions entering the atmosphere from natural sources has not been established accurately, but it is much less than anthropogenic sources.

In spite of the large additions of hydrogen sulfide and sulfur dioxide into the atmosphere, there usually are not high concentrations of sulfur compounds in the atmosphere at a given time. Reduced forms of sulfur are quickly oxidized to sulfate, and sulfur compounds are washed from the atmosphere by rain. Oxidations of sulfur compounds are represented by the following equations:

$$2H_2S + 3O_2 \rightarrow 2H_2O + 2SO_2 \qquad (13.1)$$

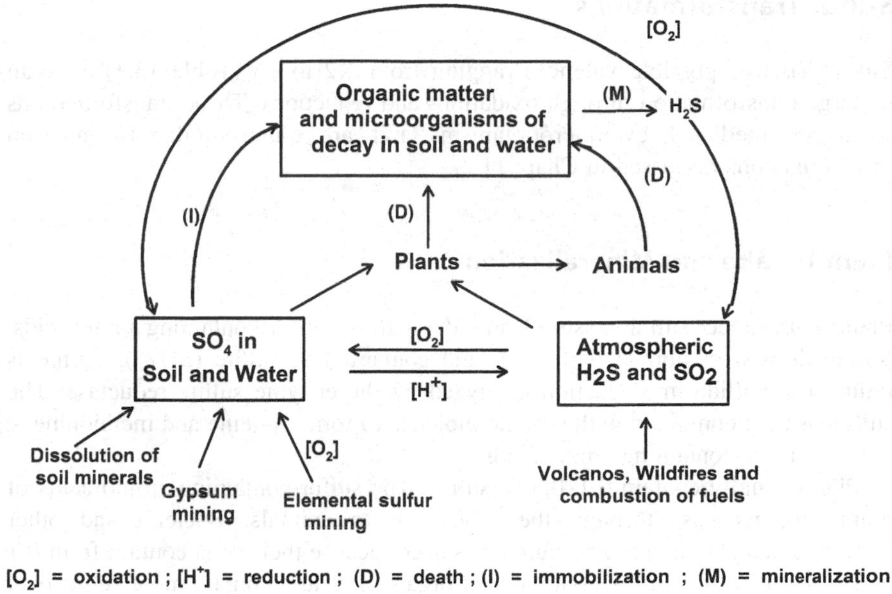

Fig. 13.1 The sulfur cycle

$$SO_2 + H_2O \rightarrow H_2SO_3 \tag{13.2}$$

$$H_2SO_3 + \frac{1}{2} O_2 \rightarrow H_2SO_4. \tag{13.3}$$

The end product of these oxidations is sulfuric acid. Rainwater may contain enough sulfuric acid to be highly acidic (pH < 5). Excessive sulfur dioxide in the air can result in toxicity to vegetation, but as mentioned above, some plants can absorb sulfur dioxide, reduce it to sulfide, and use sulfide for making sulfur-containing amino acids.

Gypsum and elemental sulfur deposits may be mined and used as industrial and agricultural sulfur. Gypsum is a major component of drywall or gypsum board widely used in building construction. Sulfuric acid is made from SO_2 derived from oxidizing sulfur or by extraction from natural gas or crude oil. Sulfur dioxide is then oxidized to H_2SO_4. Sulfuric acid has a myriad of industrial uses, but about half of it is used in fertilizer manufacturing.

Plants usually rely on sulfate as a sulfur source. Sulfur in plant biomass is passed through the food web to supply sulfur to animals. Animal excreta and remains of dead plants and animals contain sulfur that is used by bacteria. Organic sulfur in residues is mineralized by bacteria to sulfides, but in the presence of oxygen, sulfides are oxidized to sulfate. Sulfate can be reduced to sulfides by some bacteria.

Sulfur Transformations

Sulfur has five possible valences ranging from −2 to +6 (Table 13.1) and can undergo transformation through oxidations and reductions. These transformations often are mediated by microorganisms and are quite similar to nitrogen transformations discussed in Chap. 11.

Plant Uptake and Mineralization

Plants must reduce sulfate to sulfide in order to make sulfur-containing amino acids. Sulfate is first combined with ATP and converted to sulfite (SO_3^-). Sulfite is reduced to sulfide in a reaction catalyzed by the enzyme sulfite reductase. The sulfide is then combined with organic moieties to form cysteine and methionine—the two sulfur-containing amino acids.

Plants usually contain 0.1–0.3 % sulfur. The sulfur-containing amino acids of plant proteins pass through the food web to animals. Bacteria and other microorganisms must have a source of sulfur because their cells contain from 0.1 to 1.0 % sulfur on a dry weight basis. If organic residues contain more sulfur than needed by microbes of decay, sulfur will be mineralized to the environment (Ex. 13.1). Sulfur will be immobilized from the environment if there is too little sulfur in a residue to supply microbial needs.

Table 13.1 Valence states of sulfur

Valence	Name	Formula
−2	Hydrogen sulfide	H_2S
	Ferrous sulfide	FeS
	Cysteine	R-SH*
	Methionine	R-S-R*
0	Elemental sulfur	S_8
+2	Sodium thiosulfate	$Na_2S_2O_3$
	Thiosulfuric acid	$H_2S_2O_3$
+4	Sulfur dioxide	SO_2
	Sulfurous acid	H_2SO_3
	Sodium bisulfate	$NaHSO_3$
+6	Sulfuric acid	H_2SO_4
	Calcium sulfate	$CaSO_4{\cdot}2H_2O$
	Potassium sulfate	K_2SO_4

**Note*: R = organic moiety

Ex. 13.1: *Suppose that a residue containing 0.3 % sulfur and 45 % carbon (dry weight basis) is completely decomposed by bacteria. It will be assumed that there is plenty of nitrogen and other elements present to meet microbial needs. The bacteria have a carbon assimilation efficiency to 5 % and contain 0.5 % S. The sulfur mineralized from 1,000 g of residue will be estimated.*

Solution:
The carbon and sulfur content of 1,000 g residue is

$$1,000 \times 0.45 = 450 \text{ g } \textit{carbon}$$

$$1,000 \times 0.003 = 3 \text{ g } \textit{sulfur}.$$

The bacterial biomass will be

$$450 \text{ g C} \times 0.05 = 22.5 \text{ g } \textit{bacterial carbon}$$

$$\textit{22.5 g bacteria C} \div \textit{0.5} = \textit{45 g bacterial biomass}.$$

The sulfur in bacterial biomass is

$$45 \text{ g } \textit{bacteria} \times 0.005 = 0.225 \text{ g } \textit{sulfur}.$$

The residue contains 2.775 g more sulfur than needed for its decomposition, and this sulfur will be mineralized.

Usually, the C:N ratio is a more important factor controlling the rate of organic matter decomposition than is the C:S ratio.

Oxidations

Many sulfur oxidations occur chemically, but biological intervention can speed up these reactions. Bacteria that oxidize inorganic sulfur usually are obligate or facultative autotrophs. They use the energy obtained from oxidizing sulfur to convert carbon dioxide to organic carbon in their cells. This process is analogous to nitrification and results in organic matter synthesis. However, like nitrification, the process is not highly energy efficient, and the amount of organic matter produced globally by sulfur oxidizing bacteria is miniscule in comparison to photosynthesis.

The best-known sulfur-oxidizing bacteria are species of *Thiobacillus*. Some representative reactions carried out by these microorganisms are

$$Na_2S_2O_3 + 2O_2 + H_2O \rightarrow 2NaHSO_4 \tag{13.4}$$

$$5Na_2S_2O_3 + 4O_2 + H_2O \rightarrow 5Na_2SO_4 + H_2SO_4 + 4S \tag{13.5}$$

$$S + 1\frac{1}{2} O_2 + H_2O \rightarrow H_2SO_4 \quad (13.6)$$

$$5S + 6KNO_3 + 2H_2O \rightarrow K_2SO_4 + 4KHSO_4 + 3N_2. \quad (13.7)$$

In (13.7), sulfur is oxidized, but nitrogen also is released as N_2. Thus, both sulfur oxidation and denitrification are affected by the reaction.

Green and purple sulfur bacteria also can oxidize sulfur. These bacteria are highly unusual in that they are anaerobic photoautotrophs. The green sulfur bacteria include *Chlorobium* and several other genera of the family Chlorobiaceae, while the purple sulfur bacteria are of the family Chromatiaceae that includes *Thiospirillum* and several other genera. The colored, green and purple sulfur bacteria use light and energy from oxidizing sulfide to reduce carbon dioxide to carbohydrate (CH_2O) according to the following reactions:

$$CO_2 + 2H_2S \xrightarrow[\text{light}]{} CH_2O + 2S + H_2O \quad (13.8)$$

$$2CO_2 + H_2S + 2H_2O \xrightarrow[\text{light}]{} 2CH_2O + H_2SO_4. \quad (13.9)$$

Both light and anaerobic conditions are necessary for colored, photosynthetic sulfur bacteria, so their presence is greatly restricted in nature. Colored sulfur bacteria are particularly troublesome by growing on the inside tops of sewer pipes near manhole covers where light is available. The result is corrosion of the upper inner surface of the pipe near the manhole cover by sulfuric acid produced by the bacteria.

Intentional applications of elemental sulfur sometimes are used to reduce the pH in soil to promote the growth of certain acid-loving plants. The potential acidity of elemental sulfur is calculated in Ex. 13.2.

Ex. 13.2: *The amount of acidity resulting from the oxidation of 1,000 g of elemental sulfur will be estimated in terms of equivalent calcium carbonate.*

Solution:
The reaction is

$$S + 1\frac{1}{2}O_2 + H_2O \rightarrow H_2SO_4$$

and

$$\begin{array}{ccc} 1,000\ g & & x \\ S = & H_2SO_4 = & CaCO_3 \\ 32 & & 100 \end{array}$$

$$x = 3,125\ g\ CaCO_3.$$

Acid-Sulfate Soils and Acid-Mine Drainage

Sulfur oxidation can be a significant source of acidity in the environment. There are places where soils and other geologic formations contain large concentrations of metal sulfides and iron sulfide in particular. For example, sulfide-rich soils are common in existing or former coastal swamps such as salt marshes and mangrove ecosystems. In such areas, sediment contains abundant organic matter from plants, iron-bearing soil particles from river inflow, and sulfate from seawater. Decomposition of organic matter creates anaerobic conditions in the sediments of these coastal wetlands. Anaerobic bacteria use oxygen from iron oxides and sulfate, and pore water in the sediments contain ferrous iron and hydrogen sulfide. This condition in sediment is conducive for iron sulfide production as illustrated in the following unbalanced equations:

$$\underset{\text{in sediment}}{\text{Organic matter}} + \underset{\text{in sediment}}{Fe(OH)_3} + \underset{\text{in seawater}}{SO_4^{2-}} \xrightarrow[\text{anaerobic bacteria}]{} Fe(OH)_2 + H_2S + CO_2 + H_2O. \quad (13.10)$$

The reduced products of anaerobic microbial activity react forming iron sulfide

$$Fe(OH)_2 + H_2S \rightarrow FeS + 2H_2O. \quad (13.11)$$

Ferrous sulfide (13.11) is readily oxidized to iron pyrite (FeS_2) in the presence of hydrogen sulfide that may be abundant in anaerobic environments (Rickard and Luther 1997). The reaction is

$$FeS + H_2S \rightarrow FeS_2 + H_2 \uparrow. \quad (13.12)$$

Notice that in (13.12), sulfur in FeS_2 should have a valence of -1, a valence not listed for sulfur in Table 13.1. Some authors such as Schippers (2004) and Borda (2006) assign a valence of -1 to S in iron pyrite, but the -1 valence apparently causes a dilemma (Nesbitt et al. 1998). Thus, some authorities refer to the S_2 in FeS_2 as the persulfide ion with a valence of -2.

Some soils in coastal areas of South Carolina in the southern United States were reported to contain up to 5.5 % sulfide-sulfur (Fleming and Alexander 1961). Coastal areas with soil high in sulfide-sulfur concentration are common in Indonesia, the Philippines, Thailand, Malaysia, and many other countries. Metal sulfides also form in marine sediment not associated with wetlands or warm climates. For example, acid-sulfate soils apparently were first noted by Carl Linnaeus, the Swedish botanist of plant and animal taxonomy fame, in the eighteenth century in the Netherlands (Fanning 2006).

In addition to iron sulfide, copper sulfide (CuS), zinc sulfide (ZnS), manganous sulfide (MnS), and other metal sulfides form in anaerobic environments (Rickard and Luther 2006). Iron pyrite and other metal sulfides are common constituents of coal and some other mineral deposits. Mining often exposes overburden of high sulfide-sulfur content and creates spoil piles that contain sulfide-sulfur.

Soils containing sulfur concentrations of 0.75 % and greater are referred to as potential acid-sulfate soils (Soil Survey Staff 1994). As long as soils containing metal sulfides are anaerobic, these compounds are quite insoluble and have no influence on soil acidity. When such soils are exposed to oxygen, oxidation occurs and sulfuric acid forms. The oxidation of iron pyrite is a complex reaction that apparently can occur in several ways depending upon conditions existing in the sediment or soil containing pyrite (Chandra and Gerson 2011). Reactions presented by Sorensen et al. (1980) will be used here to illustrate the oxidation of iron pyrite:

$$FeS_2 + H_2O + 3.5\ O_2 \rightarrow FeSO_4 + H_2SO_4 \quad (13.13)$$

$$2FeSO_4 + 0.5\ O_2 + H_2SO_4 \rightarrow Fe_2(SO_4)_3 + H_2O \quad (13.14)$$

$$FeS_2 + 7Fe_2(SO_4)_3 + 8H_2O \rightarrow 15FeSO_4 + 8H_2SO_4. \quad (13.15)$$

The production of ferric sulfate from ferrous sulfate is greatly accelerated by the activity of bacteria of the genus *Thiobacillus*, and under acidic conditions, the oxidation of pyrite by ferric sulfate is rapid. In addition, according to Sorensen et al. (1980) ferric sulfate may hydrolyze according to the following reactions:

$$Fe_2(SO_4)_3 + 6H_2O \rightarrow 2Fe(OH)_3 + 3H_2SO_4 \quad (13.16)$$

$$Fe_2(SO_4)_3 + 2H_2O \rightarrow 2Fe(OH)SO_4 + H_2SO_4. \quad (13.17)$$

Ferric sulfate also may react with iron pyrite to form elemental sulfur and the sulfur may be oxidized to sulfuric acid by microorganisms

$$Fe_2(SO_4)_3 + FeS_2 \rightarrow 3FeSO_4 + 2S \quad (13.18)$$

$$S + 1.5\ O_2 + H_2O \rightarrow H_2SO_4. \quad (13.19)$$

Potential acid-sulfate soils are called acid-sulfate soils once oxidation of pyrite occurs.

Ferric hydroxide can react with adsorbed bases, such as potassium, in acid sulfate soils to form jarosite, a basic iron sulfate,

$$3Fe(OH)_3 + 2SO_4^{2-} + K^+ + 3H^+ \rightarrow KFe_3(SO_4)_2(OH)_6 2H_2O + H_2O. \quad (13.20)$$

Jarosite is relatively stable, but in older acid-sulfate soils where acidity has been neutralized, jarosite tends to hydrolyze

$$KFe_3(SO_4)_2(OH)_6 2H_2O + 3H_2O \rightarrow 3Fe(OH)_3 + K^+ + 2SO_4^{2-} + 3H^+ + 2H_2O. \quad (13.21)$$

Sulfuric acid dissolves aluminum, manganese, zinc, copper, and other metals from soil, and runoff from acid-sulfate soils or mine spoils not only is highly acidic but it may contain potentially toxic metallic ions. The potential for acid production in an acid-sulfate soil depends largely upon the amount and particle size of the iron pyrite, the presence or absence of exchangeable bases and carbonates within the pyrite-bearing material, the exchange of oxygen and solutes with the soil, and the abundance of *Thiobacillus*. Because the exchange of oxygen and solutes and the abundance of *Thiobacillus* are restricted with depth, the acid-sulfate condition in soils is essentially a surface problem.

Waterlogging of acid-sulfate soils restricts the availability of oxygen, and sulfuric acid production ceases when the soil becomes anaerobic. In fact, sulfate is reduced to sulfide under anaerobic conditions by bacteria of the genus *Desulfovibrio*. In natural waters, the sediment-water interface often is aerobic, so sulfuric acid production can occur slowly.

When the pyrite in spoil piles oxidizes, sulfuric acid may enter streams reducing pH. Seepage from underground mines also may be acidic. Surface coal mining was a major source of acid drainage to streams in the past. The reason was that the overburden above a coal deposit was stripped off and piled on the land surface. The last overburden removed from the coal deposit was contaminated with particles of coal that contained sulfide sulfur. This contaminated material tended to end up on top of the spoil pile where there was contact with atmospheric oxygen and oxidation of the sulfide to sulfuric acid. The acid was removed by rainwater and entered streams.

In most countries today, mining operations must separate the overburden that is contaminated with coal and stockpile it separately from the other overburden. Once the coal deposit has been mined, the contaminated overburden is placed back into the mined area and covered with the overburden that is free of coal particles. Of course liming of spoil material and development of grass cover on it also lessens the problem with acid drainage. In some cases, streams are treated with limestone to neutralize acidity.

Reductions

Sulfur reductions occur in anaerobic environments, and sulfur-reducing bacteria use oxygen from sulfate or other oxidized sulfur compounds as electron and hydrogen acceptors in respiration. The process is similar in this regard to denitrification. The predominant sulfur-reducing bacteria are species of the genus *Desulfovibrio*. Some typical reactions are

$$SO_4^{2-} + 8H^+ \rightarrow S^{2-} + 4H_2O \tag{13.22}$$

$$SO_3^{2-} + 6H^+ \rightarrow S^{2-} + 3H_2O \tag{13.23}$$

$$S_2O_3^{2-} + 8H^+ \rightarrow 2SH^- + 3H_2O. \tag{13.24}$$

The source of electrons, hydrogen, and energy for the reactions depicted in (13.22)–(13.24) are carbohydrates, organic acids, and alcohols. A representative complete reaction is

$$2CH_3CHOHCOONa + MgSO_4 \rightarrow H_2S + 2CH_3COONa + CO_2 + MgCO_3 + H_2O. \tag{13.25}$$

Sulfide

Sulfides include hydrogen sulfide and it dissociation products. Hydrogen sulfide is a diprotic acid that dissociates yielding HS^- and S^{2-}

$$H_2S = HS^- + H^+ \quad K_1 = 10^{-7.01} \tag{13.26}$$

$$HS^- = S^{2-} + H^+ \quad K_2 = 10^{-13.89}. \tag{13.27}$$

Depending on pH, the dominant sulfide species will be H_2S, HS^-, or S^{2-} (Fig. 13.2). Un-ionized hydrogen sulfide dominates in acidic environments, and in alkaline situations, HS^- will be the main form.

In highly anaerobic environments ferrous iron and sulfide are both present, and ferrous sulfide precipitates. Sulfides form with other metals, and as can be seen in Table 3.4, metal sulfides have a very small solubility product constant. Marine sediment is a particularly favorable environment for the formation of metal sulfides, because the high sulfate content of the pore water favors sulfide production.

Sulfide is a significant variable in water quality because the un-ionized form (H_2S) is highly toxic to fish and other aquatic animals. The two, ionized species (HS^- and S^{2-}) have little toxicity. The mass action expression for (13.26) can be used to estimate the proportion of sulfide in un-ionized form at a particular pH as illustrated in Ex. 13.3.

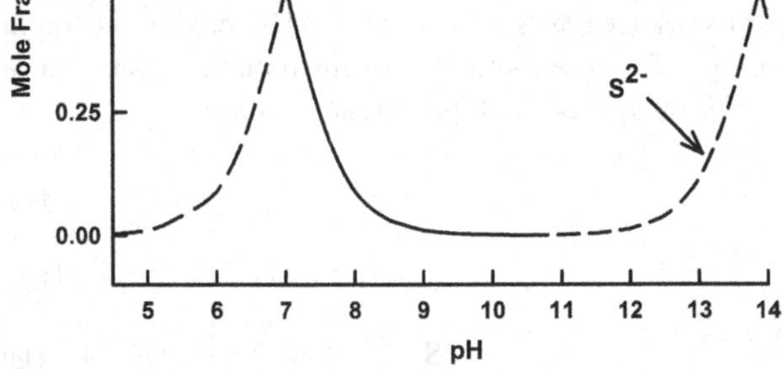

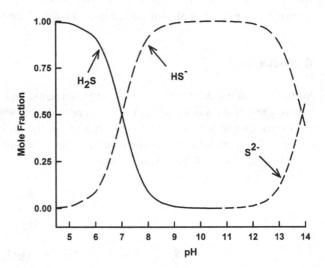

Fig. 13.2 Effects of pH on the relative proportions of sulfide species

Ex. 13.3: The percentage un-ionized hydrogen sulfide will be estimated for pH 6.

Solution:
The mass action expression of (13.26) is

$$\frac{(HS^-)(H^+)}{(H_2S)} = 10^{-7.01}$$

$$\frac{(HS^-)}{(H_2S)} = \frac{10^{-7.01}}{10^{-6}} = 10^{-1.01} = 0.098.$$

Thus, for 0.098 mol HS^-, there will be 1 mol H_2S and

$$\frac{1}{1 + 0.098} = 0.911 (or\ 91.1\%\ H_2S).$$

The first ionization constant ($H_2S = HS^- + H^+$) for hydrogen sulfide at temperatures of 0–50 °C are provided (Table 13.2). The ionization constants were used to calculate the proportions of un-ionized hydrogen sulfide at different pH values and temperatures (Table 13.3). To illustrate the use of Table 13.3, suppose

Table 13.2 First ionization constants for H_2S ($H_2S = HS^- + H^+$) at different temperatures

Temperature (°C)	K	Temperature (°C)	K
0	$10^{-7.66}$	30	$10^{-6.89}$
5	$10^{-7.51}$	35	$10^{-6.78}$
10	$10^{-7.38}$	40	$10^{-6.67}$
15	$10^{-7.25}$	45	$10^{-6.57}$
20	$10^{-7.13}$	50	$10^{-6.47}$
25	$10^{-7.01}$		

Table 13.3 Decimal fraction (proportions) of total sulfide-sulfur present as un-ionized hydrogen sulfide-sulfur (H_2S-S) at different pH values and temperatures

	Temperature (°C)							
pH	5	10	15	20	25	30	35	40
5.0	0.997	0.996	0.947	0.993	0.990	0.987	0.983	0.979
5.5	0.981	0.987	0.983	0.977	0.970	0.961	0.950	0.937
6.0	0.970	0.960	0.947	0.931	0.911	0.886	0.858	0.831
6.5	0.911	0.883	0.849	0.810	0.764	0.711	0.656	0.597
7.0	0.764	0.706	0.641	0.575	0.505	0.437	0.376	0.318
7.5	0.506	0.431	0.360	0.299	0.244	0.197	0.160	0.129
8.0	0.245	0.193	0.152	0.119	0.093	0.072	0.057	0.045
8.5	0.093	0.071	0.053	0.041	0.031	0.024	0.019	0.015
9.0	0.031	0.023	0.018	0.013	0.010	0.008	0.006	0.003

the total sulfide concentration is 0.10 mg/L at pH 8.0 and 30 °C. The factor from Table 13.3 for these conditions is 0.072; the amount of sulfur in hydrogen sulfide will be 0.1 mg/L × 0.072 = 0.0072 mg/L.

Sulfur Concentrations

Sulfate is the primary form of sulfur in natural waters. Sulfate concentrations in inland surface waters of humid regions are usually low and often only a few milligrams per liter. In arid regions, surface waters normally will have much higher sulfate concentrations with values often exceeding 50 or 100 mg/L. Ocean water averages 2,700 mg/L in sulfate (Table 4.8).

Sulfide rarely occurs in measurable concentration in aerobic surface waters. But, if the rate of sulfide release from anaerobic sediment is greater than the rate of sulfide oxidation in the water column, sulfide can accumulate in surface water. Sulfide concentration may be several milligrams per liter in hypolimnetic waters of eutrophic lakes and in certain well waters. When water containing sulfide is aerated, sulfide is rapidly oxidized to sulfate. Information on concentrations of other forms of sulfur in natural waters is scarce.

Freshwater sediments usually contain less than 0.1 % total sulfur, and marine sediments usually contain between 0.05 and 0.3 % total sulfur. In some sediments where iron pyrite and other metal sulfides have accumulated, total sulfur concentrations may range from 0.5 to 5 %. Pore water in anaerobic sediment may contain 1–5 mg/L of sulfide-sulfur.

Significance

Sulfate concentration seldom limits aquatic plant growth in aquatic ecosystems, and it is not considered a factor in eutrophication. The most common concern over sulfate concentration relates to drinking water quality. Sulfate in drinking water can impart a bitter taste. Depending upon their sensitivity, people notice the bitter taste at sulfate concentrations of 250–1,000 mg/L. Water containing sulfate also may act as a laxative to people not used to drinking it. The highest quality drinking water will not contain more than 50 mg/L sulfate, but most authorities indicate that up to 250 mg/L is acceptable for drinking water supplies. In addition, high sulfate concentration in water interferes with chlorination.

Hydrogen sulfide is extremely toxic to aquatic animals. The 96-h LC50 of hydrogen sulfide to freshwater fish ranged from 4.2 to 34.8 μg/L (Gray et al. 2002), but marine species apparently are somewhat less sensitive to hydrogen sulfide—96-h LC50s are 2–10 times greater than for freshwater fish (Bagarinao and Lantin-Olaguer 1999; Gopakumar and Kuttyamma 1996). Most authorities suggest that no more than 0.002 mg/L of sulfide should be present in natural waters and any detectable concentration is undesirable. Hydrogen sulfide toxicity to organisms results from several effects. Hydrogen sulfide inhibits the reoxidation of cytochrome a_3 by molecular oxygen, because it blocks the electron transport system and stops oxidative respiration. Blood lactate concentration also increases and anaerobic glycolysis is favored over aerobic respiration. The overall toxic effect is hypoxia, and low dissolved oxygen concentrations enhance sulfide toxicity (Boyd and Tucker 2014). The toxic form of sulfide is H_2S so sulfide toxicity also is favored by acidic conditions where most of the sulfide is un-ionized (Table 13.3). Hydrogen sulfide is rarely detectable in environments containing dissolved oxygen, and it seldom is an important factor other than in extremely polluted aquatic ecosystems.

Hydrogen sulfide also can be a contaminant in well water. Although it is not toxic to humans at concentrations normally encountered, most people can detect a bad odor at sulfide concentrations of 0.1–0.5 mg/L. At sulfide concentrations up to 1 mg/L the odor often is described as musty, but higher concentrations cause the typical "rotten egg" odor of hydrogen sulfide. In addition to the odor, elevated hydrogen sulfide concentrations are highly corrosive to water pipes and fixtures. Methods for removing hydrogen sulfide from drinking water include activated carbon filtration, aeration, and oxidation with chlorine or potassium permanganate.

Atmospheric emissions of sulfur dioxide from combustion of fossil fuels are the major cause of acid rain. Since the 1970s, many countries have worked to reduce sulfur dioxide emissions by requiring sulfur dioxide removal devices in vehicles, on smokestacks, etc. Although considerable control of sulfur dioxide emissions was achieved and global emission declined for several years, sulfur dioxide emissions are now increasing because of increased fossil fuel use by countries with rapidly developing economies such as China and India (Boyd and McNevin 2014).

References

Bagarinao T, Lantin-Olaguer I (1999) The sulfide tolerance of milkfish and tilapia in relation to fish kills in farms and natural waters in the Philippines. Hydrobiologia 382:137–150

Borda MJ (2006) Pyrite. In: Lai R (ed) Encyclopedia of soil science. Taylor & Francis, New York, pp 1385–1387

Boyd CE, McNevin AA (2015) Aquaculture, resource use, and the environment. Wiley Blackwell, Hoboken

Boyd CE, Tucker CS (2014) Handbook for aquaculture water quality. Craftmaster, Auburn

Chandra AP, Gerson AR (2011) Pyrite (FeS_2) oxidation: a sub-micron synchrotron investigation of the initial steps. Geochim Cosmochim Acta 75:6239–6254

Fanning DS (2006) Acid sulfate soils. In: Lai R (ed) Encyclopedia of soil science. Taylor & Francis, New York, pp 11–13

Fleming JF, Alexander LT (1961) Sulfur acidity in South Carolina tidal marsh soils. Soil Sci Soc Am Proc 25:94–95

Gopakumar G, Kuttyamma VJ (1996) Effect of hydrogen sulphide on two species of penaeid prawns *Penaeus indicus* (H. Milne Edwards) and *Metapenaeus dobsoni* (Miers). Bull Environ Contam Toxicol 57:824–828

Gray JS, Wu RS, Or YY (2002) Effects of hypoxia and organic enrichment on the coastal marine environment. Mar Ecol Prog Ser 238:249–279

Nesbitt HW, Bancroft GM, Pratt AR, Scaini MJ (1998) Sulfur and iron surface states on fractured pyrite surfaces. Am Mineral 83:1067–1076

Rickard D, Luther GW III (1997) Kinetics of pyrite formation by the H_2S oxidation of iron (II) monosulfide in aqueous solutions between 25 and 125°C: the mechanisms. Geochimica Cosmochim Acta 61:135–147

Rickard D, Luther GW III (2006) Metal sulfide complexes and clusters. Rev Min Geochem 61:421–504

Schippers A (2004) Sulfur biogeochemistry: past and present. Special Paper 379. In: Amend JP, Edwards KJ, Lyons TW (eds) Biogeochemistry of metal sulfide oxidation in mining environments, sediments, and soils. Geological Society of America, Washington, BC, pp 49–62

Smith SJ, van Aardenne J, Kilmont Z, Andres RJ, Volke A, Arias SD (2011) Anthropogenic sulfur dioxide emissions: 1850-2005. Atmos Chem Phys 11:1101–1116

Soil Survey Staff (1994) Keys to soil taxonomy, 6th edn. United States Department of Agriculture, Soil Conservation Service, Washington, DC

Sorensen DL, Knieb WA, Porcella DB, Richardson BZ (1980) Determining the lime requirement for the Blackbird Mine spoil. J Environ Qual 9:162–166

Micronutrients and Other Trace Elements 14

Abstract

The solubilities of most minerals from which trace metals in natural waters originate are favored by low pH. The concentration of the free ion of a dissolved trace element usually is much lower than is the total concentration of the trace element. This results from ion pair associations between the free trace ion and major ions, complex ion formation, hydrolysis of metal ions, and chelation of metal ions. Several trace elements—zinc, copper, iron, manganese, boron, fluorine, iodine, selenium, cadmium, cobalt, and molybdenum—are essential to plants, animals or both. A few other trace elements are suspected but not unequivocally proven to be essential. There are some reports of low micronutrient concentrations limiting the productivity of water bodies; but primary productivity in most water bodies apparently is not limited by a shortage of micronutrients. Trace elements—including the ones that are nutrients—may be toxic at high concentration to aquatic organisms. Excessive concentrations of several trace metals in drinking water also can be harmful to human health. Instances of trace element toxicity in aquatic animals and humans usually have resulted from anthropogenic pollution. Nevertheless, excessive concentrations of trace metals in drinking water sometimes occur naturally—an example is the presence of chronically-toxic concentrations of arsenic in groundwater that serves as the water supply for several million people in a few provinces of Bangladesh and adjoining India.

Keywords

Trace metal complexes • Metallic micronutrients • Non-metallic micronutrients • Toxicity of trace metals • Micronutrients and productivity

C.E. Boyd, *Water Quality*, DOI 10.1007/978-3-319-17446-4_14

Introduction

The discussion has until now focused on elements present in water in relatively large concentrations. Water also contains other elements called trace elements that are present in small concentrations. Some trace elements are essential for living things—they are called micronutrients. The most familiar micronutrients probably are iron, manganese, zinc, copper, and iodine, but there are others.

Trace elements are important in water quality. One or more micronutrients sometimes may be at such a low concentration as to limit productivity. At elevated concentrations—mostly in polluted waters—micronutrients and other trace elements may be toxic to aquatic plants and animals. Excessive concentrations of certain trace elements also can cause water to be unsuitable for human use because of toxicity or other unfavorable effects.

The purpose of this chapter is to discuss micronutrients and other trace elements. The focus is on the chemistry of trace elements, their sources, their concentrations in natural waters, and their ecological effects and importance in water supply.

Factors Controlling Trace Element Concentrations

The concentration of any given dissolved inorganic ion can be thought of as controlled by the solubility of a mineral compound as discussed in Chap. 3. For example, in an area with karst geology, concentrations of calcium, magnesium, and bicarbonate in water are in equilibrium with limestone. Because limestone is abundant in such areas, calcium, magnesium, and bicarbonate concentrations in water are appreciable. In the case of micronutrients and other trace elements, the controlling minerals usually are at low abundance in soil and other geological formations, as well as often being of low solubility. As a result, equilibrium concentrations of free ions of trace elements usually are very low. Free ions are unassociated with other ions, e.g., the free ion for iron is either Fe^{2+} or Fe^{3+} while that of zinc is Zn^{2+}.

Sometimes there may not be a well-defined controlling mineral for a particular trace element in a geological formation, and the source of this trace element in water may be from its incidental inclusion in other minerals, or decomposing organic matter. In such instances, the concentration of the trace ion will be low and not related to a controlling mineral. In the case of pollution—a trace element may be present at a higher concentration than typically encountered in natural waters. A portion of the ions of a trace element introduced into a water body by pollution often will precipitate in form of a controlling mineral. This restricts the free ion concentration to its equilibrium concentration with the controlling mineral; this phenomenon will occur even if the controlling mineral is not naturally present in the sediment.

Common controlling minerals for several trace ions are provided along with the dissolution equations and equilibrium constants in Table 14.1. However, trace ions

Table 14.1 Solubility expressions of equilibrium constants (K) for minerals that can control solubilities of selected trace elements in water

Name	Reaction	K
Barium carbonate	$BaCO_3 = Ba^{2+} + CO_3^{2-}$	$10^{-8.59}$
Barium sulfate	$BaSO_4 = Ba^{2+} + SO_4^{2-}$	$10^{-9.97}$
Cadmium carbonate	$CdCO_3 = Cd^{2+} + CO_3^{2-}$	10^{-12}
Calcium fluoride	$CaF_2 = Ca^{2+} + 2F^-$	$10^{-10.46}$
Gibbsite	$Al(OH)_3 + 3H^+ = Al^{3+} + 3H_2O$	10^9
Fe (III) hydroxide	$Fe(OH)_3 + 3H^+ = Fe^{3+} + 3H_2O$	$10^{3.54}$
Ferric arsenate	$FeAsO_4{\cdot}2H_2O = Fe^{2+} + AsO_4^{3-} + 2H_2O$	10^{-23}
Ferrous molybdate	$FeMoO_4 = Fe^{2+} + MoO_4^{2-}$	$10^{-6.91}$
Goethite	$FeOOH + 3H^+ = Fe^{3+} + 2H_2O$	$10^{3.55}$
Hematite	$Fe_2O_3 + 6H^+ = 2Fe^{3+} + 3H_2O$	$10^{-1.45}$
Siderite	$FeCO_3 + H^+ = Fe^{2+} + HCO_3^-$	$10^{-0.3}$
Iron chromite	$FeCr_2O_4 = Fe^{3+} + Cr_2O_4^{3-}$	–
Iron (II) hydroxide	$Fe(OH)_2 = Fe^{2+} + 2OH^-$	10^{-14}
Lead carbonate	$PbCO_3 = Pb^{2+} + CO_3^{2-}$	$10^{-13.13}$
Lead sulfate	$PbSO_4 = Pb^{2+} + SO_4^{2-}$	$10^{-7.6}$
Manganite	$4MnOOH + 8H^+ = 4Mn^{2+} + 6H_2O + O_2$	$10^{0.3}$
Manganese (II) hydroxide	$Mn(OH)_2 = Mn^{2+} + 2OH^-$	$10^{-12.8}$
Manganese carbonate	$MnCO_3 = Mn^{2+} + CO_3^{2-}$	$10^{-9.3}$
Malachite	$Cu_2(OH)_2CO_3 + 4H^+ = 2Cu^{2+} + 3H_2O + CO_2$	$10^{14.16}$
Nickel hydroxide	$Ni(OH_2) = Ni^{2+} + 2OH^-$	$10^{-15.26}$
Silver chloride	$AgCl = Ag^+ + Cl^-$	$10^{-9.75}$
Tenorite	$CuO + 2H^+ = Cu^{2+} + H_2O$	$10^{7.65}$
Zinc oxide	$ZnO + 2H^+ = Zn^{2+} + H_2O$	$10^{11.18}$
Zinc carbonate	$ZnCO_3 + 2H^+ = Zn^{2+} + H_2O + CO_2$	$10^{7.95}$
Cobalt carbonate	$CoCO_3 = Co^{2+} + CO_3^{2-}$	$10^{-12.8}$

interact with other ions and organic matter in water to form soluble entities that either increase the total concentration of the trace element or reduce the concentration of the free ion.

Ion Pairs

Cations and anions in a solution are mutually attracted, and some of them come together to form ion associations that behave as single entities. For example, in a potassium chloride solution a portion of the potassium and chloride ions are pulled closely together by electrical attraction to form a potassium chloride association (KCl^0) that is uncharged and soluble. Of course, some ion associations bear a charge, e.g., potassium and sulfate form KSO_4^- and calcium and bicarbonate form $CaHCO_3^+$. Ion associations form in accordance with the principle of chemical equilibrium, and there are equilibrium constants for their formation. Such ion associations are called ion pairs.

Ion pairs of major ions usually are not of great importance in water quality, because only a small fraction of the total concentration of major ions will be in ion pairs. In natural waters, trace ions are at low concentration, but major ions of opposite charge are at much higher concentration. Thus, a large fraction of the trace element can be paired with one or more major ions, and the amount of a trace ion occurring as ion pairs often exceeds the concentration of the free trace ion.

The formation for an ion pair is illustrated below for zinc and the major ion sulfate:

$$Zn^{2+} + SO_4^{2-} \rightleftharpoons ZnSO_4^0 \qquad K = 10^{2.38}. \tag{14.1}$$

The $ZnSO_4{}^0$ ion pair—like other ion pairs—is soluble and in equilibrium with the free metal ion concentration in solution. An ion pair does not diminish the concentration of the free metal ion provided the free ion is in equilibrium with a solid phase source. But, ion pairs can increase the concentrations of metal ions possible in water. Selected ion-pair reactions and their equilibrium constants are provided (Table 14.2). The calculation of ion pairs is illustrated in Ex. 14.1.

Ex. 14.1: *Concentrations of zinc ion pairs will be calculated for a water where the equilibrium concentration of the zinc ion (Zn^{2+}) is 0.050 mg/L ($10^{-6.12}$ M) and concentrations of anions of interest are: F^-, 0.25 mg/L ($10^{-4.88}$ M); Cl^-, 10 mg/L ($10^{-3.55}$ M); $SO_4{}^{2-}$, 5 mg/L ($10^{-4.28}$ M); $CO_3{}^{2-}$, 10 mg/L ($10^{-3.78}$ M).*

Solution:

Using ion pair equation and equilibrium constants found in Table 14.2, the ion pair concentrations are

$$\frac{(ZnCO_3^0)}{(Zn^{2+})(CO_3^0)} = 10^{5.0}$$

$$(ZnCO_3^0) = (10^{-6.12})(10^{-3.78})(10^{5.0}) = 10^{-4.90}\ \mathrm{M}\ (0.823\ \mathrm{mg}\ Zn/\mathrm{L})$$

$$\frac{(ZnSO_4^0)}{(Zn^{2+})(SO_4^{2-})} = 10^{2.38}$$

$$(ZnSO_4^0) = (10^{-6.12})(10^{-4.28})(10^{2.38}) = 10^{-8.02}\ \mathrm{M}\ (0.0006\ \mathrm{mg}\ Zn/\mathrm{L})$$

$$\frac{(ZnF^+)}{(Zn^{2+})(F^-)} = 10^{1.26}$$

$$(ZnF^+) = (10^{-6.12})(10^{-4.88})(10^{1.26}) = 10^{-9.74}\ \mathrm{M}\ (\text{negligible}).$$

Using the same technique illustrated above, for chloride ion pairs we obtain

Table 14.2 Equations and equilibrium constants for selected ion pairs of micronutrients and trace elements

Reaction	K	Reaction	K
$Ca^{2+} + F^- = CaF^+$	$10^{1.04}$	$Zn^{2+} + Cl^- = ZnCl^+$	$10^{0.43}$
$Fe^{3+} + F^- = FeF^{2+}$	$10^{5.17}$	$Zn^{2+} + 2Cl^- = ZnCl_2^0$	$10^{0.61}$
$Fe^{3+} + 2F^- = FeF_2^+$	$10^{9.09}$	$Zn^{2+} + 3Cl^- = ZnCl_3^-$	$10^{0.53}$
$Fe^{3+} + 3F^- = FeF_3^0$	10^{12}	$Zn^{2+} + 4Cl^- = ZnCl_4^{2-}$	$10^{0.20}$
$Fe^{3+} + Cl^- = FeCl^{2+}$	$10^{1.42}$	$Zn^{2+} + SO_4^{2-} = ZnSO_4^0$	$10^{2.38}$
$FeCl^{2+} + Cl^- = FeCl_2^+$	$10^{0.66}$	$Zn^{2+} + CO_3^{2-} = ZnCO_3^0$	10^5
$FeCl_2 + Cl^- = FeCl_3^0$	10^1	$Cd^{2+} + F^- = CdF^+$	$10^{0.46}$
$Fe^{2+} + SO_4^{2-} = FeSO_4^0$	$10^{2.7}$	$Cd^{2+} + Cl^- = CdCl^+$	10^2
$Fe^{3+} + SO_4^{2-} = FeSO_4^+$	$10^{4.15}$	$Cd^{2+} + 2Cl^- = CdCl_2^0$	$10^{2.70}$
$Cu^{2+} + F^- = CuF^+$	$10^{1.23}$	$Cd^{2+} + 3Cl^- = CdCl_3^-$	$10^{2.11}$
$Cu^{2+} + Cl^- = CuCl^+$	10^0	$Cd^{2+} + SO_4^{2-} = CdSO_4^0$	$10^{2.29}$
$Cu^{2+} + SO_4^{2-} = CuSO_4^0$	$10^{2.3}$	$Al^{3+} + F^- = AlF^{2+}$	$10^{6.13}$
$Cu^{2+} + CO_3^{2-} = CuCO_3^0$	$10^{6.77}$	$Al^{3+} + 2F^- = AlF_2^+$	$10^{11.15}$
$Cu^{2+} + 2CO_3^{2-} = Cu(CO_3)_2^{2-}$	$10^{10.01}$	$Al^{3+} + 3F^- = AlF_3^0$	10^{13}
$Zn^{2+} + F^- = ZnF^+$	$10^{1.26}$	$Al^{3+} + SO_4^{2-} = AlSO_4^-$	$10^{2.04}$

$$(ZnCl^+) = 10^{-9.24}\ \text{M (negligible)}$$

$$\left(ZnCl_2^0\right) = 10^{-12.61}\ \text{M (negligible)}$$

$$\left(ZnCl_3^-\right) = 10^{-16.24}\ \text{M (negligible)}$$

$$\left(ZnCl_4^{2-}\right) = 10^{-20.12}\ \text{M (negligible)}.$$

The zinc fluoride and zinc chloride ion pair concentrations are negligible, and the total soluble zinc concentration is

$$Zn_{Total} = Zn^{2+} + ZnCO_3^0 + ZnSO_4^0$$

$$Zn_{Total} = 0.874\ \text{mg/L}.$$

The total concentration of ion pairs is about 16 times greater than the free zinc ion concentration.

In Ex. 14.1, it should be noted that essentially all of the zinc combined in ion pairs is contained in a single association—$ZnCO_3^0$. Typically, a single ion pair will be dominant for a particular element.

Complex Ions

The Brønsted theory of acids and bases holds that an acid is a compound able to donate a proton (hydrogen ion) and a base is a compound able to accept a proton. There is another concept of acids and bases known as the Lewis theory. According

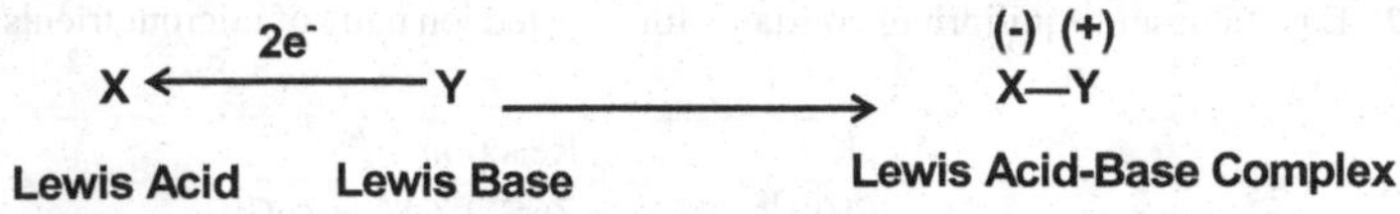

Fig. 14.1 Simple illustration of a Lewis acid-base reaction

to this theory, an acid can accept a pair of electrons and a base can donate a pair of electrons (Fig. 14.1). The Lewis theory is much more general than the Brønsted theory.

An example of a Brønsted acid is nitric acid,

$$HNO_3 \rightarrow H^+ + NO_3^-, \tag{14.2}$$

while hydroxide is an example of a Brønsted base,

$$H^+ + OH^- = H_2O. \tag{14.3}$$

An example of a Lewis acid and base is afforded by the reaction of carbon dioxide and water

$$CO_2 + H_2O = H_2CO_3 \quad K = 10^{-2.75}. \tag{14.4}$$

The reaction occurs in two steps. The oxygen atom in water donates a pair of electrons to the carbon atom in carbon dioxide, and two protons are then transferred to the negatively charged oxygen to give H_2CO_3. In this reaction, H_2O is a Lewis base and CO_2 is a Lewis acid. Other examples of Lewis acid-base reactions are

$$\underset{\text{(acid)}}{Cu^{2+}} + \underset{\text{(base)}}{4NH_3} = Cu(NH_3)_4^{2+} \quad K = 10^{12.68} \tag{14.5}$$

$$\underset{\text{(acid)}}{Fe^{2+}} + \underset{\text{(base)}}{6CN^-} = Fe(CN)_6^{3-} \quad K = 10^{42}. \tag{14.6}$$

Lewis acid-base reactions occur in natural waters, and they are important in waters with high concentrations of ammonia or cyanide or certain other pollutants as illustrated in (14.5) and (14.6). However, they are most important in metal hydrolysis and chelation to be discussed below. Boric acid, a metalloid, also forms complexes through Lewis acid-base reactions.

Hydrolysis of Metal Ions

Salts like sodium chloride and potassium nitrate dissolve in water to give neutral solutions. Other salts may form basic or acidic aqueous solutions. For example, sodium carbonate solutions are basic, while ammonium chloride solutions are acidic. Some reaction in addition to simple dissolution must be involved when such salts are put in water. This additional reaction that influences pH is hydrolysis.

The hydrolysis of ammonium from dissolution of ammonium chloride and carbonate from dissolution of sodium carbonate are shown below:

$$\begin{aligned} &NH_4^+ + H_2O = NH_3 + H_3O^+ \\ &\textit{Note: } H^+ \textit{ in water is actually in the form } H_3O^+ (\textit{see Chap. } 8). \\ &\left(\text{or simply, } NH_4^+ = NH_3 + H^+\right), \end{aligned} \tag{14.7}$$

$$\begin{aligned} &CO_3^- + H_2O = HCO_3^- + OH^- \\ &\left(\text{or simply, } CO_3^- + H^+ = HCO_3^-\right). \end{aligned} \tag{14.8}$$

Hydrolysis influences the dissociation of water. In pure water, an equal number of H^+ and OH^- are present, and this results in a neutral solution. If another reaction consumes a portion of the hydroxide ions in water, more water will dissociate in order to maintain the equilibrium constant. However, at the new equilibrium, H^+ and OH^- will not be equal and the solution will not be neutral. The anions of a weak acid capture hydrogen ions to cause a pH above neutrality. The cations of a weak base will capture hydroxide ions and cause a pH below neutrality.

Some metal ions act as weak Lewis acids by accepting pairs of electrons from the Lewis base water. The overall effect of adding a metal ion such as ferric iron (Fe^{3+}) to pure water is a reduction in pH and the formation of a complex ion containing iron and hydroxide

$$Fe^{3+} + H_2O = FeOH^{2+} + H^+. \tag{14.9}$$

The reaction depicted in (14.8) also can be written as

$$Fe^{3+} + OH^- = FeOH^{2+} \tag{14.10}$$

The OH^- that reacts with Fe^{3+} in (14.10) comes from the dissociation of water. Thus, these reactions are hydrolysis reactions.

Multiplying the mass action form of (14.10) by that for the dissociation of water gives the mass action form of (14.9)

$$\frac{(FeOH^{2+})}{(Fe^{3+})(OH^-)} \times \frac{(H^+)(OH^-)}{1} = \frac{(FeOH^{2+})(H^+)}{(Fe^{3+})}.$$

Thus, K for (14.10) times K_w equals the K for (14.9). Hydrolysis reactions for metals usually are written in the form used for (14.10).

In the case of iron, other hydrolysis reactions also occur to include

$$Fe^{3+} + 2OH^- = Fe(OH)_2^+ \tag{14.11}$$

$$Fe^{3+} + 3OH^- = Fe(OH)_3 \tag{14.12}$$

$$Fe^{3+} + 4OH^- = Fe(OH)_4^-. \tag{14.13}$$

Table 14.3 Equations and equilibrium constants for selected metal ion hydrolysis reactions

Reaction	K	Reaction	K
$Fe^{3+} + OH^- = FeOH^{2+}$	$10^{11.17}$	$Cu^{2+} + 4OH^- = Cu(OH)_4{}^{2-}$	$10^{16.1}$
$Fe^{3+} + 2OH^- = Fe(OH)_2{}^+$	$10^{22.13}$	$Zn^{2+} + OH^- = ZnOH^+$	$10^{5.04}$
$Fe^{3+} + 4OH^- = Fe(OH)_4{}^-$	$10^{34.11}$	$Zn^{2+} + 3OH^- = Zn(OH)_3{}^-$	$10^{13.9}$
$Cu^{2+} + OH^- = CuOH^+$	10^{6}	$Zn^{2+} + 4OH^- = Zn(OH)_4{}^{2-}$	$10^{15.1}$
$2Cu^{2+} + 2OH^- = Cu_2(OH)_2{}^{2+}$	10^{17}	$Cd^{2+} + OH^- = CdOH^+$	$10^{3.8}$
$Cu^{2+} + 3OH^- = Cu(OH)_3{}^-$	$10^{15.2}$		

Ferric hydroxide, $Fe(OH)_3$, is not included in the hydrolysis reactions above, because it precipitates as a highly insoluble compound in solutions with pH above 5. The other iron hydroxides, $FeOH^{2+}$, $Fe(OH)_2{}^+$, and $Fe(OH)^{4-}$ are soluble species. Assuming the concentration of free ferric iron in the aqueous phase represents the equilibrium concentration with a solid phase iron mineral, the amount of free iron removed during formation of the hydrolysis products is replaced by dissolution of more iron from the solid phase source. Thus, formation of hydrolysis products often does not affect the equilibrium concentration of free metal ions. However, they often increase the total concentration of the metal in solution.

Equations and equilibrium constants for hydrolysis of selected metals are presented in Table 14.3. The calculation of concentrations of hydrolysis products is illustrated in Ex. 14.2.

Ex. 14.2: *A water of pH 4 contains 0.0001 mg/L ($10^{-8.75}$ M) ferric iron. The concentrations of soluble iron hydroxides will be estimated.*

Solution:
From Table 14.3,

$$\frac{(FeOH^{2+})}{(Fe^{3+})(OH^-)} = 10^{11.17}$$

$$(FeOH^{2+}) = (10^{-8.75})(10^{-10})(10^{11.17}) = 10^{-7.58}\text{ M } or \text{ 0.0015 mg } Fe/\text{L}.$$

By like manner, $Fe(OH)^+$ and $Fe(OH)_4{}^-$ would have concentrations of $10^{-6.62}$ M (0.013 mg Fe/L) and $10^{-14.64}$ M (<0.001 mg Fe/L), respectively. The concentration of soluble hydroxides would total about 0.0145 mg Fe/L—145 times the concentration of free ferric iron.

Organic Complexes or Chelates

Certain organic molecules contain one or more pairs of electrons that can be shared with metal ions in solution. These molecules are known as ligands or chelating agents and the ligand-metal complex often is called a chelated metal. According to

Pagenkopf (1978), humic and fulvic acids are the most commonly occurring ligands in natural waters, but the equilibrium constants for reactions between metals and natural, organic ligands are seldom known. Even the structure of natural ligands is poorly understood. Thus, salicylic acid will be used to illustrate chelation:

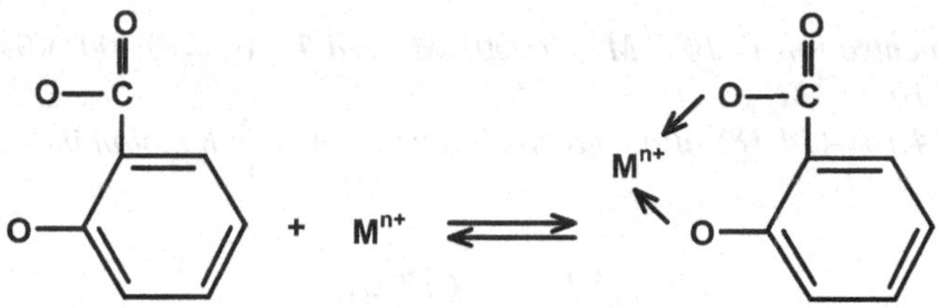

(14.14)

where M^{n+} is the metal ion.

Chelated forms behave in the same manner as hydrolysis products and ion pairs. They do not diminish the concentration of the free metal ion provided it is in equilibrium with an excess of the solid phase mineral.

In the case of organic ligands used to chelate commercial preparations of metals for use in fertilizers and other products, both the structure of the ligand and the equilibrium constant of the ligand-metal ion reaction usually are known (see Ex. 14.3). Triethanolamine (HTEA) is a common chelating agent used to chelate copper for use in water as an algicide. According to Sillén and Martell (1971), triethanolamine dissociates as follows:

$$HTEA = H^+ + TEA^- \qquad K = 10^{-8.08}. \tag{14.15}$$

The ionized form of triethanolamine can form complexes with metal ions as illustrated below for copper:

$$Cu^{2+} + TEA^- = CuTEA^+ \qquad K = 10^{4.44} \tag{14.16}$$

$$Cu^{2+} + TEA^- + OH^- = CuTEAOH^0 \qquad K = 10^{11.9} \tag{14.17}$$

$$Cu^{2+} + TEA^- + 2OH^- = CuTEA(OH)_2^- \qquad K = 10^{18.2}. \tag{14.18}$$

Ex. 14.3: *A solution 10^{-4} M in triethanolamine contains $10^{-8.35}$ M Cu^{2+} at equilibrium with cuO at pH 8.0. The chelated copper concentration will be estimated.*

Solution:

The triethanolamine concentration is given, so the TEA-concentration available to chelate copper will be computed. From (14.15),

$$\frac{(H^+)(TEA^-)}{(HTEA)} = 10^{-8.08}$$

$$\frac{(TEA^-)}{(HTEA)} = \frac{10^{-8.08}}{10^{-8}} = 10^{-0.08} = 0.83$$

$$\% \, TEA^- = \frac{0.83}{1.83} \times 100 = 45.4\%.$$

The HTEA concentration is 10^{-4} *M* = *0.0001 M, and* $TEA^- = (0.0001)(0.454) =$ *0.0000454 M* = $10^{-4.34}$ *M.*

Equations (14.16)–(14.18) allow computations of the triethanolamine-chelated copper:

$$Cu^{2+} + TEA^- = CuTEA^+$$

$$(CuTEA^+) = (Cu^{2+})(TEA^-)\ K = (10^{-8.35})(10^{-4.34})(10^{4.44}) = 10^{-8.25}\ \text{M}.$$

Following the same procedure, the concentrations of $CuTEAOH^0$ *and* $CuTEA(OH)_2^-$ *are* $10^{-6.79}$ *M and* $10^{-6.49}$ *M, respectively. The total concentration of chelated copper is* $10^{-6.49}$ *M or about two orders of magnitude greater than the* Cu^{2+} *concentration.*

Role of Hydrolysis Products, Ion Pairs, and Chelates in Solubility

Equilibrium concentrations of free trace metal ions in natural water bodies are rather low even if the controlling mineral occurs in abundance in the system. However, free metal ions can be in equilibrium with their controlling mineral and with ion pairs and inorganic and organic complexes as illustrated in Fig. 14.2 for copper ion. The presence of ion pairs and complexed copper does not alter the concentration of Cu^{2+} at equilibrium with CuO, but because of complexation, much more copper dissolves in water than would be expected from the solubility of Cu^{2+} from CuO in pure water. In pure water, only Cu^{2+} and hydrated copper would be present. The same situation is true for other metals. For example, at pH 7, the equilibrium concentration of Fe^{3+} and its hydrated forms would be so low as to be undetectable by analytical methods, but because of ion-pair and complex formation, many natural waters have 0.1–0.25 mg/L total iron. Waters that contain large amounts of humic substances may have 1 mg/L or more of chelated iron.

The major ions also form ion pairs and complexes as illustrated for calcium in Fig. 14.3. Because of the greater solubilities of their mineral forms, the complexed forms of major ions do not have as large an influence on total concentrations of major ions as do the complexed forms of micronutrient metals have on their total concentrations in water.

There is an equilibrium between the complexed forms of a metal (ion pairs and complex forms) and its free ions just as there is between the free ion and the mineral form,

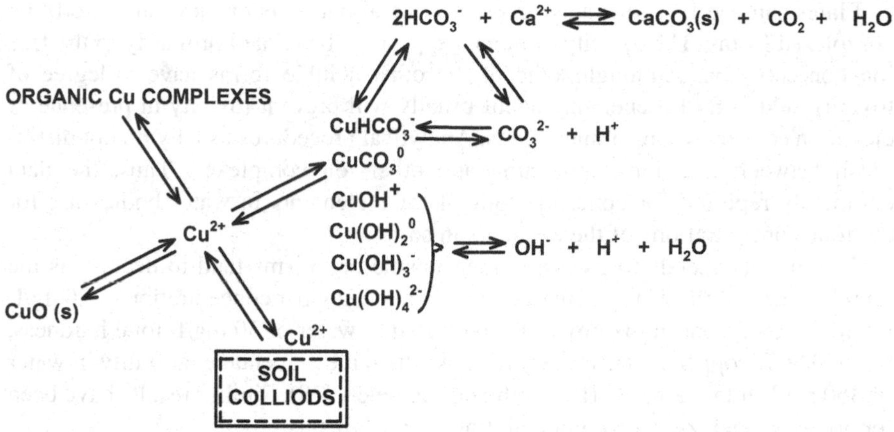

Fig. 14.2 Copper equilibria in a soil-water system containing free calcium carbonate

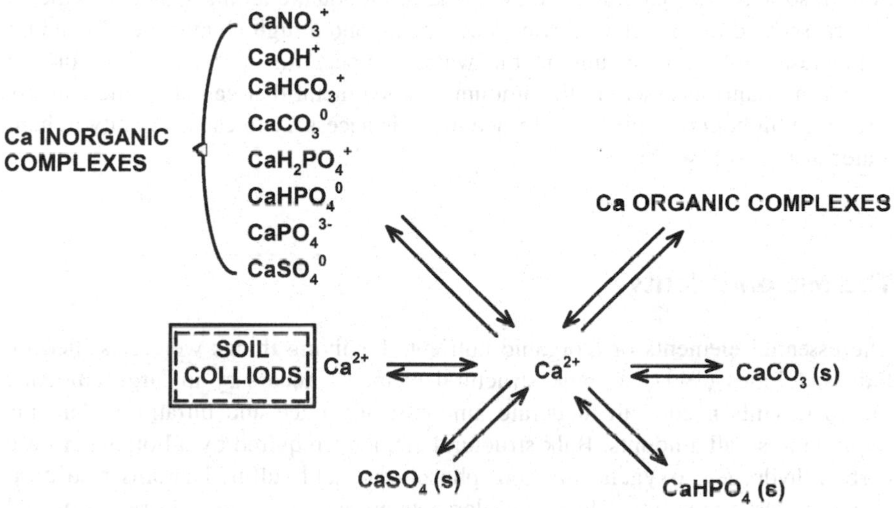

Fig. 14.3 Calcium equilibria in a soil-water system containing gypsum and calcium carbonate

$$\text{Mineral form} \rightleftharpoons \text{Free Ion} \rightleftharpoons \text{Ion pairs and complexed forms.} \qquad (14.19)$$

If some of the free ion is removed, it will be replaced by further dissolution of the mineral form or by dissociation of the complexed forms. The dissociation of the soluble ion pairs complexes can supply free ions quicker than the solid mineral forms that are in the bottom sediment. Thus, complexed metals tend to buffer metal ion concentrations in water. Metal ion buffering increases as a function of total concentrations of metal ions and ligands.

Plants can use free ions as sources of mineral nutrients or they can absorb the complexed forms. The toxicity of metals appears to be related primarily to the free ion concentration. Although some of the other soluble forms have a degree of toxicity, addition of a chelating agent usually will prevent toxicity in presence of elevation concentrations of metal ions. Analytical procedures usually do not distinguish between free ions, ion pairs, and other ion complexes. Thus, the data commonly reported for concentrations of trace elements in water bodies are for the total concentrations of the elements in solution.

The toxicity metals to fish and other aquatic organisms tend to decline as the total hardness of the water increases. For example, copper concentration of 20 μg/L killed 50 % of trout in toxicity tests conducted in water of 30 mg/L total hardness, but 520 μg/L copper was necessary to cause the same percentage mortality in water of 360 mg/L total hardness (Howarth and Sprague 1978). Similar results have been reported for lead, zinc, cadmium, and other trace metals.

Trace metals enter fish mainly through the gills. Calcium and magnesium ions—the sources of hardness in water—interfere with the transport of trace metals across the gill so that they can enter the bloodstream of aquatic animals. Trace elements are transported by an active carrier mechanism, and a high abundance of calcium (and presumably magnesium) in the water compete for absorption sites on the carrier mechanism lessening the amount of trace metal that can enter the fish. As a result, a higher concentration of trace metal is necessary to cause toxicity in hard water than in soft water.

The Micronutrients

The essential elements or inorganic nutrients for living things were classified by Pais and Jones (1997) as bulk structural elements necessary in large amounts, macroelements needed in moderate amounts, and trace and ultratrace elements required in small amounts. Bulk structural elements required by all organisms are carbon, hydrogen, oxygen, nitrogen, phosphorus, and sulfur. Diatoms also need silicon in large amounts. The macroelements are necessary for all organisms and includes calcium, magnesium, sodium, potassium, and chlorine. The trace elements, zinc, copper, iron, and manganese also are used by all organisms. The need for the ultratrace elements (Table 14.4) varies among organisms, and in some cases, there may not be a consensus opinion as to whether or not certain ultratrace elements are nutrients. Of course, all of the trace and ultratrace elements can be toxic at elevated concentrations to plants and animals.

Table 14.4 Status of ultratrace elements as nutrients

Element	Required by all or some species		Can have physiological benefits, but requirement subject to conjecture	
	Plants	Animals	Plants	Animals
Non-metals				
Arsenic (As)				X
Boron (B)	X			X
Fluorine (F)		X		
Iodine (I)		X		
Selenium (Se)	X	X		
Metals				
Cadmium (Cd)		X	X	
Chromium (Cr)			X	
Cobalt (Co)	X	X		
Lead (Pb)				X
Molybdenum (Mo)	X	X		
Nickel (Ni)			X	X
Tin (Sn)			X	X
Vanadium (V)			X	

Sixteen elements are required by both plants and animals. These elements and the 16 most abundant elements in the earth's crust are listed below in descending order of abundance:

Soil	Plant and animal biomass
O	C
Si	O
Al	H
Fe	N
C	Ca
Ca	K
K	P
Na	S
Mg	Mg
H	Na
Ti	Cl
S	Fe
Ba	Mn
Mn	Zn
N	Cu
P	B

Twelve elements are in common among the 16 most abundant elements in both the soil and in biomass. Moreover, oxygen, calcium, potassium, magnesium, and manganese are of similar order of abundance in soil and in biomass. Several elements of high abundance in soil—aluminum, titanium, and barium—are not required by plants and animals. Silicon that is highly abundant in soil is required by some higher plants and necessary in large amounts for diatoms.

Metallic Micronutrients

Some common minerals that control concentrations of common micronutrients are provided in Table 14.1. The pH greatly influences the solubility of these minerals with the concentration of the free metal ions increasing as pH decreases (Fig. 14.4). Of course, concentrations of ion pairs and other complexed forms increase as the free ion concentrations increase. A lower redox potential also increases the solubility of several micronutrients—iron and manganese in particular—and non-nutrient trace elements.

Iron and Manganese

Many enzymes such as peroxidases, catalases, and cytochrome oxidases that are important in energy transformations during respiration contain iron. In photosynthesis, iron-sulfur proteins called ferredoxins participate in the photophosphorylation process. Iron also is contained in hemoglobin that increases the capacity of the blood of vertebrates and some invertebrates to transport oxygen to their cells. The hemoglobin molecule consists of a heterocylic organic ring (porphyrin ring) with a ferrous iron ion at its center.

Manganese is mainly involved in enzyme reactions as a constituent of enzymes or as an activator. This element is particularly important as a factor in antioxidants.

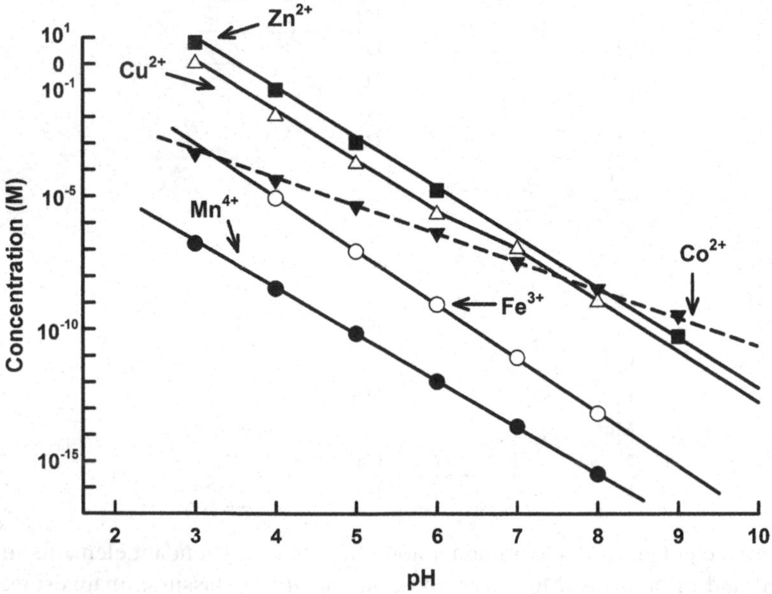

Fig. 14.4 Effects of pH on concentrations of ionic forms of trace metals

Manganese also plays a role in catalyzing the photolysis reaction that releases oxygen in photosynthesis.

Iron seldom reaches concentrations in water that are directly toxic to organisms. Nevertheless, precipitates of ferric oxide and iron humates on biological surfaces are indirectly harmful by affecting respiration, osmoregulation and other processes. These precipitates also alter benthic habitats and affect food supplies. Manganese seldom reaches toxic concentrations in water bodies. The most sensitive organisms apparently are benthic invertebrates in which excess manganese negatively impacts neuro-muscular transmission and the immune system (Pinsino et al. 2012)

The two main sources of iron ore for producing iron are hematite (Fe_2O_3) and magnetite (Fe_3O_4). Magnetitie is interesting because it is a combination of Fe^{2+} and Fe^{3+} in a 1:2 ratio—this mineral also is magnetized. Other iron oxides and hydroxides also are present—especially in soils. The principal ore of manganese is manganese dioxide (MnO_2), but there are many other oxides and hydroxides of this element, and it appears in combination with a variety of substances.

The solubility of iron and manganese under aerobic conditions is governed mainly by pH, and most surface waters will have extremely low concentrations of free iron and manganese ions. Concentrations of ferric iron and manganic manganese are very low (less than 2 μg/L) even at pH 4, but because of complex formation, natural waters usually have greater concentrations of soluble iron and manganese. In clear, inland waters in the pH range of 6–8, total iron and manganese concentrations may reach 0.25 and 0.1 mg/L, respectively. In acidic waters or in waters heavily stained with humic substances, total iron concentrations may range from 1 to 10 mg/L or more. Manganese concentrations seldom exceed 1 mg/L. Waters in humid regions will tend to be more enriched with iron and manganese than waters in arid regions. Ocean water has averages of 0.01 and 0.002 mg/L manganese, respectively (Table 4.8).

If not for the complexed forms of iron and manganese, plants could not exist in most ecosystems because of iron and manganese limitations. There is speculation that iron is a major factor limiting phytoplankton productivity in the sea. There have been serious proposals to fertilize the sea with chelated iron to increase productivity and enhance carbon dioxide removal from the atmosphere through greater photosynthesis (Nadis 1998). This would lessen the increase in atmospheric carbon dioxide concentration from anthropogenic sources. Primary productivity in some inland waters also may be limited by a shortage of iron and manganese (Hyenstrand et al. 2000).

Although iron and manganese compounds are highly insoluble in most oxygenated waters, the solubility of the controlling minerals increase with decreasing dissolved oxygen concentration and declining redox potential. Sediment pore water and hypolimnetic water usually are anaerobic and may have 20 mg/L or more of ferrous iron. Manganese concentrations also may be high in anaerobic water.

Some anaerobic bacteria can use oxygen from oxidized iron and manganese compounds as hydrogen acceptors in respiration. For example, when carbon in organic matter is mineralized to carbon dioxide by anaerobic bacteria, hydrogen ion is released,

$$CH_3COOH + 2H_2O \rightarrow 2CO_2 + 8H^+. \tag{14.20}$$

The hydrogen ion released in metabolism can then be combined with oxygen from iron or manganese compounds

$$\frac{8}{3}Fe(OH)_3(s) + 8H^+ = \frac{8}{3}Fe^{2+} + \frac{8}{3}H_2O \tag{14.21}$$

$$2MnO_2(s) + 8H^+ = 2Mn^{2+} + 4H_2O. \tag{14.22}$$

Anaerobic respiration causes the accumulation of Fe^{2+} and Mn^{2+} in sediment pore water, hypolimnetic water, and other anaerobic waters. The redox potential is around 0.2–0.3 V when Fe^{2+} and Mn^{2+} are present, and Mn^{2+} usually appears at a slightly higher redox potential than iron.

Water from some wells may contain high concentrations of iron and manganese because of low redox potential in aquifers. It is more common to find iron than manganese in well water, so the factors controlling the iron concentration in groundwater will be discussed briefly. According to Hem (1985), the iron concentration in shallow aquifers tends to be controlled by ferric hydroxide (14.21), because the water is held in aerated soil and redox potential will be comparatively high. Concentrations of ferrous iron usually will be low in shallow aquifers. Deep wells often draw water from aquifers where there is a low redox potential and iron concentrations may be controlled by iron sulfide,

$$FeS_2 = Fe^{2+} + 2S^{2-} \quad K = 10^{-26}. \tag{14.23}$$

Provided there is sufficient sulfate in the groundwater to provide enough sulfide through the activity of sulfate reducing bacteria to precipitate iron as insoluble ferric sulfide or other metallic sulfides, waters from deep aquifers will contain very little iron. Of course, they may yield waters high in sulfide. In aquifers of intermediate depth, the redox potential is lower than in shallow aquifers but higher than in deep aquifers. Ferrous carbonate (siderite) tends to be thc mineral controlling iron concentration in aquifers of intermediate depth

$$FeCO_3 + H^+ = Fe^{2+} + HCO_3^- \quad K = 10^{-0.3}. \tag{14.24}$$

Very high concentrations of iron (up to 100 mg/L or more) may occur in waters from aquifers where siderite is the iron-controlling mineral. Manganese reacts very similarly to iron, and there are some groundwaters that also are high in manganese concentration because manganese carbonate can occur in the geological matrix of aquifers.

In waterlogged soils or sediment, organic matter decomposition causes low redox potential and a high concentration of ferrous iron often results. The pH usually will be between 6 and 6.5 in such situations. This results from the reaction depicted in (14.24) and is illustrated in Ex. 14.4.

Ex. 14.4: The pH will be estimated for a sediment in which the pore water contains 20 mg/L ($10^{-3.45}$ M) Fe^{2+} and 61 mg/L (10^{-3} M) HCO_3^-.

Solution:
From (14.24),

$$\frac{(Fe^{2+})(HCO_3^-)}{H^+} = 10^{-0.3}$$

$$(H^+) = \frac{(Fe^{2+})(HCO_3^-)}{10^{-0.3}} = \frac{(10^{-3.45})(10^{-3})}{10^{-0.3}} = 10^{-6.15}$$

$$pH = 6.15.$$

When water containing Fe^{2+} or Mn^{2+} is brought in contact with oxygen, precipitation of iron and manganese compounds will occur as illustrated in the following representative reaction for iron:

$$4Fe(HCO_3)_2 + 2H_2O + O_2 = 4Fe(OH)_3 \downarrow + 8CO_2. \qquad (14.25)$$

When water of high Fe^{2+} or Mn^{2+} concentration is placed in a drinking glass in contact with air, a precipitate will soon form. Precipitation of iron and manganese can stain plumbing fixtures and kitchen utensils, and clothes laundered in such water may be permanently stained.

Iron bacteria such as *Leptothrix ochracea* and *Spirophyllum ferrugineum* obtain energy for the synthesis of organic compounds from the oxidation of ferrous salts according to the following reaction

$$4FeCO_3 + O_2 + 6H_2O = 4Fe(OH)_3 \downarrow + 4CO_2 \qquad (14.26)$$

These bacteria tend to form slimy mats known as ochre where waters containing a high concentration of iron flows or seeps onto the land surface and absorbs oxygen from the air. The oxidation of iron pyrite by *Thiobacillus* has been discussed already (see Chap. 13).

Iron also may be precipitated during normal microbial decomposition of organic matter that has a high iron content. The organic matter is decomposed, and oxidized iron compounds are precipitated. The same phenomenon can occur with manganese.

Zinc and Copper

Copper is an important cofactor in hundreds of metalloenzymes in plants, animals, and bacteria. These enzymes include those that catalyze RNA and DNA synthesis, melanin production, electron transfers in respiration, and formation of crosslinks in collagen and elastin. This element also is necessary for pigmentation in hair. In plants, copper is required for chlorophyll synthesis, root metabolism, and lignification. Zinc also is important in metalloenzymes that catalyze reactions. It is particularly important in assuring the stability of many biochemical molecules and membranes. In plants, zinc also is involved in chlorophyll synthesis.

Copper occurs in the earth's crust mainly as sulfides, carbonates, or oxides. Zinc is more abundant than copper and is present in sulfides, oxides, silicates, and carbonates. Concentrations of free zinc and copper ions usually are low because the controlling minerals have low solubilities within the pH range of most natural waters (Fig. 14.4). The solubilities of Zn^{2+} from $ZnCO_3$ and Cu^{2+} from CuO will be calculated for pH 7 in Ex. 14.5.

Ex. 14.5: *The solubilities of Zn^{2+} and Cu^{2+} in waters of pH 7 will be estimated.*

Solution:
The equilibrium constants for solubilities of $ZnCO_3$ and CuO were obtained from Table 14.1:

$$\frac{(Cu^{2+})}{(H^+)^2} = 10^{7.65}$$

$$(Cu^{2+}) = (10^{7.65})(10^{-7})^2 = 10^{-6.35}\ \text{M}\ \textit{or}\ 0.028\ \text{mg/L}$$

$$\frac{(Zn^{2+})}{(H^+)^2} = 10^{7.95}$$

$$(Zn^{2+}) = (10^{7.95})(10^{-7})^2 = 10^{-6.05}\ \text{M}\ \textit{or}\ 0.058\ \text{mg/L}.$$

The concentrations of Zn^{2+} and Cu^{2+} will be in equilibrium with inorganic complexes of zinc and copper. The possible ion pairs and inorganic complexes and reactions and equilibrium constants for their formation (Tables 14.2 and 14.3) allow the following expressions:

$$\begin{aligned}\text{Total inorganic Zn} = {} & Zn^{2+} + ZnOH^- + Zn(OH)_2^- + Zn(OH)_4^{2-} \\ & + ZnCl^+ + ZnCl_2^0 + ZnCl_3^- + ZnCl_4^{2-} + ZnF^+ \\ & + ZnSO_4^0 + ZnCO_3^0\end{aligned} \tag{14.27}$$

and

$$\begin{aligned}\text{Total inorganic Cu} = {} & Cu^{2+} + CuOH^+ + Cu_2(OH)_2^{2+} + Cu(OH)_3^- \\ & + Cu(OH_4)^{2-} + CuCl^+ + CuF^+ + CuSO_4^0 \\ & + CuCO_3^0 + Cu(CO_3)_2^{2-}.\end{aligned} \tag{14.28}$$

The concentrations of inorganic forms of copper in a water will be estimated in Ex. 14.6:

Ex. 14.6: *The concentration of total inorganic copper will be estimated for a water in equilibrium with copper oxide that contains 61 mg/L (10^{-3} M) HCO_3^-, 10 mg/L ($10^{-3.55}$ M) chloride, 5 mg/L ($10^{-4.28}$ M) sulfate, and has a pH of 8.*

Solution:
At pH 8, carbonate is not analytically detectable, but its concentration may be calculated using (8.10):

$$\frac{(H^+)(CO_3^{2-})}{(HCO_3^-)} = 10^{-10.33}$$

$$(CO_3^{2-}) = \frac{(10^{-3})(10^{-10.33})}{(10^{-8})} = 10^{-5.33}\ \text{M}.$$

Following the procedure illustrated in Ex. 14.5, the concentration of Cu^{2+} from CuO at pH 8 is $10^{-8.35}$ M. From equations and equilibrium constants in Tables 14.2 and 14.3:

$$(CuOH^+) = (10^{6})(10^{-8.35})(10^{-6}) = 10^{-8.35}\ \text{M}$$

$$\left[Cu_2(OH)_2^{2+}\right] = (10^{17})(10^{-8.35})^2(10^{-6})^2 = 10^{-11.7}\ \text{M}$$

$$\left[Cu(OH)_3^-\right] = (10^{15.2})(10^{-8.35})(10^{-6})^3 = 10^{-11.15}\ \text{M}$$

$$\left[Cu(OH)_4^{2-}\right] = (10^{16.1})(10^{-8.35})(10^{-6})^4 = 10^{-16.25}\ \text{M}$$

$$(CuCl^+) = (10^{0})(10^{-8.35})(10^{-3.55}) = 10^{-11.9}\ \text{M}$$

$$(CuSO_4^0) = (10^{2.3})(10^{-8.35})(10^{-4.28}) = 10^{-10.33}\ \text{M}$$

$$(CuCO_3^0) = (10^{6.77})(10^{-8.35})(10^{-5.33}) = 10^{-6.91}\ \text{M}$$

$$\left[Cu(CO_3)_2^{2-}\right] = (10^{10.01})(10^{-8.35})(10^{-5.33})^2 = 10^{-9.0}\ \text{M}.$$

The total inorganic copper concentration will be $10^{-6.85}$ M. The total inorganic copper concentration is 0.0090 mg/L as compared to 0.000283 mg/L of Cu^{2+}. Thus, 96.9 % of the total inorganic copper is complexed. Notice that almost all of the complexed copper is in the form of $CuCO_3^0$.

Table 14.5 Concentrations of Cu^{2+}, $CuCO_3^0$, and total copper (Cu_T) in equilibrium with tenorite (CuO) in waters of different total alkalinity concentrations at pH 8

	Mole/L × 10^{-7}		
Total alkalinity (mg/L as $CaCO_3$)	Cu^{2+}	$CuCO_3^0$	Cu_T
8	0.045	0.161	0.251
25	0.045	0.509	0.601
50	0.045	1.02	1.12
100	0.045	2.03	2.15
200	0.045	4.05	4.25

Table 14.6 Concentrations of Cu^{2+}, $CuCO_3^0$, and total copper (Cu_T) in equilibrium with tenorite (CuO) at different pH values in waters with a total alkalinity of 50 mg/L as $CaCO_3$

	Mole/L × 10^{-7}		
pH	Cu^{2+}	$CuCO_3^0$	Cu_T
7	4.5	10.2	15.2
7.5	0.45	3.21	3.81
8	0.045	1.02	1.12
8.5	0.0045	0.32	0.35
9	0.00045	0.10	0.11

It is interesting to look at the influence of total alkalinity on concentrations of Cu^{2+}, $CuCO_3^0$, and total copper in equilibrium with copper oxide at a given pH (Table 14.5). As the alkalinity increases, the total copper concentration increases, and the ratio $CuCO_3^0$:total copper increases because Cu^{2+} concentration does not change and the amount of carbonate available to form $CuCO_3^0$ rises. However, if the alkalinity is fixed and the pH varied (Table 14.6), the total copper concentration decreases with rising pH because of the simultaneous decrease in Cu^{2+}. Even though carbonate increases as pH increases, there is a decreasing amount of Cu^{2+} to react with the increasing amount of carbonate. Thus, the ratio $CuCO_3^0$:total copper is roughly the same at all pH values shown in Table 14.6.

Copper oxide (CuO) was used as the mineral controlling cupric ion concentration in Exs. 14.5 and 14.6. Copper oxide or tenorite is the controlling mineral at pH 7 and above. At lower pH values, the controlling mineral is malachite, $Cu_2(OH)_2CO_3$ (Table 14.1).

Concentrations of dissolved copper in 1,500 surface waters of the United States averaged 0.015 mg/L and the maximum concentration was 0.28 mg/L (Kopp and Kroner 1967). Seawater has an average copper concentration of 0.003 mg/L (Table 4.8). Kopp and Kroner also found a mean dissolved zinc concentration of 0.064 mg/L, and the highest concentration was 1.18 mg/L. Seawater has an average of 0.01 mg/L zinc (Table 4.8). Of course, some zinc and copper is contained on or in suspended particles in water. Anaerobic conditions do not lead to large increases in zinc and copper concentrations as were noted for iron and manganese.

Cadmium

Cadmium was found to be a nutrient for at least one species of planktonic, marine diatoms (Lee et al. 1995). The role of cadmium in increasing photosynthesis appears to be related to a beneficial influence on carbonic anhydrase activity (Lane and Morel 2000).

Cadmium is relatively rare in most geological deposits, occurring mainly in carbonate and hydroxide forms. The solubility expression for cadmium carbonate is

$$CdCO_3 = Cd^{2+} + CO_3^{2-} \quad K = 10^{-11.3}. \tag{14.29}$$

In water of pH 8 and carbonate at $10^{-5.33}$ M, the Cd^{2+} concentration at equilibrium would be about $10^{-5.97}$ M in a cadmium carbonate-water system. This is 0.12 mg/L of dissolved cadmium, and additional cadmium would be present in $CdOH^+$. In natural waters, cadmium would complex chloride as $CdCl^+$, $CdCl_2^0$, $CdCl_3^-$, and $CdCl_4^{2-}$ ion pairs.

Because of its low abundance in the earth's crust, cadmium concentrations are undetectable in most waters, and waters with detectable cadmium usually had less than 1 µg/L (Kopp and Kroner 1967). The highest concentration found by Kopp and Kroner was 120 µg/L, and most values above 1 µg/L were probably the result of pollution. Seawater has an average cadmium concentration of 0.11 µg/L (Table 4.8).

Cobalt

Cobalt is a cofactor for several enzymes, the most notable being vitamin B12 that controls red blood cell production in animals. This element also is needed for nitrogen fixation by blue-green algae and certain bacteria.

Cobalt has many properties similar to iron, but the redox potential for the cobaltic form (Co^{3+}) is greater than that for the occurrence of ferric iron, and the cobaltous (Co^{2+}) form is more common in aquatic environments. According to Hem (1985), the mineral controlling cobalt concentration in natural water likely is cobalt carbonate (Table 14.1). At pH 8 and total alkalinity of 61 mg/L, the cobalt concentration in equilibrium with cobalt carbonate would be about 2 µg/L.

Cobalt is present in soils and sediment at much lower concentrations than iron. Because of the low abundance of cobalt minerals in the earth's crust, concentrations of cobalt usually are less than expected for cobalt carbonate equilibrium in water. In fact, cobalt concentrations in inland waters often are below detection. In waters with detectable cobalt, concentrations usually are <1 µg/L (Baralkiewicz and Siepak 1999). Values above 20 µg/L have been reported in highly mineralized water. In mining areas, cobalt concentrations up to 80 µg/L were reported (Essumang 2009). Seawater averages 0.5 µg/L cobalt (Table 4.8)—about half of the cobalt in seawater is the free ion and the rest is in hydrolysis products and ion pairs (Ćosović et al. 1982).

Molybdenum

Molybdenum is an essential constituent of several enzymes that play a role in nitrogen fixation by microorganisms and nitrate reduction by plants. In animals, molybdenum is a cofactor in enzymes influencing oxidation of purines and aldehydes, protein synthesis, and metabolism of several nutritive elements.

Molybdenum is of low abundance in soils and sediment, and inland waters contain small concentrations of molybdate (MoO_4^{2-}). According to Hem (1970), average concentrations of molybdenum were 0.35 μg/L for North America rivers and 1.4 μg/L for drinking water supplies in the United States. Up to 100 μg/L has been observed in a reservoir in Colorado. Groundwater waters often contain 10 μg/L molybdenum or more. The average concentration of molybdenum in seawater is 10 μg/L (Table 4.8).

Non-metallic Micronutrients

Boron

Plants require small amounts of boron. This element in the form of boric acid bonds with molecules of pectins in plant cell walls and glycolipids in bacterial cell walls to make the cell walls more stable. There also is evidence that boron may be essential in animals.

Boron is neither a metal nor non-metal—it has properties of both and is a metalloid. Boron occurs in rocks and soils as borosilicates and borates that are relatively soluble. Its geochemical cycle is much like that of chloride, and its concentration in seawater or inland saline water is rather high compared to its concentration in freshwater.

Boron occurs in water mostly as undissociated boric acid, because this acid does not dissociate greatly at the pH of most natural waters

$$H_3BO_3 = H^+ + H_2BO_3^- \quad K = 10^{-9.24}. \tag{14.30}$$

At pH below 9.24, boron acid is more than 50 % for the total boron involved in the dissociation. For example, at pH 8, about 5 % of boric acid is dissociated.

Boron concentration in freshwater bodies in the southeastern US seldom exceeds 0.1 mg/L (Boyd and Walley 1972). A more extensive survey of 1,546 river and lake samples throughout the United States also gave a mean boron concentration of 0.1 mg/L with a maximum of 5 mg/L (Kopp and Kroner 1970). Much greater concentrations can be found in saline inland waters of arid regions; concentrations above 100 mg/L have been reported (Livingstone 1963). The concentration of boron in seawater averages 4.6 mg/L (Table 4.8). High boron concentrations in irrigation water can be toxic to plants.

Fluoride

Fluoride is considered a nutrient at optimal levels, because it is important to the integrity of bone and teeth in humans, and presumably, other animals. According to Palmer and Gilbert (2012), fluoridation of municipal water supplies for the prevention of dental cavities is recognized as the most effective dental public health measure in existence. The optimal concentration for protection against dental cavities is 0.7–1.2 mg/L of fluoride. High concentrations of fluoride can cause bone disease and mottling of tooth enamel. The current recommended limit for fluoride in drinking water is 4 mg/L for avoiding health effects, and no cosmetic effects on teeth have been reported for fluoride concentrations less than 2 mg/L. Nevertheless, there is a segment of society that opposes fluoridation of drinking water.

Fluoride, like chlorine, bromine, and iodine, is a highly electronegative element of the halogen group. Calcium fluoride is a common fluoride bearing mineral, and fluoride is contained in apatite and some other common minerals. Fluoride differs from chloride geochemically in that most fluoride is bound in rocks while the sea is a huge reservoir of chloride in dissolved form. This results from the lower solubility of fluoride compounds as compared to chloride compounds. There also is much less fluoride than chloride in the earth's crust.

The solubility of calcium fluoride may be expressed as follows:

$$CaF_2 = Ca^{2+} + 2F^- \qquad K = 10^{-10.46}. \tag{14.31}$$

The equilibrium concentration of fluoride in pure water would be about 3.9 mg/L. Most freshwaters contain less than 0.1 mg/L of fluoride, but some contain more. Some groundwaters may have several milligrams per liter of fluoride. Ocean water typically contains 1.2–1.4 mg/L of fluoride (Smith and Ekstrand 1996).

Iodine

Iodine is important for proper function of the thyroid gland that produces thyroid hormones that influence growth, development, and metabolism in humans and other vertebrates. In areas of the world where soils and waters are deficient in iodine, those who rely on local sources of food may develop iodine-deficiency disorders—the most famous of which is enlargement of the thyroid gland (goiter). The diet is an important source of iodine, and in the past when foods were produced locally, the incidence of goiter was high in areas where soils and crops produced in them were low in iodine. Dietary iodine supplements such as iodized salt were and still are a preventive measure for goiter. However, with the present day situation in which food comes from many different places, the risk of iodine deficiency in the diet is less.

Iodine always is found combined with other elements in the earth's crust, and most commonly it is present as iodide. Being a halogen, its chemistry is similar to chloride and fluoride. According to Hem (1985), rainwater typically contains 1–3 μg/L of iodine, while the range for river water is about 3–42 μg/L. The British Geological Survey (2000) reported that potable groundwater typically contained 1–70 μg/L iodide with extreme values up to 400 μg/L. Seawater has an average iodine concentration of 0.06 mg/L. There is no recommended maximum limit for iodide in potable water.

Selenium

Selenium is a component of the amino acids selenocysteine and selenomethionine. It also is a cofactor for certain peroxidases and reductases.

Selenium occurs in the earth's crust mainly as elemental selenium, ferric selenite, and calcium selenite. The chemistry of selenium is similar to that of sulfur, but it is much less common than sulfur. The ion, SeO_3^{2-} is the most stable form of selenium in water.

Selenium concentrations in groundwater and surface water seldom exceed 1 μg/L (Hem 1985). However, concentrations range from 0.06 to 400 μg/L, and there is even one report of 6,000 μg/L of selenium in groundwater (World Health Organization 2011). The average concentration of selenium in ocean water is 4 μg/L.

Those who drink water of high selenium concentration over many years may experience hair and fingernail loss, numbness in extremities, and circulation problems. In the United States, the maximum selenium concentration allowed in municipal water supplies is 50 μg/L. Selenium accumulates in the food chain, and in spite of being an essential element for aquatic animals, selenium pollution can have serious consequences on aquatic ecosystems (Hamilton 2004).

Arsenic

Arsenic is needed in small amounts—apparently as a factor in methionine metabolism—by certain animals (Uthus 1992). Arsenic oxide has several medicinal uses including treatment of a particular form of leukemia. But, arsenic is best known as a toxin; it has been used as an insecticide, fungicide, and wood preservative. It also has been used as a chemical weapon, and for intentionally poisoning (murdering) people.

Arsenic occurs in minerals such as arsenopyrite (AsFeS) realgar (AsS) and orpiment (As_2S_3), and during weathering it is released as arsenate. It also can replace phosphate in apatite (rock phosphate). The chemistry of arsenic is very similar to that of phosphorus. It usually occurs naturally at low concentrations in water, because it is not present in most soils or sediment in high concentration and the mineral forms are not very soluble.

The dissociation of arsenic acid (H_3AsO_4) is as follows:

$$H_3AsO_4 = H^+ + H_2AsO_4^- \qquad K = 10^{-2.2} \qquad (14.32)$$

$$H_2AsO_4^- = H^+ + HAsO_4^{2-} \qquad K = 10^{-7} \qquad (14.33)$$

$$HAsO_4^{2-} = H^+ + AsO_4^{3-} \qquad K = 10^{-11.5}. \qquad (14.34)$$

Below pH 2.2, H_3AsO_4 dominates, $H_2AsO_4^-$ is the dominant ion between pH 2.3 and pH 7, $HAsO_4^{2-}$ is the most abundant from pH 7.1 to pH 11.5, and above pH 11.5, AsO_4^{3-} is most abundant.

Comparison of (14.32)–(14.34) with (12.1)–(12.3) reveals that the dissociation of arsenic and orthophosphoric are essentially identical—right down to the values of the equilibrium constants. In most regards, arsenate and phosphate chemistry are quite similar. In fact, arsenate cannot be distinguished from phosphate in the usual method for determining phosphate concentration in water.

According to Kopp (1969), arsenic concentrations in river water in the United States ranged from 5 to 336 μg/L with an average of 64 μg/L. On average, seawater contains 0.004 μg/L of arsenic (Table 4.8), but concentrations of 2–3 μg/L are sometimes observed in polluted coastal waters.

Arsenic contamination of groundwater—mainly from natural sources—is a serious problem in many regions of the world—especially in south Asia. The most highly publicized incidence is in the lower Ganges River basin in nine districts of West Bengal, India and 42 districts in Bangladesh where arsenic levels in drinking water exceed the World Health Organization maximum permissible limit of 50 μg/L (Chowdhury et al. 2000). In Bangladesh, about 21 million people use well waters with more than 50 μg/L arsenic—some well waters have over 1,000 μg/L. Arsenic poisoning symptoms include lesions, keratosis, conjunctival congestion, edema of feet, and liver and spleen enlargement. In advanced cases, cancers affecting lungs, uterus, bladder, genitourinary tract often are seen. According to Chowdhury et al. (2000), of 11,180 people examined who had been drinking water with over 50 μg/L of arsenic, 24.47 % had arsenical skin lesions. An estimated 100 million people are at risk of arsenic poisoning in West Bengal and Bangladesh.

In the United States and many other countries, the emphasis on arsenic in water supplies has been related to an increased risk of cancer. The drinking water standard in the United States was 50 μg/L for many years, but in 2006 it was reduced to 10 μg/L. Some municipal water supplies have not been able to comply with the stricter standard.

Non-nutrient Trace Elements

Many non-nutrient trace elements occur in water, but only those of significance in water quality—usually because of toxicity—will be discussed here. In the interest of brevity, the effects of excessive concentrations of non-nutrient trace elements will not be discussed. But, in most cases, they interfere with certain enzymatic processes. Trace substances can reach undesirable levels naturally, but excessively high concentrations usually result from anthropogenic pollution.

Aluminum

Only oxygen and silica exceed aluminum in abundance in the earth's crust, but aluminum is not a major constituent of water except at very low pH. Aluminum occurs in many silicate rocks and as aluminum oxides such as aluminum hydroxide or gibbsite [$Al(OH)_3$] in acidic soils. Gibbsite dissolves as follows:

$$Al(OH)_3 + 3H^+ = Al^{3+} + 3H_2O \qquad K = 10^9. \tag{14.35}$$

In waters of pH 6 or less, gibbsite controls the concentration of aluminum and there is little aluminum in solution unless pH is below 5. Concentrations of Al^{3+} are: pH 4, 10^{-3} M; pH 5, 10^{-6} M; pH 6, 10^{-9} M; pH 7, 10^{-12} M.

Aluminum forms several hydroxide complexes in dilute aqueous solution, several of which are polynuclear. There are various ways of writing the equations for the formation of these complexes, and experts do not completely agree on the formulas for the complexes. The equations and equilibrium constants for aluminum hydroxide complexes given below were from Hem and Roberson (1967) and Sillén and Martell (1964):

$$Al^{3+} + H_2O = AlOH^{2+} + H^+ \qquad K = 10^{-5.02} \tag{14.36}$$

$$2Al^{3+} + 2H_2O = Al_2(OH)_2^{4+} + 2H^+ \qquad K = 10^{-6.3} \tag{14.37}$$

$$7Al^{3+} + 17H_2O = Al_7(OH)_{17}^{4+} + 17H^+ \qquad K = 10^{-48.8} \tag{14.38}$$

$$13Al^{3+} + 34H_2O = Al_{13}(OH)_{34}^{5+} + 34H^+ \qquad K = 10^{-97.4}. \tag{14.39}$$

As will be shown in Ex. 14.7, the solubility of aluminum is greatly increased by the formation of hydroxide complexes. Aluminum hydroxide acts as a base when it dissolves in acidic solution because it neutralizes hydrogen ion. Free aluminum ion hydrolyzes (14.36)–(14.39) and hydrogen ion is released, so Al^{3+} is an acid in water. The aluminum complexes increase the amount of dissolved aluminum considerably above that of the Al^{3+} concentration in an aqueous solution in equilibrium with gibbsite.

Ex. 14.7: *The concentrations of Al^{3+} and inorganic complexes of aluminum will be calculated for a water-gibbsite system at equilibrium at pH 5. At pH 5 $(Al^{3+}) = 10^{-6}$ M (see above).*

Solution:
Using (14.36)–(14.39), we obtain

$$(AlOH^{2+}) = \frac{(Al^{3+})(10^{-5.02})}{H^+} = \frac{(10^{-6})(10^{-5.02})}{10^{-5}} = 10^{-6.02}\ \text{M}$$

$$\left[Al_2(OH)_2^{4+}\right] = \frac{(Al^{3+})^2(10^{-6.3})}{(H^+)^2} = \frac{(10^{-6})^2(10^{-6.3})}{(10^{-5})^2} = 10^{-8.3}\ \text{M}$$

$$\left[Al_7(OH)_{17}^{4+}\right] = \frac{(Al^{3+})^7\,(10^{-48.8})}{(H^+)^{17}} = \frac{(10^{-6})^7\,(10^{-48.8})}{(10^{-5})^{17}} = 10^{-5.8}\ \text{M}$$

$$\left[Al_{13}(OH)_{34}^{5+}\right] = \frac{(Al^{3+})^{13}\,(10^{-97.4})}{(H^+)^{34}} = \frac{(10^{-6})^{13}\,(10^{-97.4})}{(10^{-5})^{34}} = 10^{-5.4}\ \text{M}.$$

The total aluminum concentration is $10^{-5.12}$ M or 0.204 mg/L as compared to 0.027 mg/L for Al^{3+}. In natural water, the aluminum concentration would be even higher than calculated in Ex. 14.7 because of ion pairs and other complexed forms of aluminum.

Aluminum hydroxide is amphoteric (acts as either an acid or a base). It is a base in (14.35), but an acid in the equation below:

$$Al(OH)_3 + OH^- = Al(OH)_4^- \qquad K = 10^{1.3}. \qquad (14.40)$$

At pH values above 7 where the basic reaction of gibbsite to release Al^{3+} is essentially nil, aluminum hydroxide can dissolve by reaction with hydroxide as illustrated in Ex. 14.8.

Ex. 14.8: *Concentrations of $Al(OH)_4^-$ will be estimated for pH 7, 8, 9, and 10 for a gibbsite-water system at equilibrium.*

Solution:
From (14.40),

$$\left[Al(OH)_4^-\right] = (OH^-)(10^{1.3}).$$

The concentration at pH 7 is

$$\left[Al(OH)_4^-\right] = (10^{-7})(10^{1.3}) = 10^{-5.7}\ \text{M}.$$

Repeating the computation for other pH values, we get: pH 8, $10^{-4.7}$ M; pH 9, $10^{-3.7}$ M; pH 10, $10^{-2.7}$ M.

In Ex. 14.8, the aluminum concentration resulting from $Al(OH)_4^-$ increased from 0.054 mg/L at pH 7 to 54 mg/L at pH 10. In spite of the increasing solubility of gibbsite and other aluminum minerals at pH above 7, natural waters of pH 5 and above seldom contain much aluminum because there is little aluminum in neutral to alkaline soils or sediment.

Concentrations of total aluminum in natural waters with pH values above 5 usually are very low, but occasional values of 1 mg/L or more are reported. It is assumed that these elevated concentrations result from finely-divided aluminum minerals in suspension. Waters with pH values below 4 may have extremely high aluminum concentrations up to several thousand milligrams per liter. Seawater contains an average of 0.01 mg/L aluminum (Table 4.8).

Barium

Barium occurs in the earth's crust as barite ($BaSO_4$) and witherite ($BaCO_3$). The solubilities of these compounds can be expressed as

$$BaSO_4(s) = Ba^{2+} + SO_4^{2-} \qquad K = 10^{-8.80} \tag{14.41}$$

$$BaCO_3(s) = Ba^{2+} + CO_3^{2-} \qquad K = 10^{-8.82}. \tag{14.42}$$

In pure water at 25 °C, the equilibrium concentration of Ba^{2+} from $BaSO_4$ would be $10^{-4.4}$ M (5.5 mg/L). From $BaCO_3$, the equilibrium barium concentration would be 5.3 mg/L in absence of carbon dioxide. However, as with $CaCO_3$, the solubility of $BaCO_3$ increases in presence of carbon dioxide. The hydrolysis product, $BaOH^+$, increases the solubility of barium sulfate or carbonate.

Natural waters are generally far below equilibrium concentrations with dissolved barium. Hem (1985) indicated that river waters and public water supplies in the United States had average barium concentrations of 43 and 45 μg/L, respectively, but concentrations up to 3 mg/L were found. Seawater contains an average of 30 μg/L barium (Table 4.8).

Beryllium

This element occurs in geological formations as silicates and hydroxyl-silicates. Controlling minerals are beryllium carbonate and beryllium hydroxide of very low solubilities. Concentrations at equilibrium would not exceed 1 μg/L as Be^{2+}, but complexes form and increase solubility. Natural surface waters in the United States contained 0.01–1.22 μg/L dissolved beryllium (average = 0.19 μg/L) (Kopp and Kroner 1967). Seawater contains an average of 0.05 μg/L beryllium (Table 4.8).

Chromium

Chromium is a relatively common element in the earth's crust usually found in the form of chromites such as iron chromium oxide ($FeCr_2O_3$). Chromium can exist in several oxidation states. The trivalent and hexavalent forms are common in natural waters. The concentrations of chromium in natural waters in the United States ranged from less than detection to 112 μg/L with an average of 9.7 μg/L for samples with detectable chromium (Kopp 1969). Samples with particularly high chromium concentration were from waters polluted with industrial effluents. Concentrations of chromium up to 60 μg/L were reported in well waters from an arid area in southern California (Izbicki et al. 2008). Seawater contains an average of 0.05 μg/L chromium (Table 4.8).

Cyanide

Cyanide in natural waters is almost exclusively the result of anthropogenic pollution. Cyanide is usually introduced into water as NaCN or KCN that are highly soluble salts. The CN^- ion is in equilibrium with hydrogen ion

$$HCN = H^+ + CN^- \qquad K = 10^{-9.4}. \tag{14.43}$$

Below pH 8, hydrocyanic acid (HCN) will be the predominant form. Cyanide forms Lewis acid-base complexes with metal ions such as Fe^{2+}, Cu^{2+}, Ag^+, and Cd^{2+}.

Lead

Lead occurs in several minerals to include galena (PbS), cerussite ($PbCO_3$), anglesite ($PbSO_4$), and pyromorphite (lead chlorophosphate). The controlling mineral in water is probably lead carbonate

$$PbCO_3(s) = Pb^{2+} + CO_3^{2-} \qquad K = 10^{-12.8}. \tag{14.44}$$

In pure water the lead concentration at equilibrium would be $10^{-6.4}$ M (0.08 mg/L), but carbon dioxide would increase the concentration. Lead concentration in river waters studied by Livingstone (1963) were between 1 and 10 μg/L, and Kopp and Kroner (1967) found a maximum lead concentration of 140 μg/L in surface waters in the United States. As with most other trace metals, the main reason for especially high concentrations is pollution. The ocean has an average lead concentration of 0.03 μg/L (Table 4.8).

Mercury

Mercury has a low abundance in geological formations, but it is found in many rocks and is present in coal. The most common source of mercury ore is cinnabar (HgS). In natural water, mercury concentration usually is less than 0.1 μg/L (Wershaw 1970). Elevated concentrations of mercury in water as a result of pollution—especially methylmercury—is a serious human health issue. Mercury can accumulate in organisms as organic mercury compounds such as methyl mercury and passed through the food chain to reach humans to whom it is extremely dangerous, especially to fetuses, infants, and children. Fisheries in mercury-polluted water bodies often are closed by public health authorities. Ocean waters have an average mercury concentration of 0.03 μg/L. Most waters with greater than 0.1 μg/L mercury have received mercury pollution.

Nickel

Nickel occurs in nature as oxides, sulfides, and silicates. The geochemistry of nickel is similar to that of cobalt. The predominant form in natural waters is Ni^{2+} and it forms soluble hydroxides. Nickel carbonate may be the controlling species

$$NiCO_3(s) = Ni^{2+} + CO_3^{2-} \qquad K = 10^{-8.2}. \tag{14.45}$$

Pure water in equilibrium with $NiCO_3$ would contain $10^{-4.1}$ M (4.7 mg/L) nickel, and CO_2 would increase the solubility. However, such a high concentration is not found in surface waters. North American rivers contained an average of 10 μg/L nickel (Durum and Haffty 1961), and surface waters in the United States contained up to 56 μg/L nickel (Kopp and Kroner 1967). Ocean waters contain an average nickel concentration of 2 μg/L (Table 4.8).

Silver

Silver is not abundant in most geological formations, but it sometimes occurs in deposits—often with lead—in minerals such as galena that is primarily lead sulfide. According to Hem (1970) the solubility of silver oxide is low enough to prevent high concentrations of silver at high pH, and in waters with more than 10^{-3} M chloride, silver chloride formation would prevent high silver concentrations. The reactions are

$$Ag_2O(s) + H_2O = 2Ag^+ + 2OH^- \qquad K = 10^{-7.7} \tag{14.46}$$

$$AgCl(s) = Ag^+ + Cl^- \quad K = 10^{-9.7}. \tag{14.47}$$

Median silver concentrations were 0.23 μg/L in public water supplies in the United States (Durfor and Becker 1964) and the average concentration in rivers in the United States was 0.09 μg/L (Durum and Haffty 1961).

Strontium

This element occurs in various rocks and minerals, and strontium carbonate (strontianite) and strontium sulfate (celestite) occur in sediment. The ratio of strontium to calcium is about 1:1,000 in many limestones. Strontium carbonate can be considered the controlling mineral for aqueous solutions

$$SrCO_3(s) = Sr^{2+} + CO_3^{2-} \qquad K = 10^{-9.2}. \qquad (14.48)$$

Thus, strontium is less soluble than calcium carbonate for which the solubility product is $10^{-8.2}$.

Because of its low abundance in the earth's crust, concentrations of strontium are quite low in most waters. According to Hem (1970), strontium concentrations averaged 0.06 mg/L in major North American rivers and 0.11 mg/L in public water supplies. Waters in arid regions may contain 2 mg/L or more of strontium, and ocean water contains an average of 8 mg/L (Table 4.8).

Tin

The most common ore of tin is cassiterite (SnO_2). Very little information is available on the chemistry and concentration of tin in natural waters. Livingstone (1963) reported tin concentrations as high as 100 μg/L in inland waters, but most values were much lower—usually less than 2.5 μg/L. Ocean waters contain an average of 3 μg/L tin (Table 4.8).

Vanadium

Vanadium chemistry is quite complex because this element can occur in several valence states and in both cationic and anionic forms. Under certain conditions, such as arid climate or unusually high vanadium content of rocks, vanadium concentrations may exceed 1 mg/L in surface waters (Livingstone 1963). However, most inland surface waters contain less than 1 μg/L. Ocean waters have an average vanadium concentration of 2 μg/L (Table 4.8).

Significance

Concentrations of micronutrients in freshwater bodies usually are high enough that they do not limit primary productivity. However, Goldman (1972) demonstrated that certain micronutrients limited primary productivity in water bodies located in areas with poorly developed soils such as sandy watersheds or watersheds where most of the land surface is covered by outcrops of hard, insoluble rocks. In aquaculture, fertilizers have traditionally been used to increase nutrient concentrations and stimulate primary productivity in earthen ponds to promote fish growth. There are few cases where applications of micronutrients have been necessary in aquaculture pond fertilization (Boyd and Tucker 2014). There is evidence that primary productivity of the oceans is limited by a shortage of micronutrients and especially a shortage or iron (Nadis 1998; Boyd et al. 2007).

Naturally-occurring concentrations of trace elements high enough to be toxic to aquatic life also are rare. Most incidences of trace metal toxicity result from pollution. The mining industry is a particularly important source of trace element pollution. In areas impacted by acidic rain, concentrations of heavy metals in surface waters may reach toxic levels.

Iron and manganese are particularly troublesome in water supplies. Although iron and manganese seldom are present in water at concentrations causing adverse physiological effects in humans or other animals, concentrations above 0.3 mg/L iron and 0.05 mg/L manganese may seriously affect water's usefulness for some domestic and industrial purposes. Iron and manganese in water can cause staining of plumbing fixtures, staining of clothes during laundering, incrustation of well screens, clogging of pipes, and bad taste.

Discussion of some human health issues related to excessive concentrations of trace elements has been provided above, and a summary of adverse effects of trace elements follows:

Aluminum—There is unproven concern about aluminum exposure being related to Alzheimer's disease. However, the pH of drinking water is usually above 5, and it contains little aluminum.

Arsenic—Skin damage, circulatory system problems, increased risk of cancer

Barium—Increased blood pressure

Beryllium—Intestinal lesions

Boron—Not likely to be at high enough concentration in drinking water to cause health effects

Cadmium—Kidney damage

Chromium—Allergic dermatitis; increased risk of cancer

Cobalt—Apparently, there is no concern over cobalt concentrations in drinking water.

Copper—Liver and kidney damage

Cyanide—Nerve damage and thyroid problems

Fluoride—Bone disease and mottled teeth

Iodine—There is no evidence of ill effects from drinking water containing elevated iodine concentration. Iodine tablets are sometimes used to disinfect drinking water.

Iron—No ill health effects from iron in drinking water, but iron causes taste and odor problem in water, and causes stains on laundry articles and kitchen and sanitary ware.

Lead—Developmental problems in children; kidney problems and high blood pressure in adults

Manganese—No ill health effects have been reported from manganese in drinking water, but it can cause neurologic damage upon high exposure rates. Like iron, manganese can cause taste, odors, and stains.

Mercury—The most serious effect of mercury on humans is impaired neurological development in fetuses, infants, and children. Mild exposure to mercury in adults can cause neurological issues (mood swings, weakness, twitching, headaches, etc.). More intense exposure damages the kidneys and can lead to death.

Nickel—Seldom enough nickel in water to cause health problems, but high concentrations can cause dermatitis, intestinal upset, increased red blood cells, and increased protein in urine.
Selenium—Hair and nail loss; circulatory problems
Silver—There is no evidence of ill effects from drinking water containing elevated silver concentration. If silver salts are used to disinfect water, prolonged intake could cause argyria—a condition in which the skin turns blue or bluish-gray.
Strontium—Seldom enough in water to cause health problems, but high levels can negatively affect bone and tooth development in children.
Tin—There is no evidence of ill effects from drinking water of elevated tin concentration.
Vanadium—There is some concern that long-term consumption of water with an elevated vanadium concentration can cause cancer.
Zinc—High intake over time can cause anemia and damage the pancreas.

The World Health Organization and governments in most countries have developed recommended limits on concentrations of trace elements in potable water that may or may not be enforced. The standards for concentrations of trace elements in drinking water in the United States will be discussed in Chap. 16.

Trace elements have many uses in human activities, and municipal and industrial effluents often have elevated concentrations of one or more potentially toxic trace elements. Stream waters and groundwaters in mining areas are likely to be contaminated with trace elements. Therefore, trace element toxicity in aquatic ecosystems is not uncommon.

References

Baralkiewicz D, Siepak J (1999) Chromium, nickel, and cobalt in environmental samples and existing legal norms. Polish J Environ Studies 8:201–208

Boyd CE, Tucker CS (2014) Handbook of aquaculture water quality. Craftmaster, Auburn

Boyd CE, Walley WW (1972) Studies of the biogeochemistry of boron. I. Concentrations in surface waters, rainfall, and aquatic plants. Am Midl Nat 88:1–14

Boyd PW, Jickells T, Law CS, Blain S, Boyle EA, Buesseler KO, Coale KH, Cullen JJ, DeBaar HJ, Fellows M, Harvey M, Lancelot C, Levasseur M, Owens NPJ, Pollard R, Rivkin RB, Sarmiento J, Schoemann V, Smetacek V, Takeda S, Tusuda A, Turner S, Watson AJ (2007) Mesoscale iron enrichment experiments 1993-2005: synthesis and future directions. Science 315(5812):612–617

British Geological Survey (2000) Iodine. Water quality fact sheet, London

Chowdhury UK, Biswas BK, Chowdhury TR, Samanta G, Mandal BK, Basu GC, Cahnda CR, Lodh D, Saha KC, Murkherfee SK, Roy S, Kalir S, Quamruzzaman Q, Chakraborti D (2000) Groundwater arsenic contamination in Bangladesh and West Bengal, India. Environ Health Perspect 108:393–397

Ćosović B, Degobbis D, Bilinski H, Branica M (1982) Inorganic cobalt species in seawater. Geochim Cosmochim Acta 46:151–158

Durfor CN, Becker E (1964) Public water supplies of the 100 largest cities in the United States, 1962. United States geological survey water-supply paper 1812. United States government Printing Office, Washington, DC

Durum WH, Haffty J (1961) Occurrence of minor elements in water. United States geological survey circular 445. United States Government Printing Office, Washington, DC
Essumang DK (2009) Levels of cobalt and silver in water sources in a mining area in Ghana. Int J Biol Chem Sci 3:1437–1444
Goldman CR (1972) The role of minor nutrients in limiting the productivity of aquatic ecosystems. In Likens GE (ed) Nutrients and eutrophication: the limiting-nutrients controversy. Limnology and Oceanography special symposium, vol 1, pp 21–33
Hamilton SJ (2004) Review of selenium toxicity in the aquatic food chain. Sci Total Environ 326:1–31
Hem JD (1970) Study and interpretation of the chemical characteristics of natural water. Water-supply paper 1473, United States geological survey. United States Government Printing Office, Washington, DC
Hem JD (1985) Study and interpretation of the chemical characteristics of natural water. Water-supply paper 2254, United States geological survey. United States Government Printing Office, Washington, DC
Hem JD, Roberson CE (1967) Form and stability of aluminum hydroxide complexes in dilute solution. Water-supply paper 1827-A, United States geological survey. United States Government Printing Office, Washington, DC
Howarth RS, Sprague JB (1978) Copper lethality to rainbow trout in waters of various hardness and pH. Water Res 12:455–462
Hyenstrand P, Rydin E, Gunnerhed M (2000) Response of pelagic cyanobacteria to iron additions—enclosure experiments from Lake Erken. J Plankton Res 22:1113–1126
Izbicki JA, Ball JW, Bullen TD, Sutley SJ (2008) Chromium, chromium isotopes and selected trace elements, western Mojave Desert, USA. Appl Geochem 23:1325–1352
Kopp JF (1969) The occurrence of trace elements in water. In: Hemphill DD (ed) Proceedings of the third annual conference on trace substances in environmental health. University of Missouri, Columbia, pp 59–79
Kopp JF, Kroner RC (1967) Trace metals in waters of the United States. A five year summary of trace metals in rivers and lakes of the United States (October 1, 1962 to September 30, 1967). United States Department of the Interior, Federal Water Pollution Control Administration, Cincinnati
Kopp JF, Kroner RC (1970) Trace metals in waters of the United States. Report PB-215680. Federal Water Pollution Control Administration, Cincinnati
Lane TW, Morel FMM (2000) A biological function for cadmium in marine diatoms. Proc Natl Acad Sci U S A 97:4627–4631
Lee JG, Roberts SB, Morel FMM (1995) Cadmium: a nutrient for the marine diatom *Thelassiosira weissflogii*. Limol Oceanogr 40:1056–1063
Livingstone DA (1963) Chemical composition of rivers and lakes. Professional Paper 440-G, United States geological survey. United States Government Printing Office, Washington, DC
Nadis S (1998) Fertilizing the sea. Sci Am 177:33
Pagenkopf GK (1978) Introduction to natural water chemistry. Marcel Dekker, New York
Pais I, Jones JB Jr (1997) The handbook of trace elements. Saint Lucie, Boca Raton
Palmer CA, Gilbert JA (2012) Position of the Academy of Nutrition and Dietetics: the impact of fluoride on health. J Acad Nutr Diet 112:1443–1453
Pinsino A, Matranga V, Roccheri MC (2012) Manganese: a new emerging contaminant in the environment. In: Srivastava J (ed) Environmental contamination. InTech Europe, Rijeka, pp 17–36
Sillén LG, Martell AE (1964) Stability constants for metal-ion complexes. Special publication 17. Chemical Society, London
Sillén LG, Martell AE (1971) Stability constants of metal-ion complexes. Special publication 25. Chemical Society, London
Smith FA, Ekstrand J (1996) The occurrence and chemistry of fluoride. In: Fejeskov O, Ekstrand J, Burts BA (eds) Fluoride in denistry, 2nd edn. Munksgaard, Copenhagen, pp 20–21

Uthus EO (1992) Evidence for arsenic essentiality. Environ Geochem Health 14:55–58
Wershaw RL (1970) Mercury in the environment. Geological survey professional paper 713. GPO, Washington, DC
World Health Organization (2011) Selenium in drinking water. WHO/HSE/WSH/10.01/14

Water Pollution

15

Abstract

Water pollution results from many anthropogenic activities and occurs in various forms. Soil erosion leads to contamination of natural waters with suspended soil particles and can be the cause of excessive sedimentation. Organic wastes impart a high oxygen demand often culminating in low dissolved oxygen concentrations in water bodies. Nutrient pollution of streams and lakes—primarily from nitrogen and phosphorus—results in eutrophication, deterioration of water quality and loss of biodiversity. Pesticides, synthetic organic chemicals and heavy metals from industry, and pharmaceutical compounds and their degradation products can be toxic to aquatic animals or have other adverse effects on them. Toxins in drinking water can lead to several serious illnesses to include cancer in humans. Water bodies also may be contaminated with biological agents that cause aquatic animal and human diseases. Elevated sulfur dioxide and carbon dioxide concentrations in the atmosphere as a result of air pollution can influence water quality. Wetland destruction must be considered in a discussion of water pollution, because functional wetlands are important for natural water purification. Control of water pollution from all sources is important for protecting natural resources that are becoming increasingly more scarce.

Keywords

Types of pollution • Biochemical oxygen demand • Eutrophication • Toxicity evaluation • Wetland destruction

Introduction

The human population grew slowly for many millennia and other than for areas that were most suitable for human habitation, the earth was not heavily populated. Natural ecosystems were capable of supplying the resources and services necessary

C.E. Boyd, *Water Quality*, DOI 10.1007/978-3-319-17446-4_15

to support society in a sustainable manner. The demand placed on ecosystems by humans did not cause significant damage to overall ecosystem structure and function other than in isolated, highly populated areas.

Humans initially had relatively little control over the environment, and their population was basically limited by the same factors controlling the populations of other species. The population was around ten million when agriculture was invented around 10,000 BC. This increased food availability, but population reached only 750 million by 1750 AD. However, beginning in the 1500 and 1600s, mankind began to flourish because of increasing knowledge of science and technology, and a "critical mass" was reached in the mid-eighteenth century resulting in the industrial revolution.

Since the industrial revolution, the global population has increased at a rapid rate with exponential growth since the mid-1800s (Fig. 15.1). The expanding human population has placed a huge demand on the world's ecosystems for water and other resources as well as taxing their waste assimilation capacity. One of the major impacts of the growing human population has been an increasing pollution load that has greatly deteriorated the quality of aquatic ecosystems and water supplies.

In this chapter, the major sources of pollution and their effects on natural aquatic ecosystems and water use by humans will be discussed.

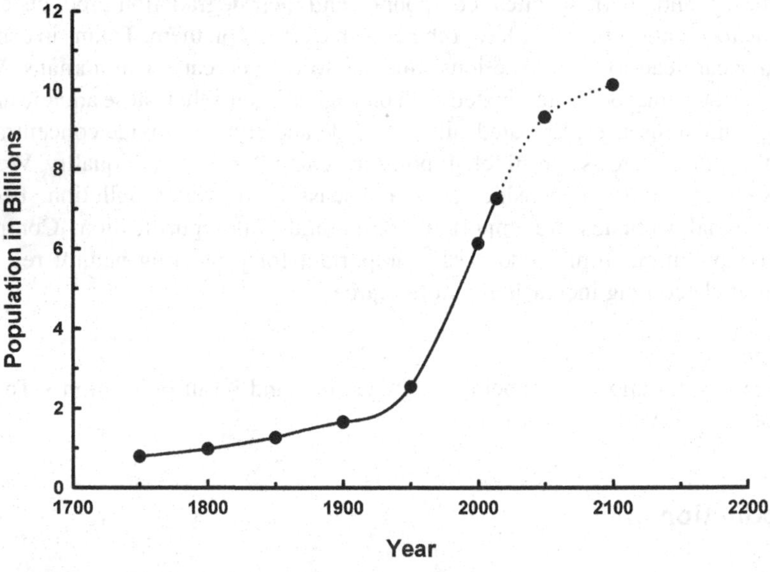

Fig. 15.1 World population 1750–2014 with projection to 2100

Overview of Water Pollution

Water pollution is separated into two broad categories—point sources and nonpoint sources. Well-defined effluent streams discharged in both wet and dry weather via a pipe, channel, or other conduit are point sources. Common point sources of pollution are industrial operations and municipal wastewater treatment plants. Storm runoff enters sewers and other effluent conduits and contributes to discharge during wet weather. Storm runoff accumulates pollutants from a broad area and is considered a nonpoint source. Runoff from urban areas, farmland, construction sites, etc., is known as nonpoint source pollution. Acidic deposition in rainfall and dry fallout also is nonpoint source of pollution.

Pollutants vary widely in their properties. Organic wastes that demand oxygen when decomposed by microorganisms are major contaminants in domestic and municipal wastewater, animal feedlot effluents, and discharges from food processing and paper manufacturing.

Suspended solids are important pollutants consisting of suspended mineral and organic particles. Solids do not settle immediately, and they can create turbidity plumes that are not aesthetically pleasing. Solids settle from water creating sediment deposits that may suffocate benthic organisms. Sediment also reduces water depth, and shallow water favors aquatic macrophyte infestations. The oxygen demand of sediment with a high organic matter content can cause anaerobic conditions in shallow areas. Suspended solids in municipal, industrial, and feedlot effluents tend to be highly organic, while effluents from agricultural land, logging operations, construction sites, and surface mining have a high proportion of inorganic, suspended solids.

Nutrient pollution results primarily from nitrogen and phosphorus in runoff or effluents. Municipal wastewater and other effluents with high concentrations of organic matter also tend to have large concentrations of nitrogen and phosphorus. Runoff from residential areas with lawns and from cropland and pastureland also contain elevated concentrations of nitrogen and phosphorus. The main concern over nitrogen and phosphorus additions in water bodies is eutrophication characterized by excessive phytoplankton productivity, and low dissolved oxygen concentration that may cause fish kills, loss of sensitive species, unsightly algal blooms, and bad odors.

A variety of chemicals used for domestic, industrial, and agricultural purposes find their way into water bodies. Toxic chemicals may be contained in effluents from normal operations or they may leak or seep from storage depots or waste storage sites.

Toxic substances may be directly harmful to aquatic life or they may accumulate in the food chain (bioconcentration) and be toxic to organisms in the food web. Bioconcentration also presents a potential food safety hazard for consumers of aquatic products. Toxins also can be hazardous in domestic drinking water or in waters used for agricultural purposes.

Runoff from surface mining and seepage from underground mines are well known sources of acidification in natural waters. Combustion of fossil fuels contaminates the air with sulfur dioxide, nitrous oxide, and other compounds that oxidize to form strong, mineral acids. Rainfall in heavily populated or industrialized areas is a major source of acidification. Nitrification also can be a significant source of acidity in surface waters. On the other hand, some effluents may be alkaline and cause an excessive pH in receiving waters.

Recent research has revealed a new water pollution concern. Many natural waters contain residues and degradation products of pharmaceutical chemicals. Some are possibly directly toxic to aquatic life, while others may have more subtle, but negative effects on the physiology of organisms. Of course, these substances also can enter the water supply for humans.

Contamination of waters with disease organisms of human origin is still a major concern in many developing nations. If human fecal material enters water, the risk of disease spread through drinking water is greatly increased.

Many industrial processes generate waste heat that may be disposed of by transfer to water. Thermally-enriched effluents may raise temperatures in streams or other water bodies to cause serious ecological perturbations.

The major contributors of several pollutants are summarized in Table 15.1 for the United States. Ten major sectors contributed more than half of the total quantities of the point sources for each pollutant. Municipal sewage plants were a leading point source of pollution. Nonpoint sources provided greater quantities of pollutants than point sources, and agriculture was responsible for more than half of the nonpoint pollution.

Desalination of seawater necessary to supplement water supply in some countries also causes pollution. The discharge water from reverse osmosis plants is of higher salinity than coastal waters. Distillation plants discharge thermally polluted cooling water, and metals from heat exchangers enter the cooling water.

Not all pollution results from contaminants contained in runoff or effluents. Sometimes, accidents may result in sudden spills of potentially toxic chemicals or other substances. For example, a highway or rail accident can result in a cargo being inadvertently spilled into a watercourse, or a ship accident can spill crude oil or other substances into the ocean or into coastal and inland waters. The most famous cases of crude oil pollution probably are the Exxon Valdez oil spill that occurred when an Exxon tanker struck a reef in Prince William Sound, Alaska in 1989, and the BP Deepwater Horizon oil spill that resulted from an accident on an off-shore oil drilling platform in the Gulf of Mexico in 2010.

Water quality can be impaired through natural processes without human intervention as illustrated by the following examples. In some coastal areas, soils contain iron pyrite. Iron pyrite oxidation in the dry season produces sulfuric acid that may leach out in the rainy season to cause acidification of surface water. Some groundwaters may be unfit for domestic or other uses because of high iron or manganese concentrations. The infamous poisoning of many inhabitants of areas in Bangladesh and India by naturally-occurring arsenic in groundwater has already been discussed (Chap. 14). Salinization has made many freshwaters too salty for most purposes.

Table 15.1 Annual contributions of selected pollutants (millions of kilograms) from each of 24 major point sources and the contribution of agriculture to total nonpoint sources in the United States as modified from van der Leedon et al. (1990)

Sector	TSS	BOD	N	P	Dissolved metals
Point sources					
Municipal sewage plants	1,746	1,723	369	33.5	4.2
Power plants	529				11.1
Pulp and paper mills	355	240			
Feedlots	191	44	18	9.9	
Iron and steel mills	113	17			3.4
Organic chemicals	65	49	19	0.6	1.6
Miscellaneous food and beverages	42	25	5	2.1	
Textiles	28	11			
Mineral mining	24				1.1
Seafood processing	23	39	5	0.6	
Cane sugar mills		23			
Miscellaneous chemicals		18			
Pharmaceuticals			40		
Meat packing			16	1.5	
Petroleum refining			7	0.7	2.7
Pesticide manufacturing			4		
Leather tanning			3		
Laundries				1.5	
Fertilizer manufacturing				1.2	
Poultry production				0.5	
Electroplating					0.2
Machinery manufacturing					0.2
Oil and gas extraction					0.2
Foundries					0.1
Major sources	3,116	2,192	486	52.1	24.8
Total	6,234	3,149	561	86.6	26.8
Nonpoint sources					
Agriculture	2,800,000	12,698	6,168	2,395	N/A
Total	4,432,000	18,730	9,107	3,519	N/A

Major Types of Pollution

Inorganic Solids and Turbidity

Inorganic solids contribute the largest weight of pollutants entering water bodies; they cause two major problems: turbidity from suspended particles and sediment accumulation when particles settle. The major source of sediment in water bodies is soil particles eroded from the land by rainfall and runoff. Falling raindrops dislodge soil particles, and the energy of flowing water further erodes the land surface and keeps the particles in suspension during transport. Factors opposing erosion are the resistance of soil to dispersion and movement, slow moving runoff because of gentle slope, vegetation that intercepts rainfall, vegetative cover to shield soil from direct raindrop impacts, roots to hold the soil in place, and organic litter from vegetation to protect the soil from direct contact with flowing water.

Erosion usually is considered to be one of three types: raindrop erosion, sheet erosion, or gully erosion. Raindrop erosion dislodges soil particles and splashes them into the air. Usually, the dislodged particles are splashed into the air many times, and because they are separated from the soil mass, they can be readily transported in runoff. Sheet erosion refers to the removal of a thin layer of soil from the surface of gently sloping land. True sheet erosion does not occur, but flowing water erodes many tiny rills in surface soil to cause more or less uniform erosion of the land surface. These rills are not seen in cultivated fields because they are removed by tillage. Gully erosion produces much larger channels than rills and these channels are visible on the landscape.

Estimating Soil Erosion Rates

The universal soil loss equation (USLE) is used widely to estimate soil loss by erosion. The initial efforts to predict soil erosion by mathematic procedures of Zingg (1940) and Smith (1941) lead to further research on the topic, and the first complete version of the USLE was published in 1965 (Wischeier and Smith 1965). The equation has been slightly revised over time, and the present form is

$$A = (R)(K)(LS)(C)(P) \tag{15.1}$$

where A = soil loss, R = a rainfall and runoff factor, K = soil erodability factor, LS = slope factor (length and steepness), C = crop and cover management factor, and P = conservation practice factor. Presentation of the instructions and tabular and graphical information for obtaining the various factors are too lengthy to include here, but there are many online sources including calculators for solving the equation.

Land surface disruption facilitates erosion; construction, logging and mining sites typically have very high rates of soil loss. Deforestation is a major concern both because of reduction in forest area and because of the serious erosion that follows. Row cropland also has a high erosion potential. The lowest rates of erosion are for watersheds that are forested or completely covered with grass. Erosion of streambeds and shorelines also can be important sources of suspended solids in water bodies.

Effects of Soil Erosion on Water Bodies

A portion of the soil particles dislodged from watersheds by erosion remain suspended in runoff when it enters streams and other water bodies. Suspended solids in waters create turbidity making the water less appealing to the eye, and less enjoyable for watersports. Turbidity also reduces light penetration into the water, and diminishes primary productivity. Moreover, suspended solids often must be removed from water to allow its use for human and industrial water supply adding to the cost of water treatment.

When turbulence in water carrying suspended solids is reduced, sedimentation occurs. Sediment creates deposits of coarse particles in areas where turbid water enters water bodies and finer particles over the entire bottom. Elevated sedimentation rates have several undesirable consequences. They make water bodies shallower, and this may lead to greater growth of rooted aquatic macrophytes. Shallower water bodies have less volume, and this may have negative ecological effects as well as reducing the volume of water that can be stored for flood control or human uses. Sediment also destroys breeding areas for fish and other species, and it can smother benthic communities.

Erosion and sedimentation are, of course, natural processes that have been operating since the earth began. The morphology of the earth's surface is the result of eons of erosion and sedimentation and other geological processes. However, natural processes tend to operate slowly allowing living organisms time to adapt. The problem today is that rates of erosion and sedimentation have been greatly accelerated by human activities, and many negative impacts are resulting.

Organic Pollution

Bacteria and other saprophytes in aquatic ecosystems remove dissolved oxygen for use in decomposing organic matter. The effect of addition of organic matter in pollutants on dissolved oxygen concentration depends upon the capacity of a water body to assimilate organic matter relative to the amount of organic matter introduced. A given organic matter load might not influence dissolved oxygen concentrations in a large body of water, but the same load might cause oxygen depletion in a smaller body of water. Likewise a rapidly flowing stream reaerates more rapidly than a sluggish stream of the same cross-sectional area, and therefore can asssimilate a greater organic matter input than the sluggish stream.

Biochemical Oxygen Demand

Assessment of the oxygen demand of a wastewater is a critical factor in water pollution control. The biochemical oxygen demand (BOD) is a measure of the amount of dissolved oxygen consumed by microscopic organisms while decomposing organic matter in a confined sample of water.

Standard 5-Day Measurement

The standard 5-day BOD determination or BOD_5 normally is measured to provide an estimate of the pollutional strength of a wastewater (Eaton et al. 2005). In the BOD_5 procedure, an aliquot of wastewater is diluted with inorganic nutrient solution and a bacterial seed added. The inorganic nutrients and bacterial seed are necessary to prevent a shortage of bacteria and inorganic nutrients that might result from dilution. Because of the possibility of oxygen demand from the bacterial seed, a blank consisting of the same quantity of bacterial seed used in the sample is introduced into nutrient solution and carried through the same incubation as the sample. To prevent photosynthetic oxygen production, samples are incubated in the dark. The incubation is continued in the dark for 5 days at 20 °C. At the beginning and end of incubation, the dissolved oxygen concentration is measured in blank and sample to permit estimation of BOD_5 as illustrated in Ex. 15.1.

Ex. 15.1: *In a BOD_5 analysis, the sample is diluted 20 times. The initial dissolved oxygen concentration is 9.01 mg/L in sample and blank. After 5 days of incubation, the dissolved oxygen concentration is 8.80 mg/L in the blank and 4.25 mg/L in the sample. The BOD will be calculated.*

Solution:

The oxygen loss caused by the bacterial seed is the blank BOD,

$$\textit{Blank BOD} = \textit{Initial DO} - \textit{Blank DO}$$

$$\textit{or } (9.01 - 8.80)\text{mg/L} = 0.21 \text{ mg/L}.$$

The oxygen consumption by the sample is

$$(\textit{Initial DO} - \textit{Final DO}) - \textit{Blank BOD}$$

$$\textit{or } (9.01 - 4.25) - 0.21 = 4.55 \text{ mg/L}.$$

The sample BOD is the oxygen consumption by the sample multiplied by a correction factor equal to the number of times the sample was diluted—20 times in this case.

$$\textit{BOD} = 4.55 \times 20 = 91 \text{ mg L}.$$

A formula for estimating BOD is

$$\text{BOD (mg/L)} = (\text{I}_{\text{DO}} - \text{F}_{\text{DO}})_s - (\text{I}_{\text{DO}} - \text{F}_{\text{DO}})_b \times \text{D} \quad (15.2)$$

where I_{DO} and F_{DO} = initial and final DO concentrations in sample bottle and blank bottle, respectively, and subscript s = sample, subscript b = blank, and D = the dilution factor.

The BOD of a sample represents the amount of dissolved oxygen that will be used up in decomposing the readily-oxidizable organic matter. Of course, if there is a lot of phytoplankton in the sample, a large portion of the BOD will represent phytoplankton respiration, because all the phytoplankton may not die during 5 days in the dark. By knowing the volume of an effluent and its BOD, the oxygen demand of the effluent can be estimated as shown in Ex. 15.2.

Ex. 15.2: *A wastewater effluent has a discharge rate of 75 m³/h and a BOD of 265 mg/L. The daily BOD load to the receiving water will be calculated.*

Solution:
The volume of effluent per day is

$$75\ \mathrm{m^3/h}\ \times 24\ \mathrm{h/day} = 1{,}800\ \mathrm{m^3/day}.$$

The daily BOD load is

$$1{,}800\ \mathrm{m^3/day}\ \times 265\ \mathrm{g/m^3}\ \times\ 10^{-3}\ \mathrm{kg/g} = 477\ \mathrm{kg/day}.$$

Thus, a BOD equal to 477 kg of dissolved oxygen will be discharged into the receiving water each day.

All of the organic matter in a sample does not decompose in 5 days as shown in Fig. 15.2. It would require many years for the complete degradation of all the organic matter in a sample. However, the rate of oxygen loss from a sample (expression of BOD) usually is exceedingly slow after 30 days, and the BOD_{30} is a good indicator of the ultimate BOD (BOD_u) of a sample.

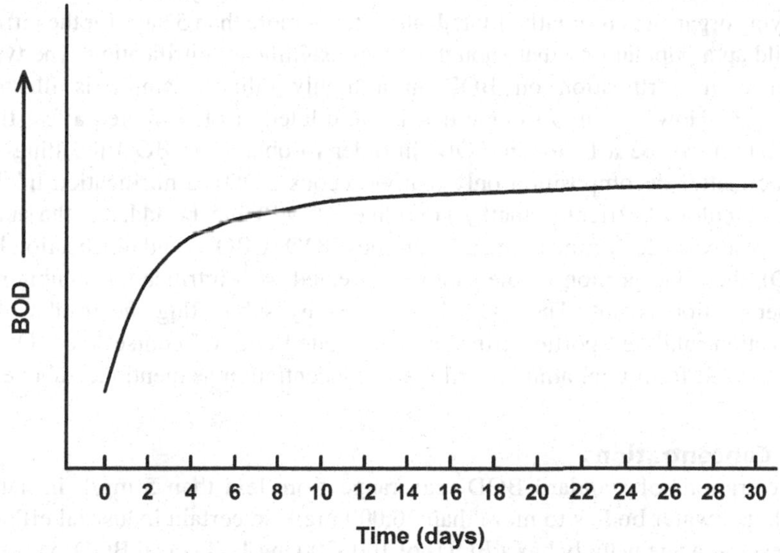

Fig. 15.2 A typical expression of carbonaceous biochemical oxygen demand (BOD) over a 30-day period

Effects of Nitrification

Many wastewaters contain appreciable ammonia nitrogen. The oxidation of ammonia to nitrate by bacteria (nitrification) consumes two moles of oxygen for each mole of ammonia nitrogen (Chap. 11) contributing to oxygen demand. Thus, the oxygen demand of ammonia nitrogen in a sample—the NOD—can be estimated from the total ammonia nitrogen concentration (Ex. 15.3).

Ex. 15.3: *A water sample contains 15 mg/L ammonia nitrogen. The BOD_{30} of the sample is determined to be 550 mg/L. The possible contribution of nitrification to the BOD will be estimated.*

Solution:

From (11.9), *it can be seen that*

$$\underset{14\ \text{g/mol}}{\overset{15\ \text{mg/L}}{N}} = \underset{64\ \text{g/mol}}{\overset{X}{2O_2}}$$

$$\text{and } X = 68.6\ \text{mg/L}.$$

Thus, nitrification could have accounted for 68.6 mg/L of BOD or 12.5 % of the BOD.

By repeating Ex. 15.3 for 1 mg/L of ammonia nitrogen, it can be seen that each milligram per liter of ammonia nitrogen nitrified by bacteria will require 4.57 mg/L O_2 or have a potential NOD of 4.57 mg/L.

In samples that are diluted several fold for BOD analysis, the abundance of nitrifying organisms is greatly diluted, and it takes more than 5 days for the nitrifiers to build up a population great enough to cause significant nitrification. The typical influence of nitrification on BOD in a highly diluted sample is illustrated (Fig. 15.3). However, in a sample that is not diluted or only diluted a few times, nitrification can be a factor in BOD. In order to obtain the BOD resulting from organic matter decomposition only (carbonaceous BOD), a nitrification inhibitor such as 2-chloro-6-(trichloromethyl) pyridine (TCMP) may be added to the sample. If it is desired to determine both carbonaceous BOD (CBOD) and nitrification BOD (NOD), then one portion of the sample is treated with nitrification inhibitor and another portion is not. The NOD is obtained by subtracting the results of the nitrification-inhibited portion from the uninhibited one. Of course, the NOD can be calculated from total ammonia nitrogen concentration as mentioned above.

BOD Concentrations

Concentrations of standard BOD_5 can range from less than 5 mg/L in natural, unpolluted water bodies to more than 20,000 mg/L in certain industrial effluents. Domestic sewage usually has a BOD_5 of 100–300 mg/L. Typical BOD_5 values for effluents from selected industries are as follows: pond aquaculture, 10–30 mg/L (Boyd and Tucker 2014); beet sugar refining, 450–2,000 mg/L; brewery, 500–1,200 mg/L; laundry, 300–1,000 mg/L; milk processing, 300–2,000 mg/L;

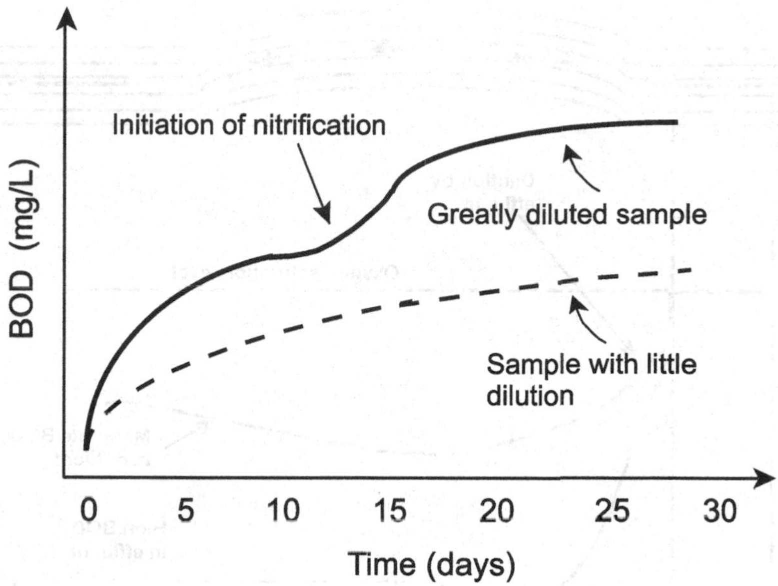

Fig. 15.3 Illustration of the expression of biochemical oxygen demand (BOD) over time in water samples that were either greatly diluted or slightly diluted with nutrient solution

meat packing, 600–2,000 mg/L; canneries, 300–4,000 mg/L; grain distilling, 15,000–20,000 mg/L (van der Leeden et al. 1990).

Effects of Wastewater Oxygen Demand

The typical response of streams to BOD loads is a dissolved oxygen sag downstream from the effluent outfall (Fig. 15.4). If the BOD load is extremely high, there may be a reach of the stream where anaerobic conditions exist. The distance downstream before dissolved oxygen concentration returns to normal depends upon the amount of BOD and the rate of stream reaeration—a topic already discussed in Chap. 6.

The rate of change of the oxygen deficit with time at a location in a stream is equal to the rate of deoxygenation caused by the BOD load minus the rate of stream reaeration (Vesilind et al. 1994). Mathematical models based on this concept are used to predict dissolved oxygen concentrations at different distances downstream from effluent outfalls. Discharge of effluents into lakes, estuaries, or the ocean also can depress oxygen concentrations in the vicinity of the outfall. The severity of this effect depends both on the BOD load and the extent to which the effluent is transported away from the outfall by water currents.

Organic wastes typically contain nitrogen and phosphorus, resulting in ammonia and phosphate being released along with carbon dioxide during decomposition. Carbon dioxide, ammonia, and phosphorus concentrations tend to increase in streams below effluent outfalls, or in the vicinity of outfalls into lakes, estuaries, and the sea. Solids in effluents settle in the vicinity of outfalls and oxygen depletion may occur in sediments with abundant organic matter.

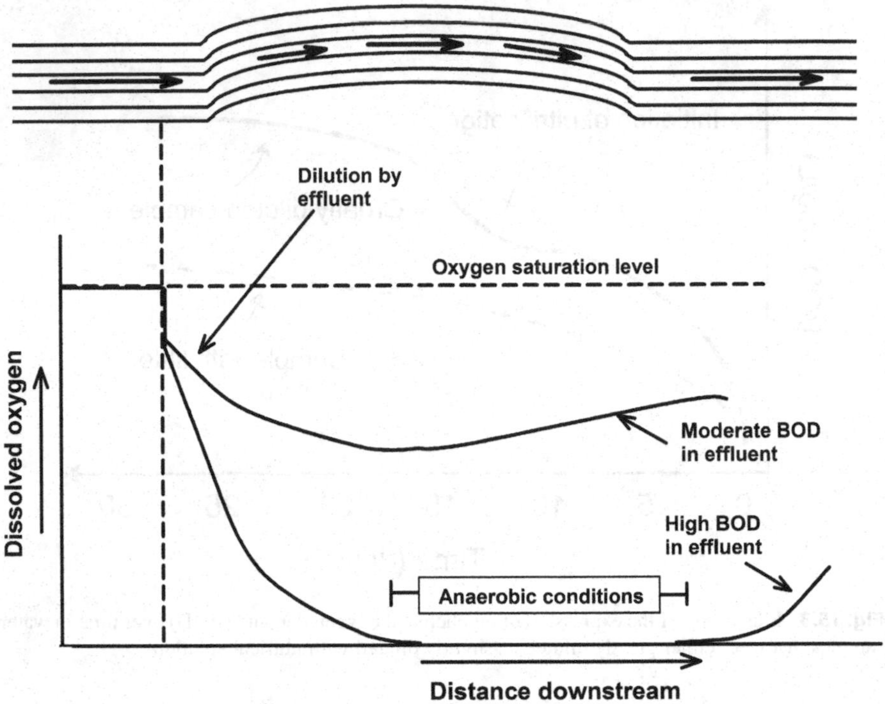

Fig. 15.4 Oxygen sag curve below an effluent outfall in a stream

There is a typical pattern in nitrogen concentrations downstream from outfalls. First, there is a large increase in organic nitrogen. Next, organic nitrogen declines and total ammonia nitrogen increases as a result of decomposition. Nitrite may also increase because of low dissolved oxygen concentration. Finally, downstream of the oxygen sag, nitrate increases and ammonia nitrogen decreases because of nitrification.

Nutrient Pollution

Nutrient pollution stimulates the growth of phytoplankton in water bodies and can lead to eutrophication. Heavy blooms of phytoplankton have a number of adverse effects. In lakes and reservoirs they can cause shallow thermal stratification with an oxygen-deficient hypolimnion. Excessive plant growth causes wide daily fluctuations in dissolved oxygen, and low dissolved oxygen concentrations that may occur at night may be harmful to aquatic animals. Blue-green algae often are abundant in eutrophic waters, and these algae are subject to sudden die-offs that can lead to dissolved oxygen depletion. Some species of blue-green algae may be toxic to other organisms, and other species may impart off-flavor to fish or crustaceans or cause taste and odor problems in drinking water. Dense algal blooms in waters used for recreational purposes are undesirable because they limit visibility into the water and can cause bad odors. Dense scums of blue-green algae can accumulate on the surface of eutrophic water bodies and detract from aesthetic value.

Large inputs of nutrients to streams can cause filamentous algae mats and encourage the growth of rooted plants in shallow water areas. Of course, dense phytoplankton blooms may occur in slow-moving streams. Nutrients released into coastal waters are mixed into seawater by currents and greatly diluted. However, in the mixing zones in estuaries, water pollution can cause eutrophic conditions with undesirable phytoplankton blooms or infestations of macrophytes in shallow water.

Certain blue-green algae, dinoflagellates, and diatom blooms in areas for mulluscan shellfish production may produce toxins that can be absorbed and stored in shell fish and later cause serious illnesses in humans.

There has been much disagreement over the actual concentrations of nitrogen and phosphorus required for eutrophication and excessive phytoplankton. Van der Leeden et al. (1990) presented data on nitrogen and phosphorus concentrations for selected lakes heavily impacted by human activity. Concentrations of total phosphorus ranged from 0.006 to 0.29 mg/L with an average of 0.043 mg/L. Total nitrogen concentrations had a range of 0.047–7.11 mg/L (average = 1.26 mg/L). Van der Leeden et al. (1990) also cited a study by the U. S. Environmental Protection Agency which indicated nearly half of 23,236 lakes surveyed in the United States were eutrophic.

In eutrophic ecosystems, the combination of widely-fluctuating, daily dissolved oxygen concentrations in the water column, low dissolved oxygen concentration in the sediment, and high ammonia concentration lessens the growth and survival of many environmentally-sensitive species of plants and animals. Eutrophic water bodies have a high abundance of a few species, and many sensitive species disappear. Thus, eutrophication tends to reduce biological diversity. Diversity is an index of how the individuals in a community are distributed among the species. There are many equations for diversity, and a common one suggested for assessing phytoplankton species diversity (Margalef 1958) is

$$\overline{H} = \frac{S - 1}{\ln(N)} \tag{15.3}$$

where S = the number of species and N = the total number of individuals. The diversity of two phytoplankton communities is calculated in Ex. 15.4.

Ex. 15.4: *Community A contains 25 species of phytoplankton and 1,000 individuals/mL while community B has 11 species and 14,000 individuals/mL. The diversity index for the two communities will be estimated by (15.3).*

Solution:

A. $\overline{H} = \frac{25 - 1}{\ln(1,000)} = \frac{24}{6.91} = 3.47$

B. $\overline{H} = \frac{11 - 1}{\ln(14,000)} = \frac{10}{9.55} = 1.05.$

In Ex. 15.4, the community with the most species relative to the total number of individuals (community A) has a much higher diversity than the other community with fewer species relative to the number of individuals. As a rule, the greater the diversity, the more stable an ecosystem (Odum 1971). Eutrophic aquatic ecosystems tend to be less stable than oligotrophic ones. This lack of stability may be reflected in sudden shifts in the abundance of species and also in sudden changes in dissolved oxygen concentrations and other water quality variables.

Toxins

Many of the chemicals used to improve human life can be toxic to humans and other organisms. There is a large number of potentially toxic chemicals, but the major classes of chemicals are petroleum products, inorganic compounds such as ammonia, heavy metals, and cyanide, pesticides and other agricultural chemicals, synthetic organic compounds used in industry, and pharmaceutical compounds that may be discarded into sewage systems or enter via human wastes.

Evaluation of the toxicity of waterborne toxins to aquatic organisms is difficult. There can be different degrees of response to a toxin. A toxin may kill aquatic organisms directly. The mortality may be rapid (acute) if a high concentration of the toxin is introduced or slow (chronic) if a lower concentration is maintained in the water. The lowest concentration at which mortality can be detected is the threshold toxic concentration. The threshold concentration also may be defined as the lowest concentration necessary to elicit some response other than death. Responses may include failure to reproduce, lesions, aberrant physiological activity, susceptibility to disease, or behavioral changes. The exposure time necessary for a toxin to produce some undesirable effect on organisms decreases with increasing concentration.

Organisms usually must absorb a certain amount of a toxin before the toxic effect occurs. The total body burden of a toxin may be calculated as illustrated in Ex. 15.5 using the following equation:

$$TBB = (DI + R) - DL \tag{15.4}$$

where TBB = total body burden, DI = daily intake of toxin, R = residual of toxin in body before exposure, and DL = daily loss of toxin from body by metabolism or excretion.

Ex. 15.5: *Suppose a 1-kg fish has a residual concentration of toxin X of 5 mg, the daily intake is 0.1 mg, and the daily loss is 0.08 mg. The total body burden will be estimated after 30 days.*

Solution:
From (15.4),

$$TBB = [30(0.1) + 5] - 30(0.08) = 5.6\ g.$$

Toxicity occurs when the body burden reaches the threshold level. If the toxin disappears from the water, organisms will eliminate the toxin, but the rate of loss usually declines as the total body burden decreases.

Bioaccumulation occurs when an organism accumulates a toxin in specific organs or tissues. For example, many pesticides are fat-soluble and tend to accumulate in fatty tissues. The term bioconcentration is used to describe the phenomenon in which a toxic substance accumulates at greater and greater concentrations as it passes through the food chain. A toxin introduced into water may be bioaccumulated by plankton. Fish eating the plankton may store this toxin in their fat and have higher body burdens than did the plankton. Birds feeding on the fish may further concentrate the toxin until a toxic body burden is reached. Obviously, bioaccumulation and bioconcentration of toxic substances by aquatic food organisms is a human food safety concern.

In synergism, the toxicity of a compound may increase as a result of the presence of another compound; the mixture of two compounds is more toxic than either compound alone. Toxicity normally increases with increasing water temperature. In the case of metals, the free ion usually is the most toxic form, and the toxicity of a metal will be less in water with high concentrations of humic substances that complex metals than in clear water. Other water quality factors such as pH, alkalinity, hardness, and dissolved oxygen concentration can affect the toxicity of substances.

Toxicity Tests

Toxicity tests are important tools of aquatic toxicology for determining the effects, including death, of different concentrations of toxins on various species. From such tests, threshold concentrations for different responses can be estimated, and the influence of exposure time and water quality conditions on toxicity can be evaluated. Based on toxicity tests, safe concentrations of pollutants can be recommended for natural waters, effluent standards for pollutants can be established, and the risk of toxicity from waterborne pollutants can be greatly reduced. Of course, some species are more sensitive than others, and the overall risk of waterborne toxins to ecosystems is extremely difficult to establish.

In acute toxicity tests, aquatic organisms are exposed for specific time periods to a concentration range of a toxicant under carefully controlled and standardized conditions in the laboratory. The mortality at each concentration is determined, and the resulting data are helpful in assessing toxicity under field conditions. Toxicity studies may be conducted as static tests in which water with toxicant is placed in chambers and organisms introduced. There may or may not be water or toxicant renewal during the exposure period. The duration of static tests seldom exceed 96 h and sometimes is shorter. Toxicity studies also may be conducted as flow-through trials in which fresh toxicant solution is continuously flushed through the test chambers. Animals may be fed in flow-through tests, and animals may be exposed to a toxicant for weeks or months. There are many sources of information on toxicity test methodology; one of the best is the *Standard Methods for the Examination of Water and Wastewater* (Eaton et al. 2005).

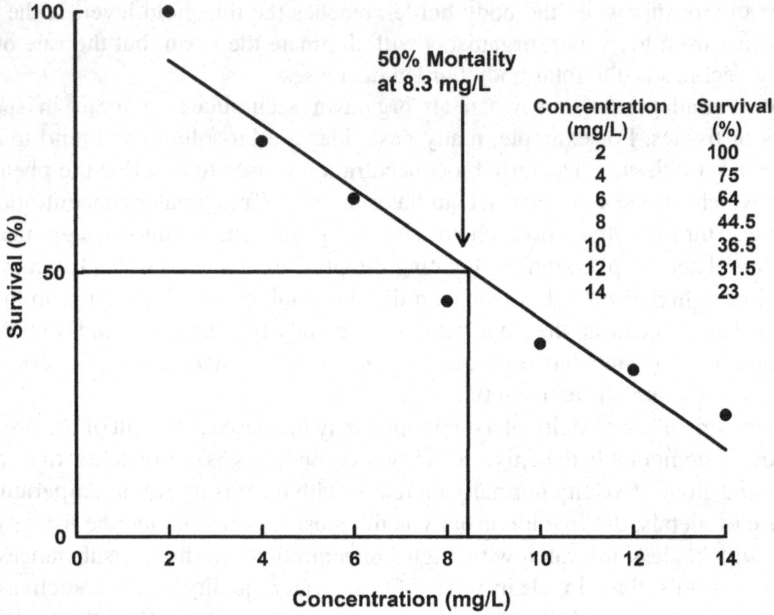

Fig. 15.5 Graphical estimation of the LC50

The most common way of analyzing results of acute toxicity tests is to calculate the percentage survival (or mortality) at each test concentration, and plot percentage survival on the ordinate against toxicant concentrations on the abscissa. Semi-log paper is normally used for preparing the graph of concentration versus mortality because the relationship is logarithmic. The concentration of toxicant that caused 50 % mortality can be estimated from the graph. The concentration of the toxicant necessary to kill 50 % of the test animals during the time that organisms were exposed to the toxicant (exposure time) is called the LC50. The exposure time of animals to toxicants usually is specified by placing the number of hours before LC50, e.g., 24-h LC50, 48 h-LC50, or 96-h LC50. The graphical estimation of the LC50 from the results of a toxicity test is illustrated (Fig. 15.5).

In addition to providing the LC50, toxicity testing can reveal the lowest concentration of a substance that causes toxicity or the highest concentration that causes no toxicity. Sometimes tests may be conducted in which the endpoint is some response other than toxicity. For example, in long-term tests, the concentration that inhibits reproduction could be measured, the concentration that produces a particular lesion, or the concentration that elicits a particular physiological or behavioral change might be ascertained.

Selected Toxicity Data

The 96-h LC50 concentrations of selected inorganic elements, pesticides, and organic chemicals for freshwater fish are provided in Tables 15.2, 15.3, and 15.4.

Table 15.2 Ranges in 96-h LC50 values for various species of fish exposed to selected inorganic elements

Inorganic substance	96-h LC50 (mg/L)	Inorganic substance	96-h LC50 (mg/L)
Aluminum	0.05–0.2	Iron	1–2
Antimony	0.3–5	Lead	0.8–542
Arsenic	0.5–0.8	Manganese	16–2,400
Barium	50–100	Mercury	0.01–0.04
Beryllium	0.16–16	Nickel	4–42
Cadmium	0.92–9	Selenium	2.1–28.5
Chromium	56–135	Silver	3.9–13.0
Copper	0.05–2	Zinc	0.43–9.2

Table 15.3 Acute toxicities of representative compounds of several classes of pesticides to fish

Trade name	96-h LC50 (μg/L)	Trade name	96-h LC50 (μg/L)
Chlorinated hydrocarbon insecticides		*Pyrethum insecticides*	
DDT	8.6	Permethrin (synthetic pyrethroid)	5.2
Endrin	0.61	Natural pyrethroid	58
Heptachlor	13	*Miscellaneous insecticides*	
Lindane	68	Diflubenzuron	>100,000
Toxaphene	2.4	Dinitrocresol	360
Aldrin	6.2	Methoprene	2,900
Organophosphate insecticides		Mirex	>100,000
Diazinon	168	Dimethoate	6,000
Ethion	210	*Herbicides*	
Malathion	103	Dicambia	>50,000
Methyl parathion	4,380	Dichlobenil	120,000
Ethyl parathion	24	Diquat	245,000
Guthion	1.1	2,4-D (phenoxy herbicide)	7,500
TEPP	640	2,4,5-T (phenoxy herbicide)	45,000
Carbamate insecticides		Paraquat	13,000
Carbofuran	240	Simazine	100,000
Carbaryl (Sevin)	6,760	*Fungicides*	
Aminocarb	100	Fenaminosulf	85,000
Propoxur	4,800	Triphenyltin hydroxide	23
Thiobencarb	1,700	Anilazine	320
		Dithianon	130
		Sulfenimide	59

Table 15.4 Acute toxicities of some representative industrial chemicals to fish

Compound	96-h LC50 (mg/L)
Acrylonitrile	7.55
Benzidine	2.5
Linear alkylate sulfonates and alkyl benzene sulfonates	0.2–10
Oil dispersants	>1,000
Dichlorobenzidine	0.5
Diphenylhydrazine	0.027–4.10
Hexachlorobutadiene	0.009–0.326
Hexachlorocyclopentadiene	0.007
Benzene	<5.30
Chlorinated benzenes	0.16
Chlorinated phenols	0.004–0.023
2,4-Dimethylphenol	2.12
Dinitrotoluenes	0.33–0.66
Ethylbenzene	0.43–14
Nitrobenzenes	6.68–117
Nitrophenols	0.23
Phenol	10
Toluene	6.3–240
Nitrosamines	5.85

Risk Assessment of Human Toxins

Toxicity tests are not conducted on human subjects, and humans are seldom exposed to a high enough concentration of a particular toxin in water to elicit an immediate response of lesions, sickness, or death. Still, over a long time, exposure to pollutants in water can adversely influence an individual's health. The USEPA has developed a system of evaluating human risk from pollutants that is used in developing drinking water quality standards that are protective of human health.

A lengthy discussion of the epidemiological techniques for assigning human health risks to pollutants is beyond the scope of this book, but a brief illustration of the general procedure is in order. The USEPA uses the concept of unit risk. For pollutants in water, a unit risk is the risk incurred by exposure to 10^{-9} g/L of the substance. A unit lifetime risk is the risk associated with exposure to 10^{-9} g/L of the pollutant for 70 years. In the case of carcinogens in water, the risk is given in terms of latent cancer fatalities. Such data must be obtained from complex epidemiological studies that have been done for various compounds, and the results are presented as the number of latent cancer fatalities (LCF) per 100,000 people.

The unit annual risk (UAR) for a compound is

$$\mathrm{UAR} = \frac{\mathrm{LCF/year}}{10^{-9}\ \mathrm{g/L}} \tag{15.5}$$

or unit lifetime risk (ULR)

$$\mathrm{ULR} = \frac{\mathrm{LCF}}{(10^{-9}\ \mathrm{g/L})(70\ \mathrm{years})}. \qquad (15.6)$$

The ULR of a hypothetical compound is calculated in Ex. 15.6.

Ex. 15.6: *A community has been drinking water for 10 years that contains 10^{-7} g/L of a compound known to be a carcinogen with a LCF of 0.2/100,000. The risk of these people developing cancer will be estimated using* (15.6).

Solution:

$$ULR = \frac{(10^{-7}\ \mathrm{g/L})(0.2\ \mathrm{LCF})(10\ \mathrm{years})}{(100{,}000)(10^{-9}\ \mathrm{g/L})(70\ \mathrm{years})} = 2.86 \times 10^{-5}\ \mathrm{LCF}.$$

About three cancer deaths per 100,000 people could be expected as a result of the community drinking the water for 10 years. Of course, individuals in the community may also develop cancer from other causes beside drinking water contaminated with the carcinogen.

Biological Pollution

The main concern over biological pollution in water arguably is the introduction of human pathogens in untreated sewage. The major waterborne diseases are gastroenteritis, typhoid, bacillary dysentery, cholera, infectious hepatitis, amebic dysentery, and giardiasis. Of course, in tropical nations, additional diseases also may be spread via the drinking water supply. The role of coliform organisms as indicators of contamination of water with human feces was discussed in Chap. 10.

Diseases of aquatic animals also can be of concern in water quality management. For example, fish and shrimp diseases are common in pond aquaculture. When the effluents from a pond containing diseased animals are released into natural waters, disease can spread (Boyd and Clay 1998).

Groundwater Pollution

Groundwater from wells is the source of drinking water for many people, and it is common for well water to be consumed without any type of treatment. Water in aquifers originates from rainfall and seeps downward until it reaches an impermeable layer. Groundwater may accumulate pollutants, because many wastes are disposed of in or on the soil. Manure of farm animals is disposed by application to fields and pastures as fertilizers, sewage is used for irrigation in some parts of the world, septic tanks seep into the soil, waste is buried in landfills, and waste is sometimes put into dumps on the land surface. Tanks that hold fuel and other chemicals are buried in the ground or placed on the land surface, and leakage or spillage is not uncommon. Pesticides are applied to fields, and a portion reaches the soil surface and more downward with infiltrating water.

In spite of the many opportunities for contamination, groundwater is protected from pollution to a large extent by the natural purifying action of the soil and other geological formations through which water must infiltrate to reach underground aquifers. The soil is composed of a mixture of particles of sand, silt, clay, and organic matter, and the underlying geological formations also contain layers of sand, silt, clay, gravel, and limestone and other fractured rocks. Water infiltrates slowly and bacteria in the soil decompose some of the dissolved organic material. Bacteria can even degrade or alter the molecular structure of agricultural and industrial chemicals to make them less toxic or render them harmless. More importantly, the soil and underlying geological formations act as a porous medium to filter harmful bacteria and other small particles from the infiltrating water. Chemical substances also are adsorbed onto the particles through which water infiltrates. Bollenbach (1975) summarized data on the movement of bacterial and chemical pollution in infiltrating water. Coliform bacteria moved through the first 1.5 m at the same rate as the water, and a few moved about 5 m in 3 days. After 2 months, a few coliforms had seeped to a depth of 10 m. Chemical pollution, however, was found to move 3 m in 4 days and to ultimately seep to a depth of 30 m. Some studies in more permeable soil showed greater rates of movement of bacteria and pollutants, but the medium through which water must seep to reach the permanently saturated groundwater zone is highly effective in filtering out particles such as bacteria and in adsorbing chemicals.

Wells themselves provide a conduit between the land surface and aquifers that can result in pollution of groundwater. A sanitary seal should be placed between the well casing and the borehole to prevent water from flowing downward in this space. Dug wells are still used in some parts of the world, and means for preventing surface runoff from entering these wells after rains should be provided. Abandoned wells should be sealed to prevent them from being conduits for groundwater contamination.

In spite of the purification of water by natural process as it seeps downward, groundwater in many aquifers is contaminated with bacteria and chemicals. This is particularly true in areas where wastes have been disposed in the soil or on its surface, and pollutants have infiltrated into shallow aquifers.

Interpretation

In many instances the only toxicity data for a substance will be the short-term LC50 on one or possibly a few species at a single temperature. Moreover, the tests usually are conducted under a specific water quality regime.

The LC50s for acute mortality (exposures up to 96 h) will be greater than for chronic mortality (long-term exposure). A plot of LC50 versus exposure time for a given toxicant will show a curvilinear decline in LC50 until the LC50 becomes asymptotic to the abscissa (Fig. 15.6). The asymptotic LC50 is the toxicant concentration above which the LC50 does not decline with greater exposure time.

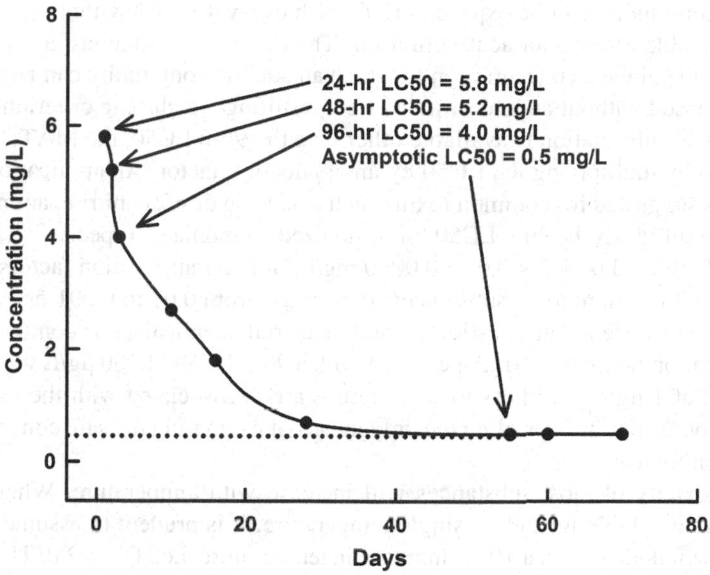

Fig. 15.6 Illustration of the asymptotic LC50

The concentration of potential toxicants in water bodies should not only be below the concentration that causes mortality, it should be less than the concentration that has adverse effects on growth and reproduction. Information on minimum lethal concentrations and no-effect levels can be obtained from full- or partial-life-cycle tests that are more difficult and expensive to conduct than short-term tests.

In life-cycle or partial-life-cycle chronic toxicity tests, the lowest toxicant concentration at which no effect is observed is called the no observed effect concentration (NOEC) and the lowest concentration that causes an effect is known as the lowest observed effect concentration (LOEC). The highest concentration of a toxicant that should not cause a negative impact on an organism is the maximum allowable toxicant concentration (MATC). The MATC is calculated as the geometric mean of the product of the NOEC and LOEC:

$$\text{MATC} = \sqrt{(\text{NOEC})(\text{LOEC})}. \tag{15.7}$$

In order to protect aquatic life, the U. S. Environmental Protection Agency (USEPA) establishes limits for pollutant concentrations that should protect aquatic communities rather than just individual species from harmful chemicals. The procedure for establishing these criteria for a pollutant will not be presented, but criteria are based on the response of very sensitive species and take into account differences in water quality that affect the toxicity of a substance. The procedures for estimating CMCs and CCCs may be found at http://water.epa.gov/learn/training/standardsacademy/aquatic_page3.cfm. The criterion maximum concentration (CMC) is the highest concentration of a substance in ambient water to which an aquatic community can be exposed briefly (1 h every 4 years) without resulting in an undesirable effect—an acute criterion. The criterion continuous concentration (CCC) is the highest concentration to which an aquatic community can be continuously exposed without an undesirable effect resulting—a chronic criterion.

When no information is available other than the 96-h LC50, the MATC often is estimated by multiplying the LC50 by an application factor. An application factor of 0.05 is suggested for common toxins such as carbon dioxide, nitrite, ammonia, or hydrogen sulfide. If the 96-h LC50 for un-ionized ammonia to a species is 1.2 mg/L, the MATC would be $1.2 \times 0.05 = 0.0060$ mg/L. Smaller application factors usually are selected for more toxic substances; they range from 0.01 to 0.001 or even less. In the case of trace metals, pesticides, and industrial chemicals, it is common to use an application factor of 0.01. A pesticide with a 96-h LC50 of 100 μg/L would have a MATC of 1 μg/L. Needless to say, there is a risk associated with the use of the application factor, but it is often the only means of estimating the safe concentration of a given toxicant.

The toxicity of most substances will increase with temperature. Where LC50 values are available for only a single temperature, it is prudent to assume that the toxicity will double with a 10 °C increase in temperature, i.e., $Q_{10} = 2.0$. Thus, if the 96-h LC50 for a pesticide at 20 °C is 0.2 mg/L, at 25 °C the 96-h LC50 could be expected to be about 0.15 mg/L. In natural ecosystems, the concentration of a toxicant will seldom be as constant as in toxicity tests. The toxin will almost never be delivered at a constant rate, and various processes will gradually or rapidly remove the toxicant from the water.

Natural environmental conditions are much different than those in toxicity tests. The toxicity of a substance may change in response to water quality conditions. To illustrate, the toxicity of nitrite to fish is much greater when dissolved oxygen concentration is low than when it is high. Animals in poor physiological condition because of environmental stress are more susceptible to most toxicants than healthy animals living in high quality water.

More than one toxicant may be present in the same water, and the two toxicants may act synergistically to produce greater effects than either will produce alone. There also could be antagonistic effects between toxicants in which the mixture is less toxic than any one of the toxicants alone.

Different sizes or life-stages of animals may have different tolerances to toxicants. Fish fingerlings usually are more susceptible to toxins than are larger fish. Coldwater species typically are more sensitive to toxins than are warmwater species.

Acidification by Atmospheric Inputs

Links exist between air pollution and water pollution that can cause changes in water quality. There are many sources of air pollution, but combustion of fossil fuels is the single largest source. Combustion of fuels releases carbon dioxide, sulfur dioxide (SO_2), nitrous oxides (NO_x), and other substances. Carbon dioxide accumulates in the atmosphere—since the beginning of the industrial revolution in the mid-1700s, atmospheric carbon dioxide concentration has risen from about 280 ppm to around 400 ppm. Sulfur dioxide and nitrous oxide do not accumulate like carbon dioxide, but the quantities emitted per year have drastically increased since the early 1900s.

More carbon dioxide in the atmosphere is thought to be the primary cause of global warming, melting of polar ice caps and glaciers, and sea-level rise. Moreover, an elevated carbon dioxide level in the atmosphere leads to a higher carbon dioxide concentration in water resulting in lower pH. The average pH of the ocean decreased from 8.12 in the late 1980s to 8.09 in 2008. The carbon dioxide concentration in the ocean averaged 0.58 mg/L in 1990; it is expected to reach 0.88 mg/L in 2065 and 1.08 mg/L in 2100. This will lead to further pH decline in the ocean. A decrease in ocean pH increases the solubility of calcium carbonate minerals such as aragonite and calcite that comprise the shells of many marine organisms—including those that make coral (Orr et al. 2005). Acidification of the ocean could have drastic effects on marine biodiversity.

Sulfur dioxide and nitrous oxides oxidize in the atmosphere as illustrated below:

$$SO_2 + OH^- = HOSO_2 \tag{15.8}$$

$$HOSO_2 + O_2 \rightarrow HO_2^- + SO_3 \tag{15.9}$$

$$SO_3 + H_2O \rightarrow H_2SO_4 \tag{15.10}$$

$$NO_x + XOH^- \rightarrow HNO_3. \tag{15.11}$$

Sulfuric and nitric acids formed in these reactions reach the earth's surface in precipitation. The result is greater acidity in freshwater bodies—especially in areas with naturally low alkalinity. There are reports of pH declines of 1–2 units since the 1940s that have adversely affected animals at all trophic levels in some water bodies (Haines 1981).

Surprisingly, recent research claims that acid rain is causing alkalization of many stream waters in the eastern United States (Kaushal et al. 2013). The reason was given as accelerated weathering of minerals in watersheds as a result of acid deposition. The increase in atmospheric carbon dioxide concentration in the atmosphere accelerates the weathering of limestone, calcium silicate, and feldspars as described in Chap. 8 causing alkalinity in surface water to increase. This process increases alkalinity, but not to the extent reported by Kaushal et al. (2013). Further investigation into the cause of stream alkalization is needed.

Wetland Destruction

Although wetlands are not a usual topic of discussion in water quality books, they are aquatic habitats, and they play an important role in water quality protection. Although wetlands are a major feature of many landscapes, a widely acceptable definition of a wetland is difficult to formulate. Cowardin et al. (1979) indicated that wetlands are covered by shallow water or the water table is at or near the land surface. According to Mitsch and Gosselink (1993) the international definition of a wetland is as follows: "wetlands, areas of marsh, fen, peatland or water, whether natural or artificial, permanent or temporary, with water that is static or flowing, fresh, brackish, or salt including areas of marine water, the depth of which at low tide does not exceed 6 m." The legal definition of a wetland in the United States is as follows: "the term wetlands means those areas that are inundated or saturated by surface or groundwater at a frequency and duration sufficient to support, and that under normal circumstances do support, a prevalence of vegetation typically adapted for life in saturated soil conditions. Wetlands generally include swamps, marshes, bogs, and similar areas."

Wetlands are highly productive ecosystems. They are nurseries and feeding grounds for aquatic animals, waterfowl, and other bird life. They also are important habitats for amphibians and some reptiles. Wetlands act as sediment and nutrient traps and provide flood protection. Riparian vegetation is a buffer zone that filters runoff and reduces the input of suspended solids and nutrients into streams. Riparian vegetation also helps control bank erosion. Coastal mangrove forests and other marine wetlands provide a similar filtration system at mouths and deltas of rivers, and mangroves reduce wave and storm damage to the coastline. As an analogy to emphasize the role of wetlands in purifying water, Mitsch and Gosselink (1993) called wetlands "the kidneys of the landscape."

The major cause of freshwater wetland loss is conversion of wetlands to agricultural land. It is estimated that 26,000,000 ha of farmland in the United States were obtained by draining wetlands (Dahl 1990). Over half of the original mangrove wetlands that occur in coastal areas of the world have been destroyed or converted to other uses (Massaut 1999). Today, wetlands are protected by law in the United States and many other countries, and their conversion to agricultural land or their filling for residential or industrial use has been essentially halted. However, in many parts of the world, wetland destruction continues.

Pollution Control

Pollution control is a broad topic and has many ramifications to include technological, political, legislative, regulatory, enforcement, business, educational, ethical, and other issues. Obviously, a thorough discussion of pollution control is beyond the scope of this book, but it is prudent to comment here on the general technological approach to pollutant removal from effluents, and some information on water quality regulations is provided in Chap. 16.

Point source pollution can be lessened if activities or operations by facilities generating effluents can be modified to lessen the amount of waste. This approach is helpful, but it will seldom be adequate. Fortunately, point source effluent can be directed through treatment units that can lessen concentrations and loads of pollutants. Some of the treatment techniques in common use are filtration and sedimentation to remove solids, activated sludge basins with aeration to rapidly oxidize organic matter, precipitation of phosphorus with ferrous chloride, aerobic reactors for nitrification and anaerobic reactors for denitrification, air-stripping of ammonia, outdoor wastewater stabilization ponds, neutralization, and precipitation of metals by chemical treatment or pH manipulation. Removal of certain substances from industrial effluents may require development of specific treatment technologies. Some effluents that contain biological pollutants must be disinfected—often by chlorination—before final discharge.

Nonpoint source effluents are not confined in a conduit, and conventional treatment techniques cannot generally be applied. The most common means of reducing nonpoint source pollution is through the use of practices that lessen the amounts of pollutants entering runoff. Such practices are called best management practices (BMPs).

Agriculture is the largest single source of nonpoint pollution, and BMPs have long been used for controlling agricultural pollution in the United States and many other countries. There are three classes of agricultural BMPs: erosion control, nutrient management, and integrated pest management. Some examples of erosion control BMPs in agriculture are cover crops, no-till farming, conservation tillage, grass-line ditches, terraces in fields on sloping land, etc. Storm water runoff management in municipalities also may consist of a system of BMPs, and BMPs are used to lessen pollution from logging, construction, and mining operations.

As will be discussed in Chap. 16, most governments have developed regulations on point source effluents that impose concentration limits, load limits, or both. Treatment of point source pollution allows compliance with the standards in the discharge regulations. Operations that discharge nonpoint source pollution may be required to use BMPs and inspections are made by the appropriate authority to verify compliance.

References

Bollenbach WM Jr (1975) Ground water and wells. Johnson Division, UOP, Saint Paul

Boyd CE, Clay J (1998) Shrimp aquaculture and the environment. Sci Am 278:42–49

Boyd CE, Tucker CS (2014) Handbook for aquaculture water quality. Craftmaster, Auburn

Cowardin LM, Carter V, Golet FC, LaRoe ET (1979) Classification of wetlands and deepwater habitats of the United States. Publication FWS/OBS-79/31, United States Fish and Wildlife Service, Washington, DC

Dahl TE (1990) Wetlands losses in the United States, 1780s to 1980s. United States Department of the Interior, Fish and Wildlife Service, Washington, DC

Eaton AD, Clesceri LS, Greenburg AE (eds) (2005) Standard methods for the examination of water and wastewater. American Public Health Association, Washington, DC

Haines TA (1981) Acid precipitation and its consequences for aquatic ecosystems: a review. Trans Am Fish Soc 110:669–707

Kaushal SS, Likens GE, Utz RM, Pace ML, Grese M, Yepsen M (2013) Increased river alkalization in the eastern U.S. Environ Sci Technol 47(18):10302–10311

Margalef R (1958) Temporal succession and spatial heterogeneity in phytoplankton. In: Buzzati-Traverso AA (ed) Perspectives in marine biology. University of California Press, Berkeley, pp 323–349

Massaut L (1999) Mangrove management and shrimp aquaculture. Department of Fisheries and Allied Aquacultures, Auburn University, Alabama

Mitsch WJ, Gosselink JG (1993) Wetlands. Van Nostrand Reinhold, New York

Odum EP (1971) Fundamentals of ecology, 3rd edn. W. B. Saunders, Philadelphia

Orr JC, Fabry VJ, Aumot O, Bopp L, Doney SC, Feely RA, Gnanadesikan A, Gruber N, Ishida A, Joos F, Key RM, Lindsay K, Maier-Reimer E, Matear R, Monfray P, Mouchet A, Najjar RG, Plattner GK, Rodgers KB, Sabine CL, Sarmiento JL, Schlitzer R, Slater RD, Totterdell IJ, Weirig MF, Yamanaka Y, Yool A (2005) Anthropogenic ocean acidification over the twenty-first century and its impact on calcifying organisms. Nature 437:681–686

Smith DD (1941) Interpretation of soil conservation data for field use. Agric Eng 21:59–64

van der Leeden F, Troise FL, Todd DK (1990) The water encyclopedia. Lewis, Chelsea

Vesilind PA, Peirce JJ, Weiner RF (1994) Environmental engineering. Butterworth-Heinemann, Boston

Wischeier WH, Smith DD (1965) Predicting rainfall erosion losses from cropland east of the Rocky Mountains. Agricultural handbook 282, United States Department of Agriculture, Beltsville

Zingg AW (1940) Degree and length of land slope as it affects soil loss in runoff. Agric Eng 22:173–175

Water Quality Regulations 16

Abstract

Water quality regulations are important for avoiding conflicts among water users, minimizing public health risks of certain chemical and biological pollutants, protecting the environment, and preventing conditions that lessen the recreational and aesthetic value of water bodies. Most countries have developed water quality regulations, but these rules vary in breadth, rigor, and enforcement. In the United States, the Clean Water Act requires that streams and other water bodies be classified according to use categories—each with specific water quality standards. The National Pollutant Discharge Elimination System (NPDES) further requires permits to discharge wastewaters into public waters. The purpose of NPDES permits is to prevent streams and other water bodies from violating their water quality standards. These permits typically have limits on concentrations of pollutants in effluents, and many times, there are limits on quantities of pollutants that may be discharged. There is a growing tendency to develop total maximum daily loads (TMDLs) that specify the total quantities of selected pollutants that can be discharged into a stream or other water body by all permit holders. Standards with which municipal water supply operations must comply also have been developed to assure safe drinking water and protect public health.

Keywords

Effluent discharge permits • Use classification of waters • Water quality standards • Water rights and water quality • Clean Water Act

Introduction

The usefulness of water declines as water quality deteriorates, and high quality water is in greater demand and has more value than low quality water. The concepts of water quality and water quantity developed simultaneously throughout human

C.E. Boyd, *Water Quality*, DOI 10.1007/978-3-319-17446-4_16

history, but until recently, few quantitative means of assessing water quality were available. Because of the importance of water in human affairs and the relative scarcity of water for many uses, conflicts over water use rights and water quality always have been common. Because of the lack of quantitative methods of assessing water quality, water rights issues historically focused on water quantity and especially on the quantity of surface water. Disputes over water rights have been settled by methods ranging from warfare to court cases to mutual agreements.

In the United States and most developed nations, water rights disputes have traditionally been settled through legal actions based on the principles of common law. In common law, court rulings are based on precedents set in previous court cases of similar nature, or if there is no precedent, the court must make a decision that will become precedent. According to Vesilind et al. (1994), common law has provided a reasonable way to deal with surface water disputes through the theories of riparian doctrine, prior appropriations doctrine, the principle of reasonable use, and the concept of prescriptive rights. Riparian doctrine holds that ownership of the land beneath or adjoining a body of surface water includes the right to use this water. Prior appropriations take the approach that water use rights are based on a "first-come, first-serve basis" and land ownership does not necessarily grant control over the use of the water associated with the land. The principle of reasonable use states that the riparian owner is entitled to make reasonable use of the water, but the court may take into account the needs of others downstream. The concept of prescriptive rights basically allows the upstream user to abuse water quantity and water quality provided the downstream riparian owner does not use the water. Thus, by lack of use, downstream owners give up their water rights.

Water Quality Regulations

Doctrines, principles, and concepts of water rights used in common law do not provide clear guidance about issues related to water quality. Of course, if a water user impairs the quality of water that another party has a right to use, the injured party may initiate a court case to seek relief. It takes years to set precedents, and current issues often are settled by ancient precedents. This does not allow good use of new technology and knowledge, and it does not provide adequate protection of environmental quality and public health. Modern governments establish rules or regulations related to water quality and water pollution abatement. A system of governmental mandates is known as statutory law. Statutory laws about water use and water quality allow a government to exert a degree of control over the quality of waters within its boundaries, minimize disputes over water use, protect aquatic ecosystems, and guard public health. In most nations, disputes over water quality and water rights can be taken to the courts, but statutory law is the major means of protecting water quality.

In the United States, federal legislation known as the Clean Water Act was passed in 1965 to provide a uniform series of procedures to deal with water pollution. The US Environmental Protection Agency (USEPA) has the responsibility for enforcing the regulations and laws mandated by the Clean Water Act, but the day to day enforcement of the Clean Water Act has been relegated mainly to the individual states. The Clean Water Act applies to governmental facilities, municipalities, industries, and private individuals. Implementation and enforcement of the Clean Water Act is obviously a tremendous task, and it has yet to be applied to all activities that affect water quality in the United States. Implementation and enforcement of the act are continuing, and water quality in the United States has improved greatly and is continuing to improve as a result. Most countries have some system of water quality legislation, but in developing nations, water quality legislation often is either poorly structured or not adequately implemented and enforced.

Effluent Discharge Permits

The initial step in water pollution control is to require a permit for discharge of pollutants from a point source into natural waters. The Clean Water Act in the United States requires a National Pollutant Discharge Elimination System (NPDES) permit for every discharge of pollutants. The holder of an NPDES permit has the right to discharge effluents containing specified amounts of pollutants at a given point usually for a period of 5 years. The NPDES does not apply to storm runoff, discharges into water treatment systems, and some other waters. Individual states and municipalities must establish regulations for discharges not covered by NPDES.

Permits such as the NPDES and similar permits in other countries typically contain effluent limitations, monitoring requirements, and reporting schedules. They also may contain other features such as use of best management practices (BMPs) to prevent or reduce the release of pollutants or to clean up spills. Permits also specify conditions related to operation of treatment systems, record keeping, inspection and entry, etc.

There is great variation among countries in requirements in water discharge permits and in the degree of implementation and enforcement of these permits. In the United States and other developed countries, water discharge permits normally are designed to protect public health and the environment and enforcement is relatively strict. In some countries, the permitting procedure is more lax. Permits may have few restrictions and requirements, and they may not be enforced to the extent necessary to protect public health and aquatic ecosystems.

Effluent Limitations

The reason for limits on the discharge of pollutants into natural waters obviously is to avoid water quality degradation. Prohibiting discharge is a possibility in some instances. But, this approach is rare, because many activities that cause water pollution are essential for society. Complete removal of pollutants by treatment of effluents before final discharge usually is both technologically and economically impossible. As a compromise, regulatory agencies usually try to achieve a balance in water quality permits to allow an activity to continue but with limits on pollutants in effluents that are technologically achievable, affordable, and protective of water quality.

Use Classification for Water Bodies

The task of assigning effluent limitations in effluent permits can be facilitated by classifying streams and other water bodies according to their anticipated maximum beneficial use such as public drinking water supplies, propagation of fish and wildlife, recreational activities, industrial and agricultural water sources, navigation, and others. Each water use requires a certain level of water quality. Thus, water bodies can be classified into use categories and each use category assigned water quality standards that must be maintained. A stream classified for agricultural and industrial use will have lower water quality standards than a stream designated for fisheries. The objective of stream classification is to force water users to limit or treat discharges to prevent streams from violating their water quality standards.

Use classification has been applied primarily to streams, but it also is used for any water body or water source. A stream classification system developed by the Alabama Department of Environmental Management in the United States is illustrated in Tables 16.1 and 16.2. This system is typical in that the stream water quality standards do not contain many quantitative guidelines. In the United States, the Clean Water Act requires that each state develop a stream classification system with criteria and standards for each use category. The standards developed by each state must attain the Clean Water Act's goal of fishable, swimmable water wherever possible and should prevent further stream degradation.

Most streams already were receiving pollution and degraded below their pristine condition before stream classification was applied. Nevertheless, by classifying streams, a government can prevent water quality in a stream or a reach of a stream from degrading further. Also, the system can be used to force improvements in surface water quality by giving streams a higher classification or disallowing the lowest classification in the system. In the United States, more streams are classified for fish and wildlife use or higher use categories than in the past. Streams seldom are downgraded in the classification system.

Stream classification can fail to provide the intended benefits of protecting public health and the aquatic environment for several reasons. The stream standards may not be adequate to protect water quality, or the effluent standards may not be strict enough to prevent the waste from causing the receiving water to violate its standard. Moreover, there are many cases where the standards simply are not enforced.

The way in which governments foster economic development also can affect water quality protection. Two general approaches are used in setting water quality standards: (1) stipulation and (2) a policy of minimum degradation of water quality.

Where the major goal of a government is to encourage economic development, subsidies can be stipulated to industry. One possible subsidy to industry is to classify steams according to low standards. From an environmental standpoint, it would be better for governments to subsidize wastewater treatment rather than to lower stream classification or relax discharge standards (Tchobanoglous and Schroeder 1985).

Water Quality Standards

Water quality standards are the quantitative or qualitative values for acceptable ranges of physical, chemical, biological, and aesthetic characteristics of water (or criteria) in stream classification standards, effluent limitation standards, drinking water standards, and other water quality standards. Examples of water quality standards can be found in Table 16.1. In Alabama, a stream classified for agricultural and industrial water supply must have a pH between 6.0 and 8.5, water

Table 16.1 Summary of Alabama stream classification system with quantitative water quality standards

Classification	pH	Temp. (°F)	Wastewater effluent limits (mg/L)	Dissolved oxygen (mg/L)	Bacteria (cfu/ 100 mL)	Turbidity increase (NTU)
Outstanding national resource waters	(No discharge permitted into these waters)					
Outstanding Alabama water	6–8.5	90° (5°R)[a]	DO: 6.0 NH_3: 3.0 BOD_5: 15.0	5.5	200	50
Swimming	6–8.5	90° (5°R)		5.0	200	50
Shellfish harvesting	6–8.5	90° (5°R)		5.0	FDA regulations	50
Public water supply	6 8.5	90° (5°R)		5.0	2,000/ 4,000 June–Sept.: 200	50
Fish and wildlife	6–8.5	90° (5°R)		5.0	1,000/ 2,000 June–Sept.: 200	50
Agricultural and industrial water supply	6–8.5	90° (5°R)		3.0		50
Industrial operations	6–8.5	90° (5°R)		3.0		50

[a]5 °F maximum rise in temperature permitted; not to exceed 90 °F

Note: Some details related to Tennessee and Cahaba Rivers and coastal waters not included

Table 16.2 Summary of qualitative narrative criteria in Alabama stream classification system

Classification	Toxicity, taste, odor, and color
Outstanding national resource waters	No discharge permitted into these waters
Outstanding Alabama Waters	Must meet all toxicity requirements, not affect propagation, palatability of fish/shellfish, or affect aesthetic values
Swimming	Must be safe for water contact, be free from toxicity, not affect fish palatability, not affect aesthetic value or impair waters for this use
Shellfish harvesting	Must be free from toxicity, not affect fish/shellfish palatability, not affect aesthetic value or impair waters for this use
Public water supply	Must be safe for water supply, free from toxicity, not have adverse aesthetic values for this use
Fish and wildlife	Must not exhibit toxicity to aquatic life of propagation, impair fish palatability, or affect aesthetic values for this use
Agricultural and industrial water supply	Must not impair agricultural irrigation, livestock watering, industrial cooling, industrial water supply, fish survival, or interfere with downstream uses. Does not protect fishing, recreational use, or use as a drinking water supply
Industrial operations	Must not impair use as industrial cooling and process water. Does not protect use as fishing, recreation, water supply for drinking or food processing

temperature of 90 °F (32.2 °C) or less, at least 3.0 mg/L dissolved oxygen, and a turbidity no greater than 50 NTU. There also are qualitative standards. In Table 16.2, streams classified for fish and wildlife propagation, the qualitative standard states: "must not exhibit toxicity to aquatic life of propagation, impair fish palatability, or effect aesthetic value for these uses."

Once stream classification systems are available, the major focus is on limiting pollutant levels in effluents so stream water quality complies with the standards for the appropriate use classification. The writer of an NPDES permit in the United States or a water discharge permit in other nations has the responsibility to protect the environment while not unduly penalizing municipalities, industries, or other users. The criteria and standards in effluent permits are selected based on experience, technical attainability, economic attainability, bioassays and other tests, ability to reliably measure the criteria, evidence of public health effects, educated guess or judgment, mathematical models, and legal enforceability (Tchobanoglous and Schroeder 1985). Considerable experience has been accumulated in setting water quality standards, and many water quality guidelines have been published, e.g., guidelines for drinking water, for protection of aquatic ecosystems, for irrigation water, for livestock watering, and for recreational waters. Still, the permit writer must decide upon the safe limits of pollutants in effluents in order to achieve the goals of these guidelines. Permits must be renewed at intervals, and obvious flaws in permits may be negotiated between the permit holder and the permitting agency. No permit is perfect, but discharge permits with water quality criteria and standards are the main tools for protecting water quality in aquatic ecosystems.

Standards on Concentrations

The simplest standards in effluent permits have criteria regarding permissible concentrations of selected water quality variables. Examples of concentration-based criteria in a water quality standard for an effluent are as follows:

Criteria	Standard
pH	6–9
Dissolved oxygen	5 mg/L or above
Biochemical oxygen demand	30 mg/L or less
Total suspended solids	25 mg/L or less

Standards of this type can prevent adverse effects on water quality in the mixing zone where the effluent mixes with the receiving water. They also put a limit on the concentration of pollutants to avoid future increases in concentration. However, if a source of water is available, the permit holder can use it to dilute the concentrations of key variables to assure compliance. Because there is no limit for discharge volume, the amounts of pollutants can be increased, and despite protection in the mixing zone, overall water quality may deteriorate in the receiving water body.

Standards on Loads

The load of a pollutant is calculated by multiplying effluent volume by the pollutant concentration, e.g., effluent volume (m^3) × pollutant concentration (g/m^3) = pollutant load (g). An example of a criterion with a load-based standard is a BOD_5 standard of no more than 100 kg/day. But, such a simple standard is unacceptable because a small discharge could have a very high BOD_5 and not exceed the standard. This scenario could lead to dissolved oxygen depletion in the mixing zone. To avoid this likelihood, a concentration limit usually is specified along with the load of a pollutant. In this case, the standard might limit BOD_5 load to 100 kg/day and prohibit a maximum daily BOD_5 concentration above 30 mg/L.

The weakness of load standards lies in the fact that the permissible effluent load is seldom known for the receiving water. Nevertheless, limiting the load can prevent the permit holder from increasing the load of a pollutant over time.

Delta Based Standards

A delta standard will specify the maximum allowable increase or delta value for one or more variables. The TSS standard could specify that TSS concentration in effluent cannot exceed the TSS concentration of intake water by more than 10 mg/L. In other cases, the standard might give the permissible effluent concentration of a variable as an increase above the receiving water concentration. For example, the turbidity cannot be more than 10 NTU greater than that of the receiving water. The delta standard is basically like a concentration standard, but it assures that the quality of the receiving water is considered.

Volume Limits in Standards

Discharge permits may require limits on volume of effluent. This is actually a way to impose a load standard without specifying the load limits provided that there are concentration limits in permits. There are situations, however, where volume limits are imposed for ecological reasons.

Total Maximum Daily Loads

The issues related to shortcomings of effluent standards discussed above have led the United States and some other countries to establish total maximum daily loads (TMDLs) for priority pollutants in the receiving water. This approach consists of calculating the maximum amount of a pollutant from all sources (natural or anthropogenic) that can be allowed without causing a water body to violate its classification standards. In some situations, the TMDL of a pollutant must be allocated among the different sources of the pollutant. To illustrate, suppose a stream reach has a TMDL for phosphorus of 500 kg/day, several industries want to discharge into the stream reach, and natural sources of phosphorus are 100 kg/day. The maximum amount of phosphorus that can be contained in effluents from the industries is 400 kg/day, and this amount normally would not be allowed in order to have a safety factor. If a safety factor of 1.5 is used, only 267 kg/day of phosphorus could be allowed. This load would have to be allocated among the different industries by some means. In other cases, the current load of a pollutant might already exceed the TMDL leading to stricter limits in permits requiring reduction in loads. Use of TMDLs allows industries to trade in pollution loads. If an industry does not need its entire assigned TMDL for a pollutant, it could sell the unneeded portion of its TMDL allocation to another industry that cannot meet its TMDL for the particular pollutant.

Toxic Chemical Standards

The USEPA has published maximum safe concentration limits for many toxic pollutants that are to be applied to avoid toxic conditions in any water of any designated use (Gallagher and Miller 1996). It is difficult, however, to establish standards for toxic chemicals in effluent permits because the toxicity of some metals and organic chemicals vary greatly with water quality conditions. To overcome this problem, there is considerable use of toxicity-based limitations that are established by whole effluent toxicity testing. The toxicity tests are conducted by exposing certain species of aquatic organisms to the effluent in question. A permit may require that toxicity testing be done to prove that the effluent is not acutely toxic to organisms in the receiving water body at the time of its discharge.

Biological Standards

Discharge permits have long contained standards for coliform organisms that can be indicators of fecal pollution and certain other microorganisms of human health concern. However, there is a growing tendency to include biological standards based on biocriteria associated with the flora and fauna of the receiving water body. Biocriteria may be used to supplement the traditional water quality criteria and standards or used as an alternative where traditional methods have not been effective. Development of biocriteria requires a reference condition (minimal impact) for each use classification, measurement of community structure and function in reference to water quality to establish biocriteria, and a protocol for determining if community structure and function has been impaired. Biocriteria are much more difficult to assess than chemical and physical criteria and standards.

Qualitative Standards and Best Management Practices

Qualitative criteria may be included in standards. There might be a provision that a visible turbidity plume at the outfall is unacceptable. Another example is to prohibit odors or foam in effluent.

Practices considered to be the best available way of avoiding pollution from a particular source—BMPs—may be included in effluent permits. One example is secondary containment around above-ground chemical storage tanks to contain the chemical if the tank leaks or ruptures. Another example is requiring proper storage of materials that could be washed into streams by rainfall.

Monitoring and Enforcement of Discharge Permits

Many times it is not possible for water discharge permit holders to comply immediately with the conditions of the permit, and a schedule for compliance may be specified. Most governments depend largely upon self-monitoring to document compliance with permit standards. The permit will specify monitoring requirements to include: frequency and method of sampling; variables and methods for monitoring; reporting schedule. Permits normally require the permit holder to immediately report when discharges are not in compliance.

Enforcement of water discharge permits normally involves administrative actions, because the permit holder must report compliance or noncompliance with the permit on a scheduled basis. In theory, the permit holder should realize when compliance is not being achieved and work to correct the problem. In the United States, federal and state enforcement of the Clean Water Act can be in the form of administrative actions or judicial actions. The administrative actions may take many forms depending upon the type of violation. An order to comply may be issued along with a compliance schedule. For serious violations, the administrative order may include a fine or other penalty. Failure to comply with administrative orders can lead to criminal prosecution. Both civil and criminal judicial enforcement is possible if administrative orders do not solve problems related to permits. The penalties resulting from judicial enforcement are more severe than those resulting from administrative orders.

Parties who feel that they have been injured by water discharges may bring civil lawsuits against permit holders if the government is not "diligently prosecuting" the violation (Gallagher and Miller 1996). The form of enforcement of water discharge permits varies greatly among nations.

Drinking Water Standards

Many countries, individual states or provinces, and international agencies have made standards for drinking water quality. For illustration, the current National Primary Drinking Water Regulations (except for radionuclides) for the United States are summarized in Tables 16.3, 16.4, and 16.5. These standards come from the Office of Groundwater and Drinking Water of the USEPA, and they are legally enforceable standards that apply to public water systems. The purpose of these primary standards is to protect drinking water quality by limiting the levels of specific contaminants that can adversely affect public health and are known or anticipated to occur in public water supply systems.

The Secondary Drinking Water Regulations for the United States are provided in Table 16.6. These are non-enforceable guidelines regulating contaminants that may cause cosmetic effects such as skin or tooth discoloration and aesthetic effects such as taste, odor, or color in drinking water.

Water Quality Guidelines

There are many lists of guidelines or criteria for five additional areas of water use: (1) public water supplies, (2) fish and wildlife, (3) agriculture, (4) recreation and aesthetics, and (5) industry. Examples of guidelines recommended for protection of aquatic ecosystems in Australia, New Zealand, and the United States are provided in Table 16.7. Guidelines such as those in Table 16.7 are not legally enforceable; they are intended as guidelines to water quality for particular uses. Of course, these guidelines may be the basis for criteria in legally enforceable effluent permits.

Table 16.3 Maximum contaminant levels (MCL) for inorganic chemicals in primary drinking water regulations in the United States (http://water.epa.gov/drink/contaminants/)

Contaminant	MCL (μg/L)	Contaminant	MCL (μg/L)
Antimony	6	Cyanide	200
Arsenic	0.01	Fluoride	4,000
Asbestos	7×10^6 fibers/L	Lead	15
Barium	2,000	Mercury	2
Beryllium	4	Nitrate (as N)	10,000
Cadmium	5	Nitrite (as N)	1,000
Chromium	100	Selenium	50
Copper	1,300	Thallium	2
		Zinc	7,400

Table 16.4 Maximum contaminant levels (MCL) for selected organic chemicals in primary drinking water regulations in the United States (http://water.epa.gov/drink/contaminants/)

Contaminant	MCL (μg/L)	Contaminant	MCL (μg/L)
Alachlor	2	Endothall	100
Atrazine	3	Endrin	2
Benzene	5	Epichlorohydrin	2,000
Benzo(a)pyrene	0.2	Ethelyne dibromide	0.05
Bromate	10	Glyphosate	700
Carbofuran	40	Haloacetic acids	60
Carbon tetrachloride	5	Heptachlor	0.4
Chloramines	4,000	Heptachlor epoxide	0.2
Chlordane	2	Hexachlorobenzene	1
Chlorine	4,000	Hexachlorocyclopentadiene	50
Chlorine dioxide	800	Lindane	0.2
Chlorite	1,000	Methoxychlor	40
Chlorobenzene	0.001	Oxamyl	200
2,4-D	70	Pentachlorophenol	1
Dalapon	0.002	Polychlorinated biphenyls (PCBs)	0.5
DBCP	0.0000002	Picloram	500
o,Dichlorobenzene	600	Simazine	4
p,Dichlorobenzene	75	Styrene	100
1,2-Dichloroethane	5	Tetrachloroethylene	5
1,1-Dichloroethylene	7	Toluene	1,000
cis-1,2-Dichloroethylene	70	Total trihalomethanes	80
trans-1,2-Dichloroethylene	100	Toxaphene	3
Dichloromethane	5	2,4,5-TP	50
1,2-Dichloropropane	5	1,2,4-Trichlorobenzene	70
Di(2-ethylhexyl)adipate	400	1,1,1-Trichloroethane	20
Di(2-ethylhexyl) phthalate	6	1,1,2-Trichloroethane	5
Dinoseb	7	Trichloroethylene	5
Dioxin	0.00003	Vinyl chloride	2
Diquat	20	Xylenes (total)	10,000

Table 16.5 Maximum contaminant levels (MCL) for microorganisms and turbidity in primary drinking water regulations for the United States (http://water.epa.gov/drink/contaminants/)

Variable	MCL
Cryptosporidium	0
Giardia lamblia	99.9 % killed or inactivated
Fecal coliform	0
Heterotrophic plate count	500 colony forming units/mL
Total coliforms	5 % samples per month can be positive
Viruses (enteric)	99.99 % killed or inactivated
Turbidity	1 NTU

Table 16.6 Secondary drinking water standards for the United States (http://water.epa.gov/drink/contaminants/)

Contaminant	Concentration	Contaminant	Concentration
Aluminum	0.05–0.2 mg/L	Iron	0.3 mg/L
Chloride	250 mg/L	Manganese	0.05 mg/L
Color	15 color units	pH	6.5–8.5
Copper	1.0 mg/L	Silver	0.1 mg/L
Corrosivity	Noncorrosive	Sulfate	250 mg/L
Fluoride	2.0 mg/L	Total dissolved solids	500 mg/L
Foaming agents	0.5 mg/L	Zinc	5 mg/L

Water quality guidelines for agriculture can protect crops and livestock from damage by poor quality water. Guidelines for industrial waters may be very important in assuring adequate water quality for various processes.

Significance

Humans must use water for many purposes, and water quality deteriorates as a result. The demand for water is increasing because of the rapidly growing human population (Fig. 15.1). Water quality deterioration has become a serious issue in many countries. Unless measures are implemented to conserve both the quantity and quality of water, the world will face a serious water shortage in the future. Some countries have developed rather elaborate systems of water quality regulations that are maintaining or improving the quality of their waters. However, many other countries have done very little to protect water quality, and serious water quality problems are occurring. It is urgent for all countries to develop water quality regulations and to enforce them rigorously. It is equally important to educate the public about the importance of protecting our fragile water supplies for future use.

Table 16.7 Guidelines for protection of freshwater aquatic ecosystems (Australian and New Zealand Environment and Conservation Council 1992; http://water.epa.gov/scitech/swguidance/standards/criteria/current/index.cfm)

	Australia/New Zealand	United States	
Variable	Concentration/level	CMC	CCC
Physico-chemical			
Color and clarity	<10 % change in compensation depth	Site specific[a]	Site specific[a]
Dissolved oxygen	>6 mg/L	Site specific[a]	Site specific[a]
pH	6.5–9.0	6.5–9.0	–
Salinity	<1,000 mg/L	–	–
Suspended particulate matter and turbidity	<10 % change in seasonal average	–	–
		Site specific[a]	Site specific[a]
Nutrients	Site specific[a]	Site specific[a]	Site specific[a]
Temperature	<2 °C increase	Site specific[a]	Site specific[a]
Inorganic (μg/L)			
Aluminum	<5 (pH < 6.5)	7.50[b]	87[b]
	<100 (pH > 6.5)	–	–
Antimony	30	–	–
Arsenic	50	340	150
Beryllium	4	–	–
Cadmium	0.2–2.0 (depends on hardness)	2.0[c]	0.25[c]
Chromium (total)	10	–	–
Chromium III	–	570	74
Chromium VI	–	16	11
Copper	2–5 (depends on hardness)	Site specific[a]	Site specific[a]
Cyanide	5	22	5.2
Iron (Fe^{3+})	1,000	–	1,000
Lead	1–5 (depends on hardness)	65[c]	2.5[c]
Mercury	0.1	1.4	0.77
Nickel	15–150 (depends on hardness)	470[b]	52[b]
Selenium	5	–	5
Silver	0.1	3.2[c]	–
Sulfide	2	–	2.0
Thallium	4	–	–
Tin (tributyltin)	0.008	0.48	0.072
Zinc	5–50 (depends on hardness)	120[c]	120[c]
Industrial organic (μg/L)			
Hexachlorobutadiene	0.1	–	–
Benzene	300	–	–
Phenol	50	–	–

(continued)

Table 16.7 (continued)

	Australia/New Zealand	United States	
Variable	Concentration/level	CMC	CCC
Toluene	300	–	–
Acrolein	0.2	3	3
di-n-butylphthalate	4	–	–
di(2-ethylexyl)phthalate	0.6	–	–
Other phthalate esters	0.2	–	–
Polychlorinated biphenyls	0.001	–	0.014
Polycyclic aromatic hydrocarbons	3	–	–
Pesticides (μg/L)			
Aldrin	0.01	3	–
Chlordane	0.004	2.4	0.0043
Chlorpyrifos	0.001	0.083	0.041
DDE	0.014	–	–
DDT	0.001	1.1	0.001
Demeton	0.1	–	0.01
Dieldrin	0.002	0.24	0.056
Endosulfan	0.01	0.22	0.050
Endrin	0.003	0.086	0.036
Guthion	0.01	–	0.01
Heptachlor	0.01	0.52	0.0038
Lindane (BHC)	0.003	0.95	–
Malathion	0.07	–	0.01
Methoxychlor	0.04	–	0.03
Mirex	0.001	–	0.001
Parathion	0.004	0.065	0.013
Toxaphene	0.008	0.73	0.0002

[a]See reference
[b]For pH 6.5–9.0
[c]For total hardness of 100 mg/L; see URL in caption for instructions
Note: In guidelines for the United States, CMC is the criterion maximum concentration and CCC is the criterion continuous concentration

References

Australian and New Zealand Environment and Conservation Council (1992) Australian water quality guidelines for fresh and marine waters. ANZECC, Canberra

Gallagher LM, Miller LA (1996) Clean water handbook. Government Institutes, Rockville

Tchobanoglous G, Schroeder ED (1985) Water quality: characteristics, modeling, modification. Adison-Wesley, Reading

Vesilind PA, Peirce JJ, Weiner RF (1994) Environmental engineering. Butterworth-Heinemann, Boston

Index

C.E. Boyd, *Water Quality*, DOI 10.1007/978-3-319-17446-4